A Dictionary of Modern War

EDWARD LUTTWAK

A Dictionary of
Modern War

Harper & Row, Publishers
New York, Evanston, San Francisco, London

Copyright © Edward Luttwak, 1971
All rights reserved. No part of this book may be used
or reproduced in any manner whatsoever without written
permission except in the case of brief quotations embodied
in critical articles and reviews. For information address
Harper & Row, Publishers, Inc., 49 East 33rd Street,
New York, N.Y. 10016.

First U.S. Edition

Library of Congress Catalog Card Number 77-159-574

Printed in Great Britain

To Zahal

Acknowledgements are due to the following for the use of photographs:

Air Portraits 44, 61
Boeing 94
British Aircraft Corporation 41, 42, 70, 74, 76
Bundesministerium der Verteidigung 59
Crown Copyright 3, 7, 8, 10, 11, 19, 20, 25, 27, 31, 33, 52, 55, 64, 65, 68
Engins Matra 73
Établissement Cinématographique et Photographique des Armées 18, 28, 29, 67, 71
Flight International 54
General Dynamics 50, 51, 75
General Dynamics/Convair 37
Hawker Siddeley 43, 66, 86, 88
Hughes Aircraft Company 63, 69
North American Rockwell 89
Philco-Ford 15, 72
Saab 48
Saabfoto 47
Shorts 84
Stato Maggiore della Marina 12
Stephen P. Peltz 39, 40, 58, 80
Tass 1, 4, 9, 16, 24, 30, 49, 53, 60, 62, 78, 79, 82, 87, 92, 93
US Air Force 56
US Army 32
US Navy 2, 5, 6, 13
Vickers 21

ACKNOWLEDGEMENTS

My very close friends, Howard and Clarice Feldman, of Washington, DC, are deeply implicated in the writing of this book. They deplore my involvement with things military, but tolerate all for the sake of friendship.

My good friends Udi and Diana Eichler, though also far from militaristic, have nevertheless helped me, as have Brian and Anne Lapping, known from Karachi to San Francisco for their hospitality.

Official regulations prevent me from thanking a number of professionals, but this applies neither to my assistant Alice Bloch nor to Miss Dahlia Ya'ari of Jerusalem. I also owe my thanks to Richard N. Perle of Los Angeles, who first introduced me to the strategic debate. A book such as this presents many difficult technical problems. My thanks are due to the editors at Penguin Books, who overcame their aversion and handled weapon technologies with care and insight.

FOREWORD

This book is a guide to the weapons and concepts of modern war. As such it is intended to cover the more important weapons currently in service as well as the terms, ideas and organizations of the military dimension of modern life. It is a very selective guide, and the choice emphasizes the 'strategic' over the 'tactical' throughout.

For obvious reasons, there is a shortage of reliable information about the more advanced weapons. The Russians publish almost nothing, and while Western arms' manufacturers do publish a great deal, many of them also cultivate 'brochuremanship', where extravagant claims are camouflaged under the pseudo-technical language fashionable in military circles. ('Advanced' means that it does not work yet; 'semi-automatic' means that everything has to be done by hand; 'anti-tank secondary kill capability' means that only a hero or a fool would approach a real-life tank armed with one of those . . .) Information about Soviet weapons comes almost entirely from intelligence reports which are leaked, selectively, by the Western military. See *Scarp* for a Soviet ICBM whose performance details were published when the Pentagon had to prod a reluctant Congress to release funds for the *Safeguard* missile defence system.

A common misconception about modern weaponry is that its operational life between production and obsolescence is very brief. In fact most successful designs tend to remain in service for a long time: the *Phantom* F-4 fighter was flying by 1958 but it will still be in service by the end of the seventies; the *Chieftain* battle tank has been operational since 1967 and it should be in use until the mid-eighties. This means that this guide will not be out of date for several years but it also means that many obsolescent weapons that are still in service have had to be excluded.

Weapons are often described in two different ways: the official designation (usually a letter/number code), and a popular name. Whenever the latter is really popular it has been used in the text, but there is a cross-reference index at the back of the book.

For those who want to read the guide, rather than to use it as a reference source, the cross-references appended to most entries provide a link from entry to entry. Most things naval are related to the general subject entry

Combat Vessels – similar general entries are *Armoured Fighting Vehicles, Anti-submarine Warfare, Missiles,* and *Combat Aircraft.* Important weapons are described in detail under individual entries such as *Mirage III, Chieftain,* and *Minuteman.* Naval weapons are entered by classes in group entries as in *Submarine, Soviet.* Miscellaneous hardware, such as *Mines* and *Torpedoes,* is discussed in single general entries with no model descriptions, since there are no meaningful design constraints. A few entries (for example *ECM, Radar,* and *Nuclear Warhead*) are rather technical: this is so since the technology involved is very important.

This guide does not cover the composition of military forces as such, except for the strategic deployments of the nuclear powers, as in the entries *NORAD* and *China, Strategic defensive forces.* The less conventional weaponry of *Chemical* and *Biological Warfare* has not been neglected, nor has *War* itself.

A NOTE ON SOURCES

JANE'S All the World's Aircraft, published by *Jane's Yearbooks,* London, is an annual survey of aircraft, missiles and rockets in production. But any one edition will not cover the very many items no longer being produced but which are still in service. Further, there is no critical assessment of the manufacturers' statements. But *JANE'S* remains the basic source, to be used with care.

The best sourcebook for armoured fighting vehicles is the *Taschenbuch der Panzer* by F. M. von Senger and Etterlin, published by J. F. Lehmanns Verlag of Munich. There is no comparable source in English.

JANE'S Fighting Ships is a comprehensive survey, but while naval guns are covered in great detail, its editors seem to be out of touch with modern missile armament.

Many modern weapons are in fact systems consisting of a platform (a land, air or naval vehicle), sensors, a delivery vehicle and a warhead, and many reference sources have become inadequate because they focus on what is in fact just one element in a system. *JANE'S Weapon Systems* attempts to overcome this drawback by focusing on the complete system, but its first editions are marred by many inaccuracies and surprising gaps.

For small arms of all kinds there is *Small Arms of the World,* edited by Joseph E. Smith and published by Stackpole Books. This is comprehensive and most reliable.

While the largest and the smallest weapons are covered by these well-known surveys, there is no handbook for middle weaponry: artillery and projectiles of all kinds. The best secondary source is the series of handbooks published by J. F. Lehmanns Verlag, Munich, Wiener's *Armeen der Neutralstaaten, Armeen der Nato-Staaten,* and *Armeen der Ostblockstaaten.*

Strategic terminology is to be used with care. Each concept is associated with a particular political and technological moment, and terms such as massive retaliation and second-strike strategy are valid only in the specific context within which they evolved. *Strategic Terminology,* by Urs Schwarz and Lázlo Hadik, published by Frederick A. Praeger and the Pall Mall Press, is an excellent trilingual glossary.

The Institute for Strategic Studies of London issues a number of publications on strategic problems, many of which are useful; but from the reference point of view the most important is the annual *The Military Balance*. This is cheap and timely, virtues not always found in some of the sources quoted above.

SECONDARY SOURCES

Weyers Flottentaschenbuch, Lehmanns Verlag, Munich.
J. W. R. Taylor: *Warplanes of the World*, Jan Allan, London.
William Green: *Combat Aircraft*.

Journals: There are hundreds of military journals. These are perhaps the most helpful:

US: *Aviation Week and Space Technology*.
  Armor.
  Air Force and Space Digest.
  Ordnance.
  US Naval Institute Proceedings.
West Germany: *Soldat und Technik*:
         Marinerundschau.
United Kingdom: *Navy*.
Switzerland: *Interavia*.   } in French, German and
         International Defense Review. } Spanish as well as English.
France: *Revue de la défense nationale* (not always reliable).

A

A. U S navy designation for attack aircraft.
◊ *Skyhawk* for A-4 and variants.
◊ *Vigilante* for A-5 and variants.
◊ *Intruder* for A-6 and variants.
◊ *Corsair II* for A-7 and variants.

A A M. Air-to-air missile. ◊ *Missile.*

Abbot F V 433. British self-propelled gun. A tracked vehicle with a squarish turret fitted with a 105-mm gun. Abbot has light armour protection (12 mm) and a ◊ *light tank* configuration with a driver in the hull and a three-man turret crew. (Two more men and extra rounds are carried in a separate vehicle.) It is sealed for both fall-out protection and wading, and there is a built-in flotation screen which can be erected to allow Abbot to swim in calm waters. The gun is effective against armour to about 1,500 yds, while in a howitzer (in-direct fire) role it has a range of 10 miles. What makes it a 'self-propelled gun' rather than a light tank is the absence of a machine-gun fired from the turret. This renders the vehicle vulnerable to infantry attack, but was considered necessary in order to abide by the armour/artillery de-marcation rules. Abbot is based on the F V 430 chassis also used in the ◊ *Trojan F V 432* ◊ *armoured personnel carrier.*
Specifications: Length: 15 ft 10 in. Width: 8 ft 8 in. Height: 6 ft 10 in. Weight: about 15 tons. Engine: Multi-fuel diesel, 240 gross bhp. Maximum road speed: 32 mph. Road range: 360 miles. The gun has a maximum range of 10 miles, and 40 rounds are carried inter-nally.
◊ *Self-propelled Artillery* for competi-tive types.
◊ *Armoured Fighting Vehicle* for general subject entry.

ABC. Abbreviation for Atomic, Bio-logical and Chemical agents used as weapons or weapon warheads. ◊ *Chemical Warfare* and *Biological Warfare.* ABC weapons are sometimes known as 'special weapons', and as 'weapons of mass destruction' in Soviet usage. ◊ *CBR* is synonymous.

ABM. Anti-Ballistic Missile (Sys-tem). A system of sensors, data-process-ing units and missiles intended to locate, intercept and destroy ◊ *ballistic missiles* or their separated warheads. The scope of an ABM is defined in terms of the extent and quality of the defence to be provided. This ranges from a full territorial defence, where enemy warheads are to be destroyed outside the atmosphere (or before they even enter it in the ascending phase of their flight), to a 'hard-point' defence, where fallout and other weapon effects over the target are acceptable so that intercepts can take place in the terminal phase of the enemy warhead's trajectory. Although boost-phase ABMs have been prospected (such as the USAF Bambi project) where the sensors (infra-red) and the interceptor missiles are satellite-borne, more realistic ABM projects, such as the US ◊ *Safeguard*, now being deployed, are based on ground radars and intercep-tors. The latter can be broadly divided into two types: atmospheric, short-range missiles with low-yield warheads, such as the US ◊ *Sprint*, and longer-range inter-ceptors with large warheads intended to kill their targets outside the atmosphere, such as the US ◊ *Spartan* and the Soviet ◊ *Galosh*.
The tasks which an ABM system has to perform include the following:

1. The detection of incoming missiles or their warheads.

2. The tracking and prediction of their course and identification of their intended target.

3. Discrimination between real warheads and other objects, including ⟡ *penetration aids* such as decoys, chaff, spent missile stages or fragmented booster parts.

4. The designation of the target through confusion ⟡ *E C M* and nuclear ⟡ *blackout*.

5. The launch and guidance control of interceptor missiles.

These missions imply an extensive data-processing capability, the protection of the ABM itself against a preliminary strike, and the destruction of enemy warheads without causing unacceptable damage to friendly or neutral territory.

Although the Safeguard ABM is based within the continental United States, seaborne and forward-area ABM systems have been proposed, with the radars on board ships or aircraft or on fixed platforms located in the Polar regions. All these systems are based on the use of very large ⟡ *phased-array radars*, which generate a large number of separate tracking beams and steer them electronically with no mechanical movement of the antenna (as opposed to the mechanically slewed single beam of classic radar).

The Safeguard programme is intended to provide a limited area defence of the US by means of the Spartan extra-atmospheric interceptor and a much thicker defence of missile silos ('hardpoints'), bomber bases and command centres (local defence). The decision to deploy it caused in 1969 a major scientific-political controversy in the United States. Safeguard's opponents argued that: (a) the system would not work, since ⟡ *MIRV*s, penetration aids and its own complexity would defeat it; (b) it would be seen by the Soviet Union as threatening their deterrent (since their warheads could not get through to deter the US). In other words, the system would not accomplish its mission and/or be thought to do so very well, thus leading to a Soviet response designed to offset Safeguard by deploying more offensive missiles. Safeguard's proponents argued that the system would

defeat lightweight penetration aids and/or force the offence to use much of the payload for more effective penetration aids, if they did want to destroy the US deterrent rather than merely deter the US by threatening its cities with retaliation in the event of an American attack. According to them, Safeguard is not intended to cope with a large, sophisticated (Soviet) attack on US cities, but merely with an attack on the ⟡ *hardened* and dispersed retaliatory forces; while the area-defence element is intended to deal only with a small unsophisticated ('Chinese') or accidental attack on US cities. Further, they argued that Soviet planners would recognize the stablizing nature of the deployment and their response would not lead to an arms race in offensive weapons.

⟡ *Assured Destruction.*

ACCHAN. Allied Command Channel. Part of the ⟡ *N A T O* command structure and parallel to, though far smaller than, ⟡ *S H A P E* and ⟡ *A C L A N T.* ACCHAN is responsible for the security of the Channel and part of the North Sea. For this purpose British, Dutch and Belgian naval forces are assigned to it. Its HQ is at Northwood, UK, and it is under the command of a British admiral (CINCHAN). There is a Channel Committee consisting of naval chiefs of staff which acts as its advisory body for general policy. The commander of ACCHAN also acts as Subordinate Commander Eastern Atlantic Area to SACLANT. ⟡ *ACLANT.*

ACE. Allied Command Europe. Command organization within ⟡ *N A T O* and subordinate to ⟡ *S H A P E.*

ACLANT. Allied Command, Atlantic. One of the three major territorial commands into which ⟡ *N A T O* is divided. Its headquarters are at Norfolk, Virginia, and it is headed by SACLANT (Supreme Allied Commander, Atlantic), who has, till now, always been an American admiral.

SACLANT is responsible for naval security in the Atlantic from the North Pole to the Tropic of Cancer, including Portuguese coastal waters. In peacetime

SACLANT controls only a small multi-national force consisting of 4 destroyers, but substantial forces are earmarked to it – in the event of war – from the British, Canadian, Danish, Dutch, Portuguese and American armed forces. ACLANT is currently oriented towards anti-submarine warfare (ASW), and it has a potentially available force of about 300 ASW vessels of all types, more than 100 'hunter-killer' submarines, 300 land-based long-range search and ASW aircraft, and a share of the US Navy's carrier-borne ASW aircraft. SACLANT also has (nominal) control over a large force of US nuclear missile submarines with strategic offensive functions.

AEC. Atomic Energy Commission. Usually used with reference to the American one, though several other countries use the same initials. The US AEC is responsible for supervising the develop-

AFCENT, AFNORTH, AFSOUTH. NATO command organizations. ⟡ *SHAPE.*

AFV. ⟡ *Armoured Fighting Vehicle.*

Aggression. The UN special committee working on a definition of Aggression has been unable to produce one after about eighteen years of deliberation. No definition is given here; everybody knows what the word means.

Airborne Early-warning System (AEW). ⟡ *Early-warning Systems.*

Aircraft Carriers, British. The British navy operates three aircraft carriers, the *Ark Royal*, *Eagle* and *Hermes*. These were originally of wartime design, with straight decks, hydraulic catapults and centreline lifts. All three have been converted to modern carrier standards with angled

British Aircraft Carriers – Specifications:

	Hermes	*Eagle* and *Ark Royal*	*Albion* and *Bulwark*
Displacement (tons)			
standard	23,000	43,000	23,300
full-load	27,800	50,000	27,300
Length (feet)	744·2	811·8	737·8
Width (hull) (feet)	90	112·8	90
Catapults	2 steam	2 steam	none
Aircraft, fixed-wing	22	34	none
helicopters	8	10	16
Missiles	Seacat	Seacat	none
No. of launchers	8 (4 twin)	24 (6 quadruple)	none
Engine power (shp)	78,000	152,000	78,000
Speed (knots)	28	31·5	28
Crew (with airmen)	2,100	1,745	1,035
Accommodation	not available	2,750	1,923–37
Year laid down	1944	1942	1944/5
Year refitted	1966	1967	1960/2

ment of nuclear energy devices, including weapons, and for formulating security regulations, in both the information and accident-prevention senses of the phrase; it is also the prime evaluator of intelligence on other countries' nuclear efforts.

decks, steam catapults, deck-edge lifts and Seacat missiles instead of anti-aircraft guns, though *Eagle* retains eight 4·5-in guns. Two more wartime carriers have been converted into 'commando carriers' (US equivalent, LPH – ⟡ *Aircraft*

Carriers, US) with accommodation for up to 900 troops and 16 helicopters. Apart from supporting 'vertical envelopment' tactics, these carriers, the *Albion* and *Bulwark*, can also double as anti-submarine carriers. The commando carriers also accommodate 4 landing craft and wheeled vehicles for the marines. (The troops are Royal Marines, organized in 'commandos' of 800–900 men each, with light vehicles and equipment.) They are fitted with 8 40-mm AA guns. British defence authorities have decided not to proceed with the construction of a new carrier. The *Ark Royal*, generally similar to the *Eagle*, has been refitted to accommodate a squadron of ◇ *Phantom* aircraft.

Aircraft Carriers, French. The French navy operates two aircraft carriers of post-

equipped with ◇ *Crusader F-8* fighters, and Étendard IV fighter-bombers as well as Alizé anti-submarine aircraft. These two carriers have the enclosed bow-angled deck, steam catapults and deck-edge elevators characteristic of modern carriers, but the elongated island, their armour and the guns are more typical of Second World War carrier configuration. The third carrier, *Arromanches*, was acquired from Britain in 1946 (initially on loan; purchased in 1951), where it had been completed in 1944 as the HMS *Colossus*. Though it has been partially modernized and now has an angled deck (though only at 4 degrees), it serves only as a training and helicopter carrier since the deck is unsuitable for modern naval fighters.

The *Jeanne d'Arc*, which has the

French Aircraft Carriers – Specifications:

	Foch and *Clemenceau*	*Arromanches*	*Jeanne d'Arc*
Displacement (tons)			
standard	22,000	14,000	10,000
full-load	32,800	19,600	12,300
Length (feet)	858·6	694·5	597·1
Width (hull) (feet)	96/104	80·2	78·7
Catapults	2 steam	1 hydraulic	none
Aircraft, fixed-wing	30	4–6	none
helicopters	none	18–20	8, heavy
Guns, type/Missiles	100-mm automatic	none	100-mm and Masurca Mk 2
Guns	8	not available	4 gun, twin launcher
Engine power (shp)	126,000	40,000	40,000
Speed (knots)	31 maximum	23·5	26·5
Radius (miles)	6,400 at 18 kts	12,000 at 14 kts	6,000 at 15 kts
Crew (with airmen)	2,150	1,019	920
Extra accommodation	none	none	700 men and light equipment

war construction, the *Foch* and the *Clemenceau*, one wartime British-built carrier (partially modernized), and one modern helicopter carrier.

The *Foch* and the *Clemenceau* are of similar design, and at 22,000 standard tons can be classified as light fleet carriers since they can carry no more than 30 aircraft. The *Foch* and the *Clemenceau* are

'commando carrier' configuration of the US LPH types and the British *Albion* and *Bulwark* (◇ *Aircraft Carriers, British*), was built as a combined anti-submarine, assault and training vessel. It can carry up to 8 heavy helicopters and a battalion of 700 marines with light equipment. For air defence, there are a twin launcher for the Masurca Mk 2 surface-to-air missile and

4 100-mm dual-purpose guns. The training role of the *Jeanne d'Arc* involves the use of lateral whaleboat implacements and certain other features which impede its use in combat, so that 'rapid' modifications will be necessary for wartime use.

2 **Aircraft Carriers, US.** The US navy operates 15 aircraft carriers of the 'attack' type which carry a full complement of fixed-wing aircraft as well as 8 antisubmarine carriers and 8 helicopter carriers. There are several other ships of the last two types in reserve. The main classes include:

Enterprise. Nuclear-powered attack carrier (CVAN). The *Enterprise* (CVAN 65), completed in 1961, is the largest aircraft carrier ever built, and the second, as yet uncompleted, CVAN, the *Nimitz* (CVAN 68), will be larger still. The *Enterprise* has a very clean deck with a block island superstructure, no funnels, no guns and 4 deck-edge lifts. It carries a complement of 70 to 100 ◊ *Phantom* fighters, ◊ *Vigilante* heavy reconnaissance planes, ◊ *Corsair II* attack fighters, A-6 ◊ *Intruder* bombers and various other ancillary and support aircraft. The absence of funnels and air-intakes allows the ship to be sealed against radiological, biological or chemical agents and has facilitated the installation of an advanced ◊ *phased-array radar* system. It is powered by 8 pressurized water-cooled nuclear reactors, with two reactors driving each of the four turbines by supplying heat to 32 heat exchangers; the total power output of 300,000 shp allows a maximum speed of 35 knots, though the rated speed is 33 knots. Even at this speed the *Enterprise* has a radius of 140,000 miles, with 400,000 miles at the economical (20 knots) cruising speed.

The *America* and *Kennedy* are two new conventionally-powered vessels (CVA) similar to the *Forrestal* class in tonnage but with a larger deck area. Like the *Forrestal* class, they carry three attack squadrons (◊ *Skyhawks* or Corsair IIs) and two fighter squadrons (Phantoms) but have a greater number of ancillary aircraft. The *America* (CVA 66) is equipped with ◊ *Terrier* missiles but the *Kennedy* (CVA 67) is to be equipped with ◊ *Tartars*: there

are no guns. These CVAs carry advanced radar and sonar systems though not the electronic sweep phased-array radar of the CVANs.

The *Forrestal* class consists of six vessels (CVAs). These were the first postwar CVAs and range in displacement (full-load) from 75,900 to 79,000 short tons. (The CVA 60, 61 and 62 of 76,000 tons are taken as representative in the specifications.) The *Forrestal* class of CVAs have been retrofitted with the SINS ◊ *inertial navigation* unit and the NTDS (Naval Tactical Data System) communication computer and display system.

Unlike the newer carriers, these CVAs were originally fitted with 2 forward and 4 aft sponson-mounted twin 5-in guns. The forward guns have been removed from the first four (CVA 59–62), while the last two, *Kitty Hawk* and *Constellation*, completed in 1961–2, were already built without the sponsons; Terrier missiles (two twin launchers) are mounted instead.

The two *Midway* class CVAs, *Coral Sea* CVA 43 and *Franklin D. Roosevelt* CVA 42, were of wartime design with straight (or 'axial') decks, centreline elevators and an open ship-like bow. They were converted during the fifties into the modern aircraft carrier configuration: angled deck, deck-edge elevators, steam catapults and enclosed bow. They can carry 50 to 80 aircraft including the 'first-line' team of Phantom and Corsair II, but carry fewer ancillary aircraft than the *Forrestal* class.

The four *Oriskany* CVAs are improved *Essex* class vessels (see below) of wartime design vintage. But unlike the other *Essex* class vessels these have been modernized to CVA standard by very extensive conversions including the angled deck and steam catapults. Since they are not suitable for the heavier Phantom fighters, they carry the obsolescent ◊ *Crusaders*.

Seven *Essex* class carriers of early wartime design have been converted into antisubmarine carriers (CVS) and helicopter assault carriers (LPH). In the antisubmarine version, they carry helicopters and fixed-wing aircraft (up to 28) and have been extensively modernized, though not all to the same standards, with some CVSs retaining straight decks, hydraulic catapults and open bows. All CVSs retain some 5-in guns. Apart from those converted to the CVA standard (the *Oriskany* type), seven *Essex* class vessels are operating as anti-submarine carriers (CVS). Three other vessels of this class were converted into helicopter assault carriers (LPH), but this role is being taken over by a new class of specially designed vessels, of which the first, the *Iwo Jima*,

Air Defence System. A system intended to prevent or oppose enemy intrusion into a defined airspace. A modern air defence system includes:

1. A chain of broad-beam radar stations to detect the presence of objects in the airspace covered; these stations may be supplemented by airborne radar (◊ *Early-warning Systems*).

2. Identification equipment to discriminate between friends, hostiles and

US Aircraft Carriers – Specifications (name-ship for each class described):

	Enterprise	America	Forrestal	Midway	Oriskany	Iwo Jima
Displacement, full-load (tons)	85,350	77,600	75,900	62,000	42,600	18,340
Length (feet)	1,123	1,048	1,039	979	904	602
Flight deck (acres)	4·5	4·5	4·1	3	2	1
Catapults	4	4	4	3	2	0
Aircraft (maximum)	100	90	90	80	70	24 helicopters
Engine power (shp)	300,000 (nuclear)	280,000	260,000	212,000	150,000	23,000
Maximum speed (knots)	35	35	33	33	33	20
Guns	none	4 5-in	4 5-in	4 5-in	4 5-in	8 3-in
Missile launchers	none	Terrier 2 twin	none	none	none	none
Crew (includes flight personnel)	4,300	4,965	3,772	3,962	3,290	528
Year completed	1961	1965	1955	1945	1950	1961

was completed in 1961. The new LPHs are intended for the US Marine Corps to support their 'vertical development' tactics, where heli-borne assault forces clear the way for amphibious landings. The new LPHs have facilities for twenty-four medium helicopters, four heavy helicopters and four light ones; more than two thousand troops can be accommodated, and there is space for a number of vehicles as well as various command and control facilities.

Many older aircraft carriers have been converted into aircraft ferries (AKV), aircraft transports (AVT), and communications relay ships (AGMR).

birds. This includes ◊ *IFF* transponders and fighter-interceptors for visual identification. In addition there may be computing and memory equipment to compare flight paths with those of known objects.

3. Narrow-beam ◊ *tracking radar* and computers to predict and follow hostile flight paths and to direct weapons against them.

4. Fighters, surface-to-air missiles and anti-aircraft artillery operating independently or by system command.

5. A human and/or computer-operated decision-making system to allocate targets to the various weapons.

6. Communication equipment to link all the parts of the system. In addition there may be electronic countermeasures (⟡ *ECM*) equipment to (a) prevent own radar and communication jamming, and (b) jam enemy radar and communications. ⟡ *NORAD, PVO Strany, CAFDA,* and *UK, Strategic Defensive Forces* for actual systems. These are intended against bomber attacks but not against ballistic-missile attacks, which require ⟡ *ABM* systems. ⟡ *Radar* for general entry on key component. Their effectiveness against ⟡ *contour-flying* aircraft and ⟡ *stand-off missiles* or rockets varies, but it is generally high for the former and low for the latter.

Air Superiority. A power is said to have air superiority within a given airspace when: (a) its aircraft of all types can operate without serious interference from the enemy, and (b) it can limit the enemy's aircraft to shallow penetrations of the relevant airspace. The possession of air superiority confers two advantages: firstly, the whole range of transport and combat aircraft in inventory, including those types which cannot sustain air-to-air combat, can be used; secondly, the side which has air superiority normally also enjoys an information superiority, since its aircraft can conduct reconnaissance missions while those of the enemy are prevented from doing so. This is of crucial importance when the conflict includes mobile land operations. Air superiority is therefore an essential objective of military operations, other than nuclear-warhead missile exchanges or revolutionary war.

The conduct of air operations in a non-nuclear conflict is therefore seen as consisting of two phases. Phase I is a struggle for air superiority in which both sides use 'first-line' fighters over the contested airspace, as well as anti-aircraft artillery and surface-to-air missiles, and perhaps bombers (and/or missiles) against airfields and perhaps the enemy's source of aircraft supplies (aircraft factories, ports, supply lines). In Phase II, the side which has gained air superiority exploits it in order to give close-support, transport and reconnaissance assistance to ground and/or naval forces. The two phases can be compressed in time but they are operationally differentiated by the fact that in Phase I only 'first-line' fighters can usually be used, whereas in Phase II all types of aircraft can be operated.

⟡ *Combat Aircraft* for a general entry on the hardware.

AK-47 (Avtomat Kalashnikov). Soviet rifle (includes AKM and RPK). The AK is a gas-operated selective-fire 'assault rifle' which in its various versions is probably the most common shoulder weapon in the world. It belongs to the 'assault rifle' category because it has optional full and semi-automatic fire and uses an 'intermediate' power cartridge. This, the 7·62-mm M. 1943, has the same diameter as the standard NATO rifle ammunition but is shorter, lighter and less powerful. Its advantages include less of a recoil 'kick', and a lighter logistic load. Its effective range is, of course, inferior to that of a full rifle-power cartridge, but far superior to that of ⟡ *sub-machine-gun* (= pistol) cartridges; it is considered adequate for most combat situations.

The AK is a short weapon of simple and reliable design which can be recognized by the separate gas return tube above the barrel and by the long 30-round curved magazine. It fires at a cyclic rate of fire of 600 rounds per minute, but well-trained infantry will use fast single shots on semi-automatic.

The AK is manufactured in several countries in two basic versions with a wooden and folding metal stock; Chinese and North Korean AKs have cruder all-round finish. The AKM is an improved AK with better sights and general finish; it will probably replace the AK in the Soviet army. The RPK is a ⟡ *light machine-gun* modification of the AKM with a simple bipod and larger 75- and 40-round magazines. It is the new standard Soviet squad support weapon.

Specifications (AK): Length: 34 in. Weight (loaded): 11·7 lb. Muzzle velocity: 2330 fps. Sights: front, post; rear, tangent at 800 metres. 30-round magazine, staggered row, detachable. Ammunition: 7·62 mm M. 1943.

⟡ *Rifle* for general entry and competitive types.

Al-Ared,   Al-Kahir,   Al-Zafir. ⟡ *Egyptian Missile Programme.*

'Alkali'. Soviet air-to-air missile. Standard equipment on interceptor-fighters. Guidance is believed to be by ◊ *semi-active radar homing*. It is 12 ft long. Its range is estimated at 4 miles.

Amatol. High explosive. ◊ *TNT*.

28 **AML. French armoured car.** The AML (Auto-Mitrailleuse Légère) is a 4-wheel ◊ *armoured car* with large-diameter wheels, independent suspension and four driven wheels. It has been produced since 1960 as a cheap substitute for the large 8-wheel ◊ *EBR* 75 armoured car, whose cost and performance are close to those of a ◊ *light tank*. Unlike the comparable British vehicle, the ◊ *Ferret*, the AML is heavily armed with alternative turrets mounting either twin machine-guns and a 60-mm mortar, or anti-tank missiles or a 90-mm smooth-bore gun for fin-stabilized rounds. Like most French ◊ *armoured fighting vehicles* it is mechanically less than satisfactory. Nevertheless, it has already sold in certain less discriminating export markets, such as Iraq and Saudi Arabia as well as South Africa.

Specifications: Length: 12 ft 5 in. Width: 6 ft 5 in. Height: 6 ft 6 in. Weight: 4·17 tons. Engine: air-cooled petrol 89 bhp. Road speed: 56 mph. Road range: 370 miles. Ground clearance: 1 ft 1 in. Turrets: HE 60: 2 machine-guns and a 60-mm breech-loading mortar; H.90: medium velocity 90-mm MECAR anti-tank gun. There is also an ◊ *ENTAC* missile adaptation. Crew: 3.

A.M.S.A. (B-1A). US strategic bomber project. Ever since the B-70 bomber project was cancelled in 1962, the USAF has maintained the Advanced Manned Strategic Aircraft on a research and development basis in order to keep open the strategic bomber option in future deployment programmes. The B-70 was cancelled because its cost and vulnerability were not offset by any significant operational advantage as compared to ◊ *Minuteman* missiles. The RS-70, a modified version, had greater flexibility (it was intended to search for un-destroyed missile silos and other targets) but was still too vulnerable, both on the ground and against sophisticated air defences, to justify its very considerable capital and operating costs. (Owing to the limited ECM payload it was perhaps more vulnerable than the B-52 ◊ *Strato-fortresses*.)

In 1969 the A.M.S.A. reappeared under yet another form. This project is intended to follow up a 'damage-limiting' counter-force attack aimed at the enemy's 'hardened' missile sites. It is therefore supposed to find missile sites which have survived the prior missile attack and, having done so, to destroy them. At the same time, in order to survive against sophisticated air defences (◊ *PVO Strany*) it must fly very low, in the ◊ *contour-flying* ('hedge-hopping') mode, by using a terrain-avoidance radar, a computer and an autopilot. Automatic flying at supersonic speed inside an air-craft which continually rises, falls and swerves to avoid ground features will not be pleasant for the crew. More important, it will be very difficult for them to find undestroyed missile sites. The RS-70 lacked this capability while flying at a high-visibility 70,000-ft course, but the A.M.S.A. is supposed to do this at 300 ft altitude or so, i.e. with a very near radar horizon. (This assumes that no instant intelligence is available from observation satellites.) Further, an aircraft flying very low needs a great deal more fuel than one moving at the same speed at high alti-tudes, so that the ◊ *avionics* or the weapon payload would have to be small. (◊ *AWACS* project.)

◊ *US, Strategic Offensive Forces.*

AMX 13. French light tank/tank destroyer. Originally intended as an air-transportable tank, the AMX 13 became a tank destroyer, and when it first ap-peared in 1949 its 75-mm (long) gun was adequate for this role. Apart from being very light at 15 tons, the AMX 13 has many original features: the driver sits in the hull alongside the engine; and there is only a two-man turret crew since the loader is replaced by an automatic loading device. This is made possible by the use of a trunnion-mounted turret which turns normally but where the whole upper turret elevates, instead of the gun elevating inside a fixed turret. This oscillating turret is higher and heavier than a conventional one, but it removes many design constraints and permits automatic loading since the gun breech is in a fixed position relative to the loading device.

The AMX 13 has been produced in a number of versions mounting different turrets including a 'colonial' (now 'counter-insurgency') turret with a short 75-mm gun; and with an ultra-long 75-mm. These, as well as the original FL 10 turret and gun, have been replaced by a 105-mm gun turret. This is the standard tank gun calibre but it fires lower-velocity HEAT rounds, to reduce stress on the very light (now 15-ton) chassis. The AMX chassis has also been used in a whole family of vehicles: the AMX-VTT ⟡ *armoured personnel carrier*; a twin 30-mm anti-aircraft ⟡ *self-propelled gun*; a 105-mm howitzer, self-propelled; a recovery vehicle; and others.

Specifications: Length: 16 ft 0 in. Width: 8 ft 2 in. Height: 7 ft 6 in. Weight: 15 tons. Engine: petrol 245 bhp. Speed: 37 mph. Road range: 220 miles. Ground pressure: 10·8 lb sq in. Ground clearance: 1 ft 2 in. Gun: 105 mm, 32 HEAT rounds carried. Secondary weapons: 2 MGs. Crew: 3.

⟡ *Light Tank* and *Tank Destroyer* for general entries.

18   AMX 30. French tank. The new French main battle tank (MBT), and the first MBT brought to the production stage in France since the Second World War. It is of conventional shape and layout, with three men in the turret and a driver in the hull; like other up-to-date MBTs it has a diesel engine, giving a good road range and speed. The AMX 30 is at 33 tons the lightest of the new generation of MBTs, reflecting French military opinion on the limited value of armour. The AMX 30 has little ballistic protection beyond that needed to cope with 20-mm automatic cannon. On the other hand, the armour envelope is well shaped, with no shot traps or vertical plates. Another good feature of the AMX 30 is a 20-mm automatic cannon, which is useful against slower aircraft as well as ground targets. But the main 105-mm gun fires expensive

and complicated 'compensated ⟡ *HEAT*' rounds, which are mounted on ball-bearings so that the shell spins – as in all high-velocity rounds – but the core does not (in order to avoid the degradation of performance which affects HEAT spin-stabilized rounds). Other MBT guns use simpler and more conventional ⟡ *AP* and ⟡ *APDS* rounds, but perhaps the AMX 30 is too light to bear the recoil stresses. There appears to be some doubt about the reliability and cost-effectiveness of this complex and very expensive tank.

Specifications: Length: 20 ft 11 in. Width: 10 ft 2 in. Height: 8 ft 3 in. Weight: 33 tons. Engine: diesel 720 bhp. Road speed: 40 mph. Road range: 300 miles. Ground pressure: 10·8 lb sq in. Ground clearance: 1 ft 6 in. Main gun calibre: 105 mm, with HEAT rounds; 56 rounds carried. Coincidence optical range-finder and night-vision aids. Crew: 4.

⟡ *Main Battle Tank* for competitive types.

⟡ *Armoured Fighting Vehicle* for general entry.

AMX VTT/VTP. French armoured   29 personnel carrier. The AMX VTT (*véhicule transport de troupe*), also designated AMX VTP, is a combat vehicle for the infantry derived from the chassis of the AMX 13 light tank; there are also mortar carrier, engineer and command vehicle versions. Unlike the standard US and British carriers (⟡ *M.113* and ⟡ *Trojan*) this is a true combat vehicle and not merely a 'battle taxi': 8 of the 12 troops carried are provided with lateral firing/observation ports, and there is a machine-gun turret on the superstructure. The armour, of variable thickness with a maximum of 40 mm in the front provides ballistic protection against various battle-field threats, up to and including 20-mm cannon.

Specifications: Length: 18 ft 2 in. Width: 8 ft 2 in. Height: up to 7 ft 1 in. Weight: 12·5 tons. Road speed: up to 40 mph. Road range: 250 miles. 7·5-mm MG mounted in standard French version, 7·62-mm or 12·7-mm MG in other versions. Crew: driver and 12 men. (The VTT is used by the Belgian and Dutch armies as well as the French army.)

AN-12. ◊ *Cub.*

AN-22. ◊ *Cock.*

'Anab'. Soviet air-to-air missile. An air-to-air missile with alternative ◊ *infra-red* and ◊ *semi-active radar homing* warheads. It equips the ◊ *Firebar* and ◊ *Fishpot* interceptors of the Soviet Air Defence Command (◊ *PVO Strany*). It is 11·5 ft long, and the range is estimated at 6 miles.

ANF. ◊ *Atlantic Nuclear Force.*

Antarctic Treaty: Demilitarization of Antarctica. The Antarctic Treaty, which came into force in June 1961 for a period of thirty years (renewable), was signed in 1959 by the representatives of Argentina, Australia, Belgium, Chile, France, Japan, New Zealand, Norway, South Africa, the UK, the USA, and the Soviet Union. It includes a prohibition on all military activities and the detonation (or storage) of nuclear explosive devices on Antarctica. Unrestrained mutual inspection by any or all parties was accepted by all signatories. The treaty covers all land south of the 60th Parallel. ◊ *Demilitarized Zone.*

Anti-submarine Warfare (ASW). Submarines are expensive and complex weapons, and there are not many of them around. At the present time no more than 500 submarines of all types are operational, and the number is declining as diesel/electric submarines are being replaced by fewer, albeit more effective, nuclear-powered boats. But the modern submarine is also extraordinarily effective by virtue of the fact that in an age of global surveillance by satellite the submarine can provide a concealed platform for weapons of all kinds, and especially for long-range missiles with nuclear warheads.

As explained in the ◊ *submarine* entry, only the nuclear-powered vessels now being deployed by Britain, France, the United States and Russia are true submarines, the older diesel/electric boats being in fact submersibles. While the nuclear-powered submarines are capable of cruising at high speeds over long ranges under water, the diesel/electrics can dive only for short and rather slow submerged cruising; unless, that is, they use snorkel tubes, in which case they are vulnerable to detection, and detection, or rather location, is of course the name of the game in anti-submarine warfare. Once a submarine is detected it can be easily destroyed, since its pressure hull is very vulnerable to explosive shock waves under water. When a submarine is on the surface it is of course just another small vessel and just as easily destroyed. But submarines are not supposed to meet an attack on the surface, and finding a submerged submarine is generally very difficult.

The main ASW detection method is by ◊ *sonar.* This works on a principle similar to radar but does so in a far less favourable medium. Temperature layers, distance and the presence of living creatures of all kinds interfere with its operation. Large passive sonar sets (intended to receive sound waves emitted by submarines) are fitted to the hulls of most warships and all ASW vessels. Such vessels also carry active sonar (where the set emits a signal and receives the echo) either as fixed installations or in the variable-depth form. Helicopters can dip sonar sets into the water while hovering overhead. Long-range maritime-patrol aircraft, and ASW fixed-wing aircraft in general, cannot of course suspend a sonar set down into the water while flying overhead. Such aircraft drop buoys containing active or passive sonar sets, which are fitted with a radio transmitter to relay the data back to the aircraft. These devices, known as sonabuoys, can be submerged to pre-set depths. Sonabuoys are hardly ever recoverable and rather expensive.

Hunter-killer submarines too have to depend on sonar to locate their target, and of course such boats carry the most elaborate active and passive sets. The sonar technique is also used in detection chains, laid on the sea bed itself, such as the US C-Systems, which transmit the information back to the home base by wire links or possibly by radio. But they are vulnerable since there is nothing to prevent one country from removing or destroying them so long as they are outside territorial waters. But no sonar device will locate a submarine which is cruising very slowly or remaining motionless deep under water, and which has a skilful crew, capable of using thermal layers and colonies of marine life for its own concealment. ◇ *MAD* (magnetic anomaly detection) is an even more problematic method of detection. The idea is to spot the minute changes in the earth's magnetic fields produced by a submerged submarine. The technique is unreliable and in any case works only at the shortest ranges. ASW aircraft nevertheless carry these in a long tail boom designed to offset the magnetometer from all the magnetic gear in the aircraft.

'Sniffer' devices, also known as ◇ *Autolycus*, which detect the presence of diesel fumes are also limited to short ranges, and of course are entirely ineffective against nuclear submarines, the main threat. Yet another technique is the observation of the ripples caused by a submarine near the surface, but it is not useful when the plane from which the observation is to be made is flying several thousand feet above the sea, let alone at night or in bad weather.

But submarines of all kinds have to communicate with their home base, and they do so by radio. Radio direction-finding is an old-established and well-developed technique. Submerged submarines broadcast in the very low frequencies (VLF) which allow the (long) radio waves to travel through water. Interested parties therefore keep a continuous watch over those frequencies. A submarine on the surface of course be detected by land, sea or airborne radars, but modern submarines are fitted with ◇ *ECM* equipment which enables them to detect others' radar emissions. And nuclear submarines can of course remain fully submerged for an entire cruise.

The inadequacy of all these techniques explains the recourse to hundreds of ASW vessels and aircraft of all kinds in the hope that numbers will make up for the deficiencies of the equipment. Detection then becomes far easier, but accurate position-fixing still remains a problem. Anti-submarine weaponry includes depth charges released from aircraft or vessels directly overhead, depth charges projected by rockets or mortars (◇ *Hedgehog*), and ◇ *torpedoes* of all kinds launched from ships, aircraft or other submarines. Newer naval weapons are the ASW missiles such as ◇ *Asroc*, *Ikara* and Malafon. These, ballistic or not, are intended to fly towards the sea

area where a submarine has been detected and then to release a homing torpedo or depth charge which can be nuclear. Hunter-killer submarines use homing, magnetic, time or impact-fused torpedoes with ranges of up to several thousand yards.

Only the US navy's hunter-killer submarines have a long-range ASW weapon. This, the ◇ *Subroc*, is not a simple weapon. It is in fact a ballistic missile which is fired under water torpedo-fashion and then leaves the water and flies a ballistic path to the target area. It then releases a nuclear depth charge which penetrates the water before exploding.

All these sensors and platforms are used by the Western powers in order to keep track of the 370 or so Russian submarines (◇ *Submarines, Soviet*). To do the job, the Western powers deploy some 400 ASW vessels, 150 or so hunter-killer submarines of which a third are nuclear-powered, more than 500 long-range maritime patrol aircraft and some 1,200 carrier-borne ASW aircraft, including helicopters. These systematically patrol the oceans with linear-programming techniques being used to optimize the efficiency of the patrol grids.

But the easiest way of keeping track of enemy submarines is to pick them up as they leave their base and simply follow them. Only another submarine can do this, because aircraft are liable to lose track quite easily, while surface vessels lack the speed. Hydrofoils and hovercraft may eventually be useful for ASW work, but none of those developed so far have sufficient range and stability for this work. In any case, in ASW silence is the thing. While submarines minimize sonar-detectable noises by moving slowly, the hunters too cannot use high speeds in or on the water without degrading the performance of their sonars. Hence the hunter-killer submarine remains the best platform for submarine tracking. Russian submarines are picked up and trailed as they leave port or when they cross narrow places such as the Dardanelles. The Russians play the same game on the *Polaris* vessels. But comprehensive tracking by these means requires many (nuclear) hunter-killers, and neither side has enough to do the job reliably.

ANZUS. Popular name of the ◊ *Pacific Security Treaty*, which is a security treaty of indefinite duration signed in September 1951 by representatives of Australia, New Zealand, and the United States. Under the terms of the Treaty, which was offered by the United States to the other parties as a substitute to their (rejected) application to join NATO, each party undertakes to offer (unspecified) help to any other party in the event of an armed attack on their territories or military forces anywhere in the Pacific. This is, however, conditional on US Congressional approval of such aid. Australian and New Zealand forces participated in the Vietnam war alongside the United States, though their contribution was rather small. The United Kingdom government expressed an interest in joining the Treaty, but did not do so.

AP (Armour Piercing). The letters AP before an ammunition specification imply that it has some armour-penetration capability. Though machine-gun and rifle-calibre ammunition is also made in AP versions, AP usually refers to artillery-calibre (20 mm/0·8 in and above) ammunition used in high-velocity guns. The armour-piercing performance of AP shot varies, for a given metallic composition, with the muzzle velocity and the diameter. For a given velocity, the larger the diameter the greater the thickness of the armour which can be perforated. Thus there is a choice between using large-calibre ammunition at relatively low velocities and smaller-calibre ammunition at higher velocities. The former requires larger and heavier guns and mounts or tank turrets, the latter requires more expensive – and more delicate – high-velocity guns, although the higher the muzzle velocity the greater is the fall-off in performance due to aerodynamic drag.

In addition to steel 'shot' monoblock AP rounds, there is other armour-piercing ammunition, for example APCR shot, which is composite with a hard high-density core and a soft light jacket. Since it is lighter than an AP projectile of the same calibre, its muzzle velocity is higher and it can penetrate relatively thicker armour because the available energy is concentrated on a smaller ('sub-calibre') area of the target.

'APCBC' describes conventional monoblock AP shot with an ultra-hard cap. Comparing APCR with APCBC rounds fired from the same gun, the performance of APCR is superior at all but extreme ranges. APCR is used in the 100-mm gun of the Soviet T54 and, under the designation HVAP, in the 90-mm gun of the US ◊ *Patton* M.48 tank. APHE rounds are specifically designed for 'soft' targets within thinner armour envelopes: there is an explosive charge within the shot which explodes once the APHE has penetrated the envelope. Thus APHE rounds can be very lethal, but their penetration is inferior to that of equivalent AP rounds.

◊ *APDS* for another type of AP shot. ◊ *HEAT* for an altogether different method of penetrating armour, and *HESH* for still another but less successful method.

◊ *Armour,* for the intended victim of all these.

APC. ◊ *Armoured Personnel Carrier.*

APCBC (Armour-Piercing Ballistic Capped). An armour-piercing shot. ◊ *AP.*

APCR (Armour-Piercing Reduced (Calibre)). An armour-piercing shot. ◊ *AP.*

APDS (Armour-Piercing Discarding-Sabot). A type of gun ammunition developed in Britain and used in high-velocity guns. APDS, which was first used in 1944, is a composite round which has an inner core and outer 'sabots' which separate as soon as the shot leaves the muzzle. Since the round is lighter than conventional (full-bore monobloc) steel shot, the muzzle velocity is higher, and it penetrates better because its kinetic energy is applied to a smaller area of the target. The APDS also has a longer effective range, since the frontal area – and the corresponding aerodynamic drag – is smaller than with normal ammunition.

Though the dispersal of the sabots causes problems, i.e. if the gun is fired over the heads of friendly troops, APDS has been widely accepted. It is used with the 105-mm gun fitted to the US M.60 (◊ *Patton*), the German ◊ *Leopard* and the

Swiss Pz.61 as well as the British ◊
Centurion, and for the 120-mm gun of the
◊ *Chieftain*.

A typical APDS shot will penetrate
armour seven times as thick as the diameter
of the core or four times the calibre of the
gun if fired at 4800 fps and at point-blank
range; operationally, the 105-mm APDS
is effective against most armour at ranges
up to 2,000 yds.

◊ *AP* and *HEAT* for other primary
tank gun rounds; ◊ *HESH* for a secondary (i.e. general-purpose rather than anti-
armour) ammunition.

**APHE (Armour-Piercing High Ex-
plosive).** A type of ammunition which
contains an explosive capsule within the
solid shot envelope. It is highly effective
against armoured targets which contain
human or other 'soft' equipment, but the
penetration power is correspondingly reduced vis-à-vis solid shot. ◊ *AP*.

Armour. A material intended to protect
men, installations or equipment. Aluminium, titanium and magnesium alloys,
plastic laminates, nylon fabrics and mastic-
stone composites have all been used, but
the most common form of armour is heat-
treated alloy steel. This comes mainly in
two forms: homogeneous and face-
hardened. The first has constant ballistic
and mechanical qualities throughout,
which in modern tank armour means high
tensile strength (of the order of 150,000
lb sq in) and medium hardness (i.e.
Brinnell N° 300); face-hardened armour
consists of a very hard first layer – which
resists penetration but lacks toughness –
and a more elastic core to support the
face. Naval deck armour and modern tank
armour are homogeneous, while face-
hardened armour is still used for some
light armoured vehicles and side protection for naval vessels. Steel armour usually
means rolled or wrought plates, but
castings are used for tank turrets and
even for entire tank hulls, as in the M.48
and M.60 US tanks. (Steel armour is
usually a nickel-chrome-molybdenum
alloy, but other alloys including man-
ganese-silicon-chromium-nickel steel have
been used, especially in Soviet tank
armour.)

Sloping increases effective protection

against high-velocity projectiles by more
than the horizontal thickness achieved,
Thus 100-mm plate inclined at 60 degrees
has a horizontal thickness of 200 mm but
gives the ballistic protection of 300-mm
vertical plates.

Aluminium has been used in a variety
of vehicles, including the ◊ *M.113* US
◊ *armoured personnel carrier*, but it does
not save much weight when substituted
for steel to give equal ballistic protection.
Armour is used to protect sensitive parts
and pilots in ground attack aircraft, in aircraft carrier decks and sides, and of course
for armoured fighting vehicles. In the case
of the latter, new projectile warheads, including shaped charge or *HEAT* and
Squash Head (◊ *HESH*) ammunition,
have complicated the game. In both cases
the thickness and sloping effective against
high-velocity rounds (◊ *AP*) does not help
much, and spaced armour is required: a
thinly spaced small detonating screen
suffices for *HESH*, but not for *HEAT*
rounds, which need about a foot.

Body armour was revived during the
First World War, and the steel helmet is a
basic piece of military gear, but actual body
armour, effective against small arms or
more commonly splinters, is also used in
the form of vests composed of nylon-
fibreglass and laminated resins. Those
currently issued to pilots and some ground
troops can reduce casualties considerably.

◊ *Armoured Fighting Vehicle* for the use
of armour.

Armour-piercing. Ammunition type. ◊
AP.

Armoured Car (AC). A wheeled ◊
armoured fighting vehicle used in reconnaissance, patrol and police-security roles.
Early ACs, which were conversions based
on car or truck chassis, had very limited
armour protection, machine-gun armament and poor cross-country mobility.
Later, specially developed and better-
armoured vehicles appeared with all-
wheel drive, independent suspensions and
gun armament. Two classes of armoured
cars have emerged: (a) small 4 by 4 light
ACs for the liaison, patrol, internal
security and light anti-tank role, such as
the British ◊ *Ferret* series, the Soviet

BRDM, and the French ◊ *AML* series; (b) heavy, 10 tons or more multi-wheel ACs for the heavy reconnaissance role, such as the British ◊ *Saladin*, a 6 by 6, and the French 8 by 8 ◊ *EBR* series. For various reasons, the deployment of ACs is largely confined to armies with colonial or similar experience, while the heavy ACs are used only in such armies. The US army has not used ACs since the Second World War, but the Cadillac Gage Commando 4 by 4 may be introduced in limited numbers as a result of the Vietnam experience. The German army does not use ACs at all, though the border police have some, including Saladins.

◊ *Armour* for details of protection. ◊ *Light Tank* for the competitive concept. ◊ *Armoured Forces* for tactical concepts.

Armoured Fighting Vehicle (AFV). General term for military vehicles with cross-country mobility and armour protection. The first AFVs were developed as specialized instruments (i.e. the tank to defeat trench systems) intended to assist traditional infantry, artillery and cavalry. Later, the concept of all-AFV forces was gradually accepted following the success of the German Panzer divisions, which demonstrated that mixed forces of artillery, infantry and engineers carried in AFVs and combined with tanks could make a creative use of battlefield mobility. Independent all-AFV forces need a variety of different vehicles, and since the Second World War AFV design-patterns have been developed to fulfil the various roles (see list below). The amount of armour protection varies widely, and so do the engines, though water-cooled diesels power most recent designs. Since weight has to be widely distributed to assure mobility over unpaved surfaces, most AFVs are tracked, though multi-wheel systems – as used in armoured cars – are adequate for lighter vehicles. Tracked drive is expensive and sometimes mechanically fragile, gives poor fuel economy and imposes a short and wide shape. Although longer, narrower vehicles would be preferable in many cases, tracked drive is essential to achieve low ground pressures per square inch in heavy vehicles, and offers far more ballistic protection than rubber wheels.

A functional system of AFV classification is followed in the text: For tracked, heavily armoured vehicles with turret-mounted armour-defeating general-purpose weapons usable on the move, ◊ *Main Battle Tank*. For similar vehicles below 25 tons weight, ◊ *Light Tank*. For armoured vehicles intended to carry infantry into battle (i.e. where the crew exceeds the number needed to man the vehicle), ◊ *Armoured Personnel Carrier*. For AFVs, other than those with turret-mounted guns, specifically intended to fight tanks, ◊ *Tank Destroyer*. For AFVs mounting guns or howitzers, ◊ *Self-Propelled Artillery* and *Artillery*. For all wheeled AFVs other than APCs, ◊ *Armoured Car*.

Apart from these main categories, which may overlap, there are many special-purpose conversions, such as command vehicles, weapon carriers, supply carriers, recovery vehicles, engineer vehicles, demolition and flamethrower vehicles, and missile carriers based on all types of AFV chassis.

◊ *Armour* for details of the nature of protection.

◊ *Armoured Forces* for basic tactical concepts.

◊ *Missile* (anti-tank) and *HEAT*, *HESH* and *AP* for armour defeating ammunition, as well as '*Bazooka*', *Recoilless Weapons* and *Artillery* for the means of delivery.

Armoured Forces. A general term used to describe military forces partly or wholly equipped with armoured fighting vehicles (AFVs). Since the development of the first AFVs just before the First World War, these have been used in a variety of ways. At first, they were regarded as supporting weapons for the then conventional arms, the infantry and cavalry. Hence the early use of tanks to support infantry attacks and of armoured cars for scouting. By the beginning of the Second World War, three methods of use had been evolved:

(a) Tanks and other AFVs as supporting weapons for the infantry advancing at infantry speeds and spread amongst the latter.

(b) All-tank forces for limited penetrations in the cavalry manner, which were to exploit breakthroughs achieved by the infantry and artillery.

(c) Large mixed forces of AFVs combining all arms and operating in depth through all phases of the fighting. British doctrine until the later part of the Second World War and to some extent even afterwards was a combination of (a) and (b), and different tanks were developed for the two roles. French doctrine was essentially (a). American doctrine tended towards (c), which was the main German and later Russian doctrine throughout the war. This has since been accepted by almost all mechanized armies, though it is uncertain whether some armies, such as the French, have mastered the complex command and control requirements such a use of armour entails. Israeli practice in 1967 was quite original: some units of all-tank forces operated in depth; others operated according to the third doctrine, while even (a) was used in places.

In modern armies the infantry is now carried in ◊ *armoured personnel carriers*, while the artillery and ancillary services are mechanized so that the type (c) use of armour has blurred the traditional distinctions between the branches of the armed forces. In terms of organizational tables therefore, 'infantry' divisions simply have fewer tanks and more APCs than 'armoured divisions'. Apart from internal security operations, any major land fighting would now involve the use of mixed forces in role (c), unless it is all conducted in city centres or on mountain peaks. Given the equipment available, type (c) use of forces means variable 'mixes' of tanks with infantry, artillery, engineers and support units all carried in armoured vehicles; and the traditional infantry-armour differentiation is now limited to poor countries.

For a classification of relevant equipment ◊ *Armoured Fighting Vehicle*. ◊ *Indirect Approach, Blitzkrieg, Hedgehog,* and *Defence in Depth* for some of the relevant tactical concepts.

Armoured Personnel Carrier (APC).

An armoured cross-country vehicle designed to carry infantry into battle. The APC is the vehicle of the modern mechanized infantry, and it evolved from early conversions of commercial trucks and war-

time half-tracked vehicles such as the German Sd. Kfz. 251/10 and the US M.3.

The APC concept originated in the need for infantry to accompany tanks in order to protect them from enemy infantry. APCs have to be (at least) lightly armoured and need cross-country mobility for movement across the battlefield. Since the Second World War, fully-tracked and armoured vehicles have been developed for this role: the US ◊ *M.113*, the Soviet ◊ *BTR-50(P)*, the German HS 30 and ◊ *Spz. Neu*, the British ◊ *Trojan FV 432*, and French ◊ *AMX VTT*, and several others. Though the role of mechanized infantry is now no longer confined to the protection of tanks, the design of the American and British APCs confines them to this function, since they have no firing ports to enable their crews to fight from within. Both the Trojan and the M.113 have thin armour, limited armament and high (and hence vulnerable) silhouettes. This reflects British and American doctrines according to which the crew 'rides to battle' in its APC and then 'dismounts' to fight on foot. Soviet, French, German, Swiss and Swedish APCs on the other hand are designed so that the crew can fight on the move, with firing ports, good frontal armour and turret-mounted machine-guns and/or 20-mm cannon. The reason for this difference lies in different appreciations of the value of aimed fire, the Anglo-American idea being that aimed (i.e. man-killing) fire requires – and justifies – the abandonment of expensive APC protection. Recent combat experience in Vietnam appears, however, to have converted the US military from the 'battle-taxi' to the combat vehicle concept.

Tracked vehicles are expensive, have poor fuel economies and a design limit of $1\cdot1:1$ to $1\cdot8:1$ in length-to-width imposed by the need to steer by slewing. For this and other reasons wheeled APCs – such as the British ◊ *Saracen*, the French *EBR-VTT* and the Soviet ◊ *BTR-60* – have been used in some armies. Some wheeled APCs based on truck conversions are also in use, such as the British FV 1611, the Soviet ◊ *BTR-152* and the Dutch DAF YP 408.

A basic variable in APC design is the degree of amphibious mobility. The US M.113 is a wader-swimmer propelled by its track in water; the British Trojan FV 432 is 4 tons heavier so that it requires an expensive and vulnerable flotation screen. The Soviet BTR-50(P) is derived from the boat-shaped ◊ PT-76 amphibian tank and swims well with water-jet propulsion; the German Spz. Neu has sealing and a snorkel so that it can submerge down to 16 ft.

◊ *Armoured Fighting Vehicle* for general entry.

◊ *Armoured Forces* for tactical concepts.

Arms Control. A general term which encompasses subsidiary concepts such as ◊ *disarmament* and ◊ *arms limitation*. It covers any arrangement, between two or more states, which limits the number or the performance characteristics of any weapon system (including related ancillary equipment); or the number, organization, deployment or use of armed forces of the parties. This arrangement may or may not be explicit, and any mutual restraint which is recognized as intentional falls within the concept.

Arms Limitation. An explicit or implicit agreement between states designed to prevent a qualitative or quantitative increase in determined weaponry or military personnel. The US–British–Japanese naval agreements and the Anglo–German naval agreement of the inter-war period, which established mutually agreed tonnage ratios for combat vessels, are examples of the application of this concept. The Test Ban Treaty is also specific to a particular class of weapons, but instead of being limited to given signatories was intended as a universal treaty.

◊ *Arms Control* and *Disarmament* for wider concepts. The more 'limited' concept of arms limitation appears to be a more workable one in an ◊ *arms race* environment, and it is more likely to lead to concrete results, particularly when an implicit agreement is acceptable to the parties involved.

Arms Race. An increase in the quantity or quality of the weaponry or military personnel of two or more states, which is both cumulative and recognized as mutually competitive by the parties involved. There appears to be a low correlation between arms races and war.

An arms race starts when one party is seen as increasing its military strength. This increase may be a deliberate attempt to improve the relative capability of the party concerned; it may be the result of purely internal pressures reflecting bureaucratic or private vested interests; it may be the result of normal replacement or recruitment policies in a situation of technological or demographic change; or the increase may be only apparent, a matter of faulty perceptions on the part of the other parties. If the parties to an arms race seek to maintain certain specific relative levels of armaments, no stable relationship may be possible. And even if they accept parity the long lead-time required to develop weapon systems in the presence of technological change will impede the establishment of a stable relationship. Unless predictions about weapon-system delivery-dates, performance and output are exactly right, imbalances – and competitive adjustments – will result. A static stability agreement is usually unacceptable, even assuming accepted and fool-proof inspection, since other parties will then 'catch up' with the parties to such an agreement.

◊ *Arms Control.*

Artillery. A general term for assorted weaponry exceeding small-arm calibre (i.e. 0·6 in/15 mm or more); also used to describe the branch of the armed forces equipped with these weapons. But this definition, in terms of barrel diameter, has been confused by the introduction of new types of weapons which have larger calibres but are not classic full-recoil tube artillery: ◊ *recoilless weapons*, rocket-launchers and ◊ *grenade*-launchers. Tactical rockets and missiles are also often described as 'artillery' but are here treated separately. (◊ *Rocket* and *Missile.*)

There remain the guns, cannon, howitzer and mortars of classic tube artillery. In all of these a warhead is projected through a barrel by the expansion gases produced by the explosion of the propelling charge. With the exception of mortars, modern artillery is breech-loading, with a breech block which slides or screws into the rear of the barrel. Again except for mortars, a recoil-absorbing mechanism is used to maintain the stability of the weapon, and this is attached to the chassis assembly, the carrying vehicle or fixed mounting. The range and accuracy of the weapon depend on the power of the explosive cartridge and the length of the barrel, all other things being equal. The projectiles are stabilized by the spinning effect induced by the rifling of the barrel, though fin-stabilized ammunition fired through smooth barrels is also used for some special purposes.

Weapons firing ammunition at a high muzzle velocity through long barrels have relatively flat trajectories; these direct-fire weapons are properly described as guns or cannon. Anti-aircraft, anti-tank, naval and tank weapons are all 'guns'. Small-calibre anti-aircraft and aircraft guns are usually described as 'automatic cannon' – these generally range in calibre from 20 to 40 mm and are – as the name implies – fully automatic. Anti-tank guns are primarily intended to fire armour-piercing ammunition (such as APDS, AP and APCR) and are usually mounted on low-silhouette two-wheel chassis with small frontal shields. The same type of gun is also used on tanks (◊ *Main Battle Tank*) and ◊ *tank destroyers*. 'Field artillery' comprises all mobile medium and large calibre guns, and also howitzers.

Howitzers are also breech-loading with a rifled barrel (i.e. spin-stabilized), but are intended for indirect fire. This means relatively short barrels and lower velocities for parabolic trajectories with the weapon being fired in high elevation. A howitzer – all other things being equal – is lighter and cheaper than a gun of equivalent calibre, but it fires the same weight of shell over a correspondingly shorter range. Thus a typical 155-mm gun has a maximum range of 26,000 yds, while a comparable 155-mm howitzer has a 16,000-yd range. In effect, howitzers are intended to deliver high-explosive shells in plunging fire with either (or both) weapon and target protected by frontal cover.

Mortars are usually smooth-bore, muzzle-loaded weapons with short barrels and are intended for firing at very high elevation. Since recoil forces then operate almost vertically downwards, they can be absorbed by cheap and simple base-plates, instead of the complex and expensive recoil-absorbing gear needed in guns and howitzers. Thus mortars can deliver a given weight of shell at a fraction of the cost and weapon weight of the other two weapon types, but at the expense of range and precision. Breech-loading mortars are necessarily more complex than the far more common muzzle-loaders but they overcome the 'venting' problem. The diameter difference between bomb and barrel (required for loading) leads to the loss of part of the expansion gases, affecting range and precision. The 60-mm mortar which equips the HS 60 version of the French ◊ *AML* armoured car is a breech-loader, as are the Soviet 160-mm and 240-mm heavy mortars.

Ground artillery used to be classed into mobile and static, but this old distinction is no longer meaningful since few static (in fact coastal) artillery pieces survive; but there is a newer distinction between towed and ◊ *self-propelled artillery*. The latter ranges from anti-tank guns mounted on simple motorized four-wheel chassis, such as the German 90-mm Pak, to tracked and armoured turreted vehicles, such as the British ◊ *Abbot FV 433*. While artillery in modern Western armies is increasingly self-propelled, Soviet artillery is still largely towed, except for some anti-aircraft and airborne assault guns (◊ *ASU*-57, *ASU*-85). This difference in approach appears to reflect both doctrinal and economic considerations. The Soviet military still appear to believe in the value of massive artillery barrages, and this renders uneconomic the use of expensive tracked and fully enclosed vehicles, since their ammunition-carrying capacity is necessarily limited (20–60 rounds). Sustained fire would require extra ammunition carried in 'soft' vehicles, and this invalidates the use of cross-country armoured vehicles since their capabilities would remain under-utilized. Western armies, on the other hand, have adopted enclosed armoured vehicles to satisfy a requirement for small scale and precise fire delivered from weapons immune to most battlefield hazards. Any 'barrage' would in fact be nuclear, and the large scale concentrations of conventional artillery which the Soviets still appear to believe in are considered to be obsolete by most Western military planners.

◊ *Artillery, British, French, Italian, Soviet, U S, West German.*

Artillery, British. After a long period of technical leadership, exemplified in the 25-pounder gun-howitzer still widely used, Britain has not developed any artillery weapons in recent years, except for a 105-mm light gun, the ◊ *Abbot* 105-mm self-propelled gun and an undistinguished 81-mm mortar (this in collaboration with the Canadian forces). At present the British army uses the Italian 105-mm howitzer (◊ *Artillery, Italian*) and the M.44 U S built 155-mm self-propelled howitzer.

Artillery, French. Unlike Britain and West Germany which rely largely on U S or cooperatively produced equipment, the French army deploys a full range of domestically produced equipment. Much of it is moderately efficient, but it is plagued by high costs and maintenance problems.

Field artillery:
O B 155 M.1950 A large and very complicated towed howitzer of 155-mm calibre weighing 9 tons. It fires a 95-lb shell up to 19,400 yds and has a multi-axis wheeled chassis.

Self-propelled artillery:
O B 155 50/63 (AMX 155) A 155-mm howitzer mounted on the stripped-down chassis of the ◊ *AMX 13* light tank. There are two variants, one with the characteristics of the OB 155 M.1950 above and another with a longer barrel which improves the range to 23,600 yds. The arrangements weigh about 19 tons, and there are large frame spades to absorb the recoil. The 12-man crew is carried in an ◊ *AMX VTT* transporter, except for 2 driving crew in the vehicle.
O B 105, AMX 105 A and B Two variants of the AMX 13 light tank. Both mount a 105-mm howitzer, the first in a casemate and the second in a rotating turret. The A model has a 4-man crew and mounts a shielded weapon with a maximum range of 15,500 yds. The B model is less of an improvised arrangement and has a 5-man crew and a more modern weapon with a maximum range

of 16,500 yds. The B has not yet been deployed.

Anti-aircraft:
AMX 30 mm D C A Another conversion of the AMX 13 light tank. This mounts twin radar-controlled 30-mm guns with a combined rate of fire of 1,200 rpm. There is a 5-man crew.

Artillery, Italian. The Italian army is equipped largely with U S-made weapons, both modern and antiquated; but one outstanding Italian artillery weapon is widely used, and it has also been supplied to the British and West German armies:

Obice da 105/14 (105-mm Italian pack howitzer). This is an ultra-light and exceedingly ingenious design based on a classic shielded two-wheel chassis. The large muzzle brakes allow the use of high-velocity ammunition, and the howitzer can be converted into an anti-tank gun by lowering the assembly. The weapon can be dismantled into 12 loads for (relatively) easy manhandling. It weighs only 2,925 lb (as opposed to 4,400–5,500 lb for a standard 105-mm) and fires standard 105-mm ammunition (shells weighing about 33 lb, which in this weapon have a maximum range of 11,200 yds).

Rocket-launcher:
1.A.100.R An experimental 24-barrel rocket-launcher mounted on an Italian jeep-like vehicle.

Artillery, Soviet. The artillery is, by tradition, a very important branch of the Soviet armed forces. Its equipment is modern and diversified, and independent artillery units still exist at brigade level. But Soviet artillery is still largely mounted on unprotected towing chassis, while Western artillery is increasingly mounted in armoured and tracked vehicles. In view of the otherwise extensive mechanization of the Soviet army, the old-fashioned configuration of much of the artillery is probably the result of a deliberate choice, rather than of an economic necessity. It appears in fact to reflect Soviet belief in the value of large-scale 'barrage' non-nuclear fire; when a great deal of ammunition is needed, it must necessarily be carried in 'soft' vehicles, and there is obviously no point in providing a given level of protection to only part of a team which must stay together.

Anti-aircraft artillery includes:
ZPU (Zenito Pulemetnaja Ustanovka): Heavy machine-guns mounted on simple wheeled chassis for towing. These come in three main versions: twin 14·5 mm, quadruple 14·5 mm, and the more modern twin 23 mm. All of these have manual control and box magazines. Variants with the firing platform only (and

fire of 1,000 rounds per minute; each 57-mm gun has an even more theoretical rate of fire of 120 rounds per minute. Both have excellent mobility, being mounted on vehicles derived from the ◊ *P T-76* (the ZSU-23-4) and the ◊ *T 54* tank chassis.

Anti-tank artillery includes:
Three basic anti-tank guns mounted on two-wheel chassis with shields in the 57-mm (M-43) the 85-mm (M-45) and 100-mm (M-55) calibres. Their specifications are in the table.

Soviet Anti-tank Guns – Specifications:

	Weight (lb)	Rate of fire (rpm)	mm of armour at 60° penetrated
M–43	2,500	25	100 at 550 yds
M–45	3,750	20	113 at 550 yds
M–55	5,940	10	110 at 1,100 yds

no wheels) are mounted on the ◊ *BTR-152* and other armoured vehicles. A larger-calibre M-38/M-39 single 37 mm has a similar configuration with a four-wheel chassis which doubles as a static platform; this too has manual controls and although obsolete in the USSR is still operational with allied armies. Other towed larger-calibre guns include the M-50 57 mm, the M-44 85 mm, the M-49 100 mm, and the M-55 130 mm. These are all adapted for radar control and have a dual anti-aircraft, anti-tank capability.

But the main anti-aircraft weapons

There is a newer Czech 85-mm anti-tank gun designated M-52 which penetrates 123 mm of armour – at 90 degrees – over a range of 1,100 yds. And the M-55 motorized anti-tank gun, which is basically a 57-mm gun mounted on an open four-wheel chassis. (This gives no armour protection and has only short-range mobility.)

Field artillery includes:
Two conventional field howitzers, the 122-mm M-38 and the 152-mm M-43, and the M-55 152-mm gun-howitzer; their specifications are shown below.

Soviet Field Artillery – Specifications:

	Weight (lb)	Range (yds)	Weight of shell (lb)
M–38	5,400	12,980	47·9
M–43	7,900	13,690	88
M–55	13,000	2,780	105 (and this can also penetrate 101 mm of armour at 1,100 yds)

deployed with first-line units are the ZSU (Zenitnaja Samochodnaja Ustanovka) self-propelled and armoured vehicles. The ZSU-23-4 mounts four 23-mm cannon and the ◊ *SU-57-2* mounts twin 57-mm guns. Each 23-mm gun has a cyclic rate of

The M-55 122-mm gun, which weighs 11,500 lb and fires a 56-lb shell over a range of 23,000 yds, or penetrates 129 mm of armour at 1,100 yds; some units have the older M-37 122-mm gun, whose capabilities are similar but which weighs 4,800 lb more. Now an even more modern 122-mm gun, the D-30, is being introduced. But the outstanding Soviet field artillery piece is the ultra-modern M-54 130-mm gun. This can fire six 80-lb rounds per minute over a range of 29,700 yds, or, when used in an anti-tank role, can penetrate 250 mm of armour at 1,100 yds. The heaviest gun deployed with the Soviet artillery is the M-55 203-mm gun, which weighs more than 22 tons and fires a 340-lb shell over a range of 31,900 yds. Since the introduction of the M-54 this gun has become less important in the Soviet artillery.

Rocket artillery:
Apart from the nuclear warhead rockets of the ◊ *Frog* series, which are the Soviet counterparts of Western weapons such as the ◊ *Honest John*, the Soviet army also uses extensively small-calibre rockets fired from multi-barrel launchers. These, developed from the rockets used with the wartime 'Stalin Organs', come in several different configurations mounted on trucks or trailers. The unguided spin-stabilized rockets are 'aimed' by pointing the whole multi-barrel assembly in the general direction of the target. Although inaccurate, they are effective for area fire and have a devastating psychological effect. The main rockets are: BM-14, a 14-cm calibre rocket weighing 90 lb with a range of 9,900 yds, which is used with 8-, 16- and 17-barrel launchers mounted on trucks or jeep-sized trailers; the BM-

BMD-25, which weighs 990 lb and has a range of 22,000 yds.
◊ *Artillery* for general subject entry.
◊ *ASU* for Soviet *Tank Destroyers.*

Artillery, US. Unlike Soviet artillery, 32 which is classic in form, i.e. towed, US artillery is largely self-propelled. In fact, apart from the lighter weapons such as mortars and 105-mm howitzers, all US artillery is to be self-propelled, though some towed 155-mm howitzers will probably remain in service. The US army no longer deploys anti-tank guns, i.e. low-slung high-velocity guns fitted with shields, and anti-tank tasks are to be performed by other tanks or by light weapons such as the ◊ *TOW* missile, ◊ *recoilless weapons* or ◊ *bazookas.* Many of the weapons entered below are in service with ◊ *NATO,* ◊ *CENTO* and ◊ *SEATO* armies as well as with the US army and Marine Corps.

Mortars:
The US army relies less on mortars than most other modern armies. Presumably the intention is to use 105-mm howitzers in this role, offsetting their greater weight by air transport right into the battlefield. This could turn out to be a costly policy. Again unlike other armies, US forces do not deploy a 120-mm mortar, and the 107-mm weapon used instead is not very effective. The 81-mm mortars in service are much like those found in almost all armies.

US Mortars – Specifications:

Model	Calibre	Weight (lb)	Weight of bomb (lb)	Range (yds)	Rate of fire (skilled crew)
M.29	81 mm	107	7·3	4,250	25 bombs per minute
M.1	81 mm	132	7·3	3,300	30 bombs per minute
M.30	107 mm	638	26·2	6,000	25 bombs per minute

24, a rocket which weighs 250 lb, has a 7,700-yd range, and is used with 12-barrel launchers mounted on trucks or tractors; the new 11·5-cm calibre rocket, which weighs 132 lb has a 16,500-yd range, and is mounted on a 40-barrel launcher. The largest of the group is the

Anti-aircraft:
The M.55, a quadruple 0·50 machine-gun arrangement mounted on a light towable chassis, is obsolete but still widely used outside the US. Similarly obsolete are the radar-assisted M.51 75-mm AA gun and the M2A2 90-mm gun. The for-

mer is inferior to the widely used Bofors 40-mm, while the latter is an old-style AA gun unsuitable for radar fire control. Both the M.51 and the M2A2 are still used outside the US. The M.42A1 is a twin 40-mm AA weapon mounted on a tracked and lightly armoured vehicle with an open turret. The two guns have a combined rate of fire of 240 rpm, and 400 rounds of ammunition are carried. This weapon is inferior to the Soviet ⚹ SU-57-2, though both are obsolescent since they lack radar fire control.

Current AA weapons are the ⚹ Chaparral missile-launcher and the M.163, which consists of an ⚹ M.113 personnel carrier modified to mount an M.61 multi-barrel 20-mm cannon. This fires at a variable rate of 1,000–3,000 rounds per minute and has radar ranging and optical fire control. The vehicle has a two-man crew and can carry only 1,800 rounds of ammunition. This weapon is likely to be effective against slow aircraft, especially helicopters, while the Chaparral, with which the M.163 is to serve in the divisional AA battalions, is intended to take care of faster air targets.

Field artillery:

The M.101 is the standard 105 howitzer used for front-line fire power by the US army and Marine Corps. Though rather

otherwise it would not hold the recoil. The M.114 is the standard 155-mm howitzer now being replaced by self-propelled models, at least in part. The M.59 155-mm gun, better known under its popular name of 'Long Tom', is no longer in service with US forces, but as a cheap piece of long-range field artillery it is still widely used outside the US, as is its companion piece, the M.115 8-in (203-mm) howitzer. Both are mounted on rather large multi-wheel towable chassis, but neither is very mobile. Both have long since been replaced by self-propelled artillery in US service.

Self-propelled artillery:

The standard equipment now in US service is based on two chassis arrangements. A fully enclosed light armoured vehicle is used for the 105- and 155-mm howitzers, while a stripped chassis mounting the weapon and little else is used for the 203-mm howitzer and the 175-mm gun. The M.108 and M.109 are essentially the same vehicles fitted with the 105 and 155-mm howitzer respectively. Both are built out of light alloys, and in both the weapon is fitted into an enclosed turret. The M.107 and M.110 are mounted on another rather good chassis. There is no turret or fixed casemate, and the weapon is fitted on a turntable. The gun crews and

US Field Artillery – Specifications:

Model	Calibre	Weight	Weight of shell (lb)	Range (yds)	Service status
M.101	105 mm	4,466 lb	33	12,300	standard light howitzer
M.102	105 mm	3,000 lb	30/33	12,300	modern replacement of M.101
M.114	155 mm	12,670 lb	95	16,500	standard medium howitzer
M.59	155 mm	14 tons	96	25,800	outside the US only
M.115	203 mm	13·5 tons	200	18,000	outside the US only

old, it is still very effective. It is now being replaced by the ultra-modern M.102, which is much lighter and which has a clever cylinder recoil grip instead of the classic spades. This arrangement allows fast swivelling, since the cylinder moves freely sideways though not straight back –

ammunition are carried in separate vehicles.

There is also an older set of vehicles now phased out from US service but still widely used outside the US. The lightest is the M.52 105-mm howitzer mounted on a then standard tracked chassis with a fully enclosed turret. The M.44 155-mm howitzer is a cruder arrangement with the weapon mounted in a fixed casemate. Another pair of weapons mounted on a standardized chassis are the M.53 155-mm gun and the M.55 203-mm howitzer, now being replaced by the M.107 175-mm gun and the M.110 203-mm howitzer. All four have atomic shells available.

added shield and a longer barrel, and other modifications have been made. It weighs 5,750 lb and fires a 33-lb shell at ranges up to 15,500 yds.

90-mm anti-tank This is a motorized 4-wheel chassis mounting a shielded 90-mm anti-tank gun. In essence it is a hybrid between the towed and the self-propelled, which aims at the economy

US Self-propelled Artillery – Specifications:

Model	Calibre	Gun (G) or Howitzer (H)	Vehicle Total weight (tons)	Road range (miles)	Ammunition capacity	Shell weight (lb)	Fire range (yds)	Service status US	Elsewhere
M.108	105	H	22	230	86	—	—	trials	Belgium
M.109	155	H	25	230	28	95	16,000	in service	NATO
M.107	175	G	31	400	none	132	36,000	in service	NATO
M.110	203	H	29	400	none	198	18,600	in service	NATO
M.52	105	H	27	100	102	33	12,500	obsolescent	NATO
M.44	155	H	31	100	24	95	16,500	obsolescent	NATO
M.53	155	G	54	170	10	96	26,000	obsolescent	NATO
M.55	203	H	54	170	10	200	18,000	obsolescent	NATO

Rockets:
The US army and several NATO members deploy the ◊ *Honest John* and the ◊ *Little John* unguided nuclear-warhead rockets, intended for the bombardment of rear areas. In addition to these, the ◊ *Lance,* ◊ *Pershing* and ◊ *Sergeant* missiles are also available. While the Lance is intended to replace the un-guided rockets of the John series for tactical use, the Pershing and the Sergeant are far more than substitutes for long-range artillery and could clearly have a strategic capability.

Artillery, West German. The West German artillery is now in the process of being re-equipped with domestically pro-duced weapons, replacing the older – and not so old – US weapons with which it was hurriedly equipped following the re-armament decision.

Field artillery:
IFH 105-mm L This is a German re-make of the M-101 US 105-mm howitzer (◊ *Artillery, US*). There is an

of the former with at least some of the mobility of the latter. The big difference of course is the lack of an enclosed and armoured turret or casemate and the limited traverse and elevation of the gun. ◊ *Jagdpanzer.*

Asdic. ◊ *Sonar.*

'Ash'. Soviet air-to-air missile. A large Soviet air-to-air missile with cruciform wings and tail surfaces. Radar homing and possibly also infra-red homing. Ash equips the new 'Fiddler' heavyweight fighter. It is about 17 ft long.

ASM. Air-to-surface missile. ◊ *Missile.*

Asroc RUR-5A. Anti-submarine missile. Asroc is an anti-submarine weapon operational since 1961 and deployed on many US navy vessels; it is also available to some US allies. It consists of a system of sonar detectors, a fire-control computer (which tracks the target and aims the missile), a launcher device for 8 missiles, plus the missile itself. The missile is a ballistic solid-fuel rocket carrying a war-head which is either an acoustic homing torpedo or a nuclear depth charge. After a sonar detection, the computer tracks the target, and the missile-launcher turns to-wards the target; a choice of warhead is then made. After launching, the missile follows a ballistic trajectory, sheds the rocket motor and then the airframe. When the payload is a homing torpedo, a parachute opens to slow the seaward plunge; when a nuclear depth charge is used, it falls to and sinks into the water to explode at a set depth.

Specifications: Length: 15 ft 0 in. Diameter: 1 ft 0 in. Span of fins: 2 ft 6 in. Weight: 1,000 lb. Range: About 6 miles maximum and 1 mile minimum.

◊ *Subroc* and *Ikara* for other anti-submarine missiles.

Assured Destruction. A concept as-sociated with the strategic policies of US Secretary McNamara and usually defined as 'the capability of deterring deliberate nuclear attack upon the United States and its allies, by maintaining, continuously, a highly reliable ability to inflict an un-

Union) has been published: 'about one fifth to one third of the population and one half to two-thirds of the industrial capacity'. This requirement, translated in terms of nuclear warheads delivered on target, means that about 400 megaton-equivalents are required to achieve the minimum population destruction figure (much less is required to satisfy the in-dustry constraint, which is therefore ignored). This does not mean that 400 delivery vehicles with one-megaton war-heads are required, since on the one hand a megaton's worth of damage can be done by two or more smaller warheads adding up to a sub-megaton yield (◊ *MIRV*), while on the other hand the number of delivery vehicles has to take into account maintenance, reliability and the effects of enemy activity. These 'degradation fac-tors' have been spelled out (in a highly simplified form) by semi-official sources, so that the number of missiles which the US needs in order to maintain its 'assured destruction' capability can be derived by dividing the 400 one-megaton-equivalents by the total estimated 'degra-dation factors':

Degradation factors, stated as fractions	Estimated Values	
	Nominal	*Conservative*
Missiles available on station	0·9	0·8
Missiles ready to fire	0·9	0·8
Missiles which launch correctly	0·9	0·8
Missiles which fly correctly	0·9	0·8
Total fraction surviving all engineering hazards	*0·66*	*0·41*
Fraction surviving a Soviet first strike, assuming the latter destroys 0·75/0·67 of all missiles, including faulty ones	*0·49*	*0·27*
Fraction which penetrates Soviet ◊*ABM* defences (of 1972)	*0·37*	*0·18*

acceptable degree of damage upon any single aggressor, or combination of ag-gressors, at any time during the course of a strategic nuclear exchange, even after absorbing a surprise "first strike" '. Per-haps surprisingly, a quantitative measure of the damage which is thought to be un-acceptable to the enemy (the Soviet

In other words, out of 1,000 missiles de-ployed about 370 would reach their in-tended target according to the nominal estimate and only 180 according to the conservative estimate. Since 400 megaton equivalents are held to be required for deterring the Soviet Union, the force to be deployed works out to something between 1,100 and 2,200 megatons, according to these calculations.

This way of looking at American strategic deployment policies provided a logical framework for limiting weapon requirements to a given number of offensive weapons. It is implicitly an offence-only formula since the effects of an ABM, in terms of casualty reduction, are not taken into account, while an ABM to protect the deterrent is not envisaged. In fact, the advocates of assured destruction campaigned vigorously against ABM development in general, since it seemed to them that the 'arms race' between the two super-powers could be stabilized when *both* sides had an assured destruction capability. They therefore all but encouraged the Soviets to catch up with the US in offensive missile deployments while they deplored the Soviet deployment of an ABM. The stability of a system where both sides have an assured destruction capability, which fascinated the planners after years of policy without logical guidelines, in fact rested on a transient combination of technological conditions (◊ *MIRV*). The new rules of the game – accurate warheads which have high probabilities of killing hard silos – make an offence-only strategy obsolete, since only an ABM can absorb a substantial enemy strike potential without leading to an arms race. It is important to remember that the policy of assured destruction did lead to a stabilization of US deployments: the total US strategic budget declined from 13·3 billion dollars in 1959 to 8·3 billion in 1970. If there has been an 'arms race', the running has all been on one side.

◊ *US, Strategic Offensive Forces*.

ASU. Soviet designation for 'airborne' tank-destroyers. ASU (Aviadesantnaja Samochodnaja Ustanovka) means 'air-transported self-propelled artillery', but both ASU vehicles are in fact tank destroyers, specifically intended for fighting other armoured vehicles.

The ASU-57 is a small tracked vehicle with an open box-like fighting compartment, from the frontal plate of which emerges the long 57-mm gun (M-55 Pak, ◊ *Artillery, Soviet*). The ASU-57 has been operational since 1957 with the Soviet armed forces, and it has also been

supplied to the Polish army. Apart from the gun, the vehicle mounts a machine-gun which is also fired through the frontal plate. The ASU weighs about 6 tons and has a 3-man crew and frontal armour 12-mm thick. The road speed is thought to exceed 35 mph.

The ASU-85 is an altogether more elaborate vehicle, with all-round armour protection, the very mobile chassis of the ◊ *PT-76* light tank, and a long 85-mm gun. The gun is still mounted in the frontal plate, but a more effective sponson allows a wider side-to-side coverage. The roof is fully enclosed and the frontal plate has an armour thickness estimated at 40-mm. The 3-man crew sits alongside (driver) and behind the gun. With its 18-ton weight, the ASU-85 is probably intended as a 'follow-up' weapon to be landed from transport aircraft rather than as a paradropped assault vehicle. Unlike the PT-76 on whose chassis it is based, the ASU-85 is not amphibious.

◊ *Tank Destroyer* for the general subject entry. ◊ the US *Sheridan M.551* for its direct competitor.

ASW. ◊ *Anti-submarine Warfare*.

Atlantic Nuclear Force (ANF). An abortive British proposal for a combined mixed-manned and national nuclear force which would have extended the ◊ *MLF* plan to include a member (the UK) which did not want to invest in that sea-borne force, but whose government did want to create a NATO multilateral deterrent force. The proposed line-up of nuclear delivery systems would have added the British ◊ *Polaris* and V-bomber force to the fleet of surface ships equipped with Polaris missiles envisaged under the MLF scheme.

'Atoll'. Soviet air-to-air missile. A   **49** first-generation air-to-air missile similar to the US ◊ *Sidewinder*. It is 9 ft 2 in long, and less than 5 in in diameter. Guidance is by infra-red homing. This weapon is effective only in fine weather and if fired dead astern, in a pursuit course. It equips the ◊ *Fishbed Mig-21* fighters.

Attack Aircraft. ◊ *Fighter*.

Autolycus. Submarine detection device. This device, known in the USA as the 'sniffer', is intended to discover the recent presence of a submarine by detecting the diesel fumes it leaves behind. Its effective range is very short, and it is of course useless against nuclear-powered submarines. ◊ *Anti-submarine Warfare.*

Avionics. A general term for the electronic and other devices which perform navigation, target detection, weapon-control and communication functions in aircraft and missiles. The quantity and sophistication of the avionics payload is a major determinant of aircraft effectiveness, often more important than speed performance, and it is the type and quantity of the avionics which determine the 'mission-orientation' of ◊ *fighter* aircraft. 'Multi-mission' aircraft such as the US ◊ *Phantom F-4* and the Swedish ◊ *Draken* and *Viggen* achieve this status because of their large avionics payload, or rather because of their large and flexible fuel/weapons/avionics payloads.

The main constituents of avionics are:

1. Radio equipment, for communications and navigation. All aircraft have some, but the range, choice of frequencies and sophistication of the equipment varies considerably.

2. Radar. Fighters with air-to-air radar detection, tracking and ranging equipment are usually described as 'all-weather'; other radar equipment for ground-speed, altitude, and aircraft recognition (IFF) is used on many military – and civilian – aircraft. These and specifically military systems are described under ◊ *radar.*

3. Inertial platforms. A sophisticated navigation aid or independent guidance unit (as on modern ballistic missiles) as yet deployed only on the more advanced aircraft. ◊ *Inertial Guidance.*

4. Electronic Countermeasures (◊ *ECM*). ECM and ECCM (counter-ECM) equipment is increasingly important on all aircraft systems, but particularly so on bombers; aircraft with a primarily ECM payload have been developed (see EA-6A/B under *Intruder*).

5. Computers. These are part of most

of the systems described above; and the overall development of avionics is largely a function of the progressive increase in the output-to-weight ratios of digital computers.

Other avionics systems include television, optical and sight-stabilizer applications.

AWACS. Airborne Warning and Control System. A projected US ◊ *radar* and control system intended to replace the present ground networks within the US–Canadian NORAD air-defence organization. AWACS, which reached contract definition in late 1969, would consist of a force of converted transport aircraft equipped with long-range surveillance and tracking radars, including 'downward-looking' tracking radar that can identify intruders flying at very low altitudes. (But the technology of separating contour-flying aircraft from ground 'clutter' still presents problems.) The AWACS is also intended as a decision-making control system for active weapons, the ◊ *Delta Dart F-106X* interceptor and surface-to-air missiles; its main attribute in this role is its low vulnerability to ballistic missile attack.

The complete system would also include ground-based OTH long-range radar for distant early warning as well as the F-106X and the AWACS aircraft. As a bomber attack is detected by the OTH or surveillance radars of the AWACS, these would search for and acquire the 'bandits'; simultaneously, the communications and control unit in the aircraft would alert the fighter and missile units and prepare them for action, and when the 'bandits' come within range, the AWACS would direct the defensive effort.

◊ *NORAD* for the present ground-based system.

'Awl'. Soviet air-to-air missile. A large Soviet air-to-air missile. About 16 ft long with radar homing guidance. It may equip the ◊ *Fiddler* first-line interceptors of the Soviet Air Defence Command, ◊ *PVO Strany.*

B

B. USAF designation for strategic bombers.
◊ *Stratofortress* for B-52.
◊ *Hustler* for B-58.

'Badger' Tu-16. Soviet medium bomber. The twin-jet Badger of which more than 2,000 have been built, has been the basic Soviet medium bomber. About 750 remain in service, but many have been converted to reconnaissance and naval roles, supplied to foreign countries, or scrapped.

Badger A, supplied also to Iraq and Egypt, is the basic bomber with a glazed nose, air-to-ground scanner radar, and a maximum bomb load of about 20,000 lb.

Badger B, supplied also to the UAR and Indonesia, is a naval version which has been adapted to carry two ◊ *Kennel* anti-shipping missiles under the wings.

Badger C is a missile-carrier for the ◊ *Kipper* and *Kelt* stand-off missiles designed to deliver nuclear warheads while the carrier aircraft stays out of air-defence range. This Badger C has a larger nose radar and other electronics for missile control.

All the Badgers have a crew of seven plus and carry three sets of twin 23-mm cannon turrets in ventral, dorsal and tail positions. The tail set has automatic radar ranging. A seventh cannon is fixed firing forward from the nose. This armament, and the Mach 0·87 top speed, are characteristic of a bomber which has been operational since 1955, and its effectiveness against modern air defences is low; but it is still a formidable weapon in certain environments – and versions replete with electronics are used in long-range naval reconnaissance over US and British naval forces. (The Badger has been succeeded by the Tu-22 ◊ *Blinder*.)

Specifications: Twin-jet, swept-wing' multi-seat medium bomber. Length: 120 ft. Height: 35 ft 6 in. Wing span: 110 ft. Gross wing area: 1,820 sq ft. Normal take-off weight: 150,000 lb. Maximum level speed in optimal conditions (35,000 ft height) loaded: 587 mph. Range at 480 mph cruising speed and 6,600 lb of bombs: 3,975 miles.
◊ *Bomber* for general subject entry.

Balance of Terror. An equilibrium between powers based only on each side's possession of particular weapons which allow each party to cause unacceptable damage to the other(s). This is as opposed to the classical concept of the 'balance of power', where all sources of strength and weakness are assumed to produce an equilibrium which does not rely on particular items of equipment, and is thus stable, and operative at all levels of the relationship. The 'balance of terror' is seen as unstable, because the introduction of new weapon systems - or the expectation on the part of others that this is or will be taking place - can shatter the equilibrium. The 'balance of terror' is usually thought of as being effective only in extreme situations, as when vital interests are challenged. The strategy of ◊ *massive retaliation* aimed at equating the concepts of 'balance of power' and 'balance of terror', since it sought to achieve a global *status quo* on the basis of nuclear delivery systems only – partly in order to save the costs of maintaining a global military presence and large non-nuclear forces. The term 'balance of terror' may have originated with J. Robert Oppenheimer, who described the USA–USSR relationship as that between two scorpions in a bottle, each able and unwilling to sting the other fatally. Albert Wohlstetter has pointed out that if both the super-powers

were ever able to destroy each other completely 'it would be extraordinarily risky for one side *not* to destroy the other, or to delay doing so . . . since this is the sole way it can reasonably emerge at all [from the conflict].'
◊ *Assured Destruction, Deterrence.*

Ballistic Missiles (includes ICBM, IRBM, MRBM, SRBM). Ballistic missiles are unmanned rocket-propelled vehicles which fly to their target in a high parabolic trajectory which in intercontinental ballistic missiles (ICBMs) lies mostly outside the atmosphere. Ballistic missiles are guided only during the ascending phase of the trajectory with the altitude, bearing and velocity setting the missile in a predictable course up to the apogee and then in a descending curve to target. Ballistic missiles consist of a booster stage, one or more 'sustainer' propulsion

well as swivelling nozzles, but the latest form of control affects the direction of the rocket thrust by means of fluid injection – this is the method used on the ◊ *Polaris A3*, ◊ *Minuteman 3* and other more advanced ICBMs. ◊ *Missile* for propulsion methods.

Ballistic missiles are conventionally classified into the range and launching method categories shown in the tables below.

ICBMs, IRBMs and MRBMs are generally 'strategic' weapons intended against homeland targets. SRBMs and unguided rockets such as the US Honest John and the Soviet Frog series are regarded as tactical since they are intended against battlefield concentrations and fixed-point facilities in the rear of the combat zone.

The conventional classification in terms of launching methods is given in the table overleaf.

			Range spectrum *(nautical miles)*
ICBM	Intercontinental ballistic missile		at least 5,000
IRBM	Intermediate-range ballistic missile		at least 1,500
MRBM	Medium-range ballistic missile		at least 500
SRBM	Short-range ballistic missile		up to 500

stages, and a re-entry vehicle which in long range missiles consists of a heat shield (to protect the missile during re-entry into the atmosphere) and the nuclear warhead as well as ◊ *penetration aids.* ◊ *MIRV* systems, where two or more warheads are separately targetted, may include a manoeuvring 'dispenser' stage.

The first operational ballistic missile, the wartime German V-2, was propelled by a liquid-fuel rocket motor and guided by gyroscopic stabilizers which actuated graphite control surfaces through a servo-system. Early ICBMs, like the US Atlas, were set on their ballistic course by radio command, but modern ICBMs use pure ◊ *inertial guidance,* which is immune to electronic countermeasures (◊ *ECM*) as well as lightweight, accurate and reliable. Ballistic missiles use control surfaces as

Ballistic missiles are usually programmed to follow a minimum-energy path which is fixed and predictable between any two points; but at some sacrifice in payload a depressed or elevated trajectory can be obtained instead. This is available as an ICBM penetration tactic intended to defeat ◊ *ABM* (anti-ballistic missile) defences such as the American ◊ *Safeguard.*

◊ *Nuclear Warhead* for the usual payload (though some SRBMs have chemical warheads as an alternative; ◊ *Chemical Warfare*), as well as *Penetration Aids.*

Ballistic Missile Defence (System). ◊ *ABM.*

Ballistic   Missile   Early-warning System. ◊ *BMEWS.*

BAR. Browning Automatic Rifle. One of the first light machine-guns, first produced in the US in 1918, the BAR is still used by the armies of many countries, although it is no longer issued by the US army and Marine Corps.

There are several variants of the BAR produced in different calibres, but the most common is the US M.1918 A2 Cal. 30 with a simple bipod, a 20-round

1,000 men. Soviet, Warsaw Pact and Chinese battalions tend towards the lower figure.

'Bazooka' M.20 BI/M.72. Anti-tank rocket-launchers. 'Bazooka' is the popular name for anti-tank rocket-launchers, originally applied to the war-time US 2·36-in rocket-launcher, but now used for all weapons of this type. These

Ballistic Missiles: Classification by method of launch

	US designation	
Silo-launched	LGM	Where a silo is a ◊*hardened* site; see the US *Minuteman*.
Under-water-launched (But SLBM is often used for UGMs as well)	UGM	Where the platform is a (submerged) submarine, as in the case of the US ◊*Polaris* or the Soviet ◊*Serb*.
Ship (or submarine) launched (But often used for all submarine-launched missiles)	SLBM	Remedial weapons deployed on board surface ships (none operational) or submarines, but fired only on the surface, such as the Soviet ◊*Sark*.
Mobile-launched	MGM	There are no mobile ICBMs (but see the Soviet ◊*Scrooge*), but the improvement in ICBM accuracies, the development of MIRV payloads and other factors may lead to the deployment of MGM strategic missiles as a form of protection.

magazine, a length of 47·8 in, and a weight of 19·4 lb. The cyclic rate of fire is variable between 300 and 650 rounds per minute. Because of the weight and optionally slow rate of fire, it is very accurate – more so than the current equivalents.

◊ *Light Machine-gun* for general entry and competitive types.

Battalion. An army formation formally subordinate to a ◊ *regiment* or ◊ *brigade,* or in some cases directly to the ◊ *division,* and comprising a number of ◊ *companies*. In the US army the ROAD division comprises about 10 battalions as well as ancillary and support units attached to the divisional level. A battalion may comprise anything between 300 and

are lightweight tube-shaped launchers for rockets fitted with ◊ *HEAT* (shaped or hollow charge) anti-tank warheads. In general these are cheap short-range weapons with low accuracies. This and the 20–30-yd-long gas blast, which is highly visible and dangerous to friendly troops, make 'Bazookas' little more than desperation weapons against tanks. ◊ *Recoilless weapons* and anti-tank ◊ *missiles* are replacing rocket-launchers, though not in the lowest echelons.

The US M.20 BI has a 3·5-in calibre, weighs about 12 lb, and launches a 8·9-lb fin-stabilized rocket. The HEAT warhead weighs 1·8 lb, and the weapon is effective to about 150 yds, though shorter ranges are usually achieved. The M.20 has been extensively copied and purchased and is widely used outside the US. The US army uses a newer 3-in (66-mm) rocket-launcher, designated M 72 (LAW), which weighs only 4½ lb with one rocket. The disposable launcher-rocket package is only 25 in long but extends to 34 in for firing.

52 **'Bear' Tu-20 (TU-95). Soviet long-range bomber / reconnaissance aircraft.** A very large, swept-wing strategic bomber of 1956 operational vintage, the Tu-20 has appeared in three main versions:

Bear A. The basic bomber with air-to-ground radar and gun turrets in ventral, dorsal and tail positions.

Bear B. A developed bomber version with in-flight refuelling, new radar and avionics as well as provisions for the Kangaroo stand-off missile. This version is credited with a 7,800-mile maximum range (with a 25,000-lb payload) and a top speed of 500 mph at optimum height (41,000 ft). In a maximum payload configuration it can carry up to 40,000 lb over unspecified ranges.

Bear C. A naval reconnaissance version which is reported to have extra avionics and cameras.

All Bears are powered by four turbo-props with an estimated rating of 14,795 shp each and have a maximum take-off weight of about 360,000 lb.

About 90 Bear Bs are still serving in the ◊ *USSR strategic offensive forces*, but their penetration capabilities against sophisticated air defence are not thought to be very high.

◊ *Bomber* for general subject entry.

Beretta. An Italian concern which manufactures, amongst other things, distinctive small arms, some of which are widely used outside Italy; like ◊ *Bofors*, ◊ *FN*, ◊ *Hispano-Suiza*, and the Oerlikon/Contraves concerns, it is a supplier of opportunity to the (legal) arms market.

The Beretta M.1951 9-mm pistol is used in several smaller armies. It is a heavy automatic firing the standard Parabellum cartridge, and it has a distinctive and very well designed slanted grip.

The M.38/49 is a sub-machine-gun widely used in all Italian military and

security forces. The wooden stock and the two-trigger arrangement (one for semi- and one for full-automatic) are unusual features; the rest is standard, with a rate of fire of 550 rpm, a Parabellum 9-mm cartridge, and a stated useful range of 110 yds.

◊ *Sub-machine-gun* for general subject entry.

Biological Warfare (BW). The use of disease-producing micro-organisms as weapons or weapon warheads, as forbidden by the 1925 ◊ *Geneva Convention* (reaffirmed by a UN General Assembly motion of 1966 passed with 91 votes in favour and 4 abstentions). In anticipation of possible use, the large BW establishments of the USSR and USA – as well as those of many smaller powers – are (a) developing suitable strains of disease-causing organisms and (b) preparing suitable distribution systems. Bacteria, viri and rickettsia produce about 160 known infectious diseases, but each exists in very many different strains, each of which is a separate agent from the BW point of view. Thus, for example, *Pasteurella pestis*, the plague bacterium, has more than 140 strains. Good military micro-organisms are selected from the very large number of strains on the basis of six main criteria:

(a) virulence, i.e. the damage produced by the infection. This must be severe, though not necessarily permanent,

(b) infectivity, i.e. the size of dose required to start a continuing infection. This must be low to ensure ease of distribution;

(c) stability, i.e. the survival capability of the strain on its journey to the 'host';

(d) the degree of natural immunity, which must be low to ensure general impact on the target population;

(e) the availability of vaccines. Vaccines must be available, for otherwise there is a backlash risk, but they should not be commonly available to the enemy;

(f) the ease or otherwise of therapy. Clearly the strain must resist commonly available cures like antibiotics. Epidemicity, i.e. the ease of inter-host communication, must be low, since the idea is to neutralize a given population rather than to start a world-wide pandemic.

Since criteria (a), (b) and (c) can be changed in the laboratory – by producing 'artificial' strains – a great deal of BW research is conducted in this field.

The basic attractions of BW are that the agents are relatively cheap and the weight of an effective warhead is low. BW is thus especially attractive to poor countries, but BW agents still need delivery systems, and these make up a large proportion of modern weapon costs.

Lethal agents with BW potential include: *Poxvirus variolae* (smallpox); *Pasteurella pestis* (plague); *Whitmorella pseudomallei* (melioidosis); *Malleomyces mallei* (glanders); *Bacillus anthracis* (anthrax). And the most lethal of all, *Botulinum*, which is the product of an organism but strictly speaking is not biological at all, since it is a poison and not an organism.

Where no animal intermediary, usually a rodent or biting insect, is required, the most efficient medium of distribution for BW agents is aerosol, where the organisms are in a pressurized liquid released through nozzles into the atmosphere and enter the host by inhalation. The aerosol package can be built into artillery shells or missile warheads or – with certain precautions – be used in aerial bombing. There is a dilemma between accuracy of aim and infectivity, since really small particles (1–5 microns), which are very infective – since they penetrate the lung wall – descend so slowly that winds may carry them right off the target area. Sabotage action, such as putting BW agents in reservoirs and ventilation systems can be very effective on a small scale, but public water facilities are regularly checked, and the agent would be detected. The Pentagon has its own guarded water-purification facility.

From the military point of view, a lethal agent has no significant advantage over an incapacitating agent which temporarily neutralizes a population. Incapacitating agents with military potential include: *Pasteurella tularensis* (Tularaemia); *Brucella melitensis* (Brucellosis); Dengue virus family (breakbone fever); *Coxiella burnetii* (Q fever). This last has the distinction of being very infective, since a single albeit optimally distributed ounce is sufficient to infect 28 billion people.

For other ABC/CBR weapons ◊ *Chemical Warfare*.

'Bison' Mya-4. Soviet long-range **53** **bomber.** The Myasishchev Bison is a four-jet swept-wing long-range bomber which, together with the ◊ *Bear Tu-20*, is operational in the USSR strategic offensive forces. The maximum range is estimated at 6,050 miles, and it can carry up to 20,000 lb of ordnance over somewhat shorter ranges. Its maximum take-off weight has been estimated at between 250,000 and 350,000 lb. Its length is 162 ft, and it has a wing span of 170 ft. Like other bombers of its generation (operationally about 1956), some versions of the Bison have gun turrets for onboard defence, as well as standard air-to-ground radar. A maritime version, for long-range reconnaissance, has also been seen, with extensive radar, electronic countermeasures and communications equipment. Though about 100 Bisons are still operational, their penetration capabilities against US defences are likely to be low. With air re-fuelling the Bison is said to have a 7,000-mile range at 520 mph with 10,000 lb of bombs. The maximum speed at high altitude is estimated at Mach 0·87. ◊ *Bomber* for general subject entry.

Blackout. One of the effects of nuclear explosions is the emission of intense heat which ionizes air particles in the atmosphere. Beta-ray emission from the radioactive debris of the explosion has the same effect. Ionized air distorts and absorbs electromagnetic radiation, thus affecting the operation of ◊ *radar*, which is based on the use of this form of energy. 'Blackout' detonations have been widely described as effective countermeasures to ballistic missile defence systems based on radar (◊ *ABM*). This penetration tactic envisages the use of 'precursor' ICBMs to black out the defence radars so that subsequent ICBMs can get a 'free ride'.

The intensity of the blackout effect depends on the yield and altitude of the detonation, the respective positions of the radar and the detonation, and on the frequency of the former. The effect on a system of radars, such as used by the ◊ *Safeguard* ABM, depends on the number of 'precursors' and the precision with which they are coordinated, in terms of accuracy both in timing and in space. After making due allowances for errors (and for 'precursors' intercepted by the defence prior to their detonation), the neutralization of an ABM by means of this tactic requires a large number of ICBMs, since an 'window' left in the blackout will still allow the radar to observe the incoming warheads. This is therefore by no means the cheap and highly reliable ABM countermeasure which it has been claimed to be.

◊ *Penetration Aids* for a listing of other ballistic-missile defence countermeasures.

'Blinder' Tu-22. Soviet supersonic medium bomber. The Blinder, which has been operational since 1962, is an advanced twin-jet bomber which can carry a bomb-load of 12,000 lb and has a top speed of Mach 1·5 and a range exceeding 2,500 miles. First seen in 1961, the standard version, Blinder B, has a 2- or 3-man crew and carries the ◊ *Kitchen* stand-off missile for nuclear delivery at long range. There are at least three other versions in service. Visible features of Blinder B include a large nose radar, automatic tail cannon which fires under radar control, a refuelling probe in the nose and camera windows under the nose; these may also be used for visual navigation and weapon delivery. The maximum take-off weight is estimated at 175,000 lb, the wing span is about 80 ft and the overall length 130 ft.

◊ *USSR Strategic Offensive Forces.*
◊ *Bomber* for general subject entry.

Blitzkrieg. German, literally 'lightning war'. A tactical method used by the German armed forces during the first stages of the Second World War. It consisted of the use of mobile and armoured forces to achieve deep penetrations of the enemy front in order to cut supply lines, and envelop the by then disorganized defence. The initial breakthrough was to be achieved with coordinated fighter-bomber attacks, concentrated artillery fire and sheer attrition. Once the forces had broken through, engagements were minimized in order to maintain momentum. Obstacles were either avoided – whence

the need for cross-country vehicles – or dealt with by air attack; as the column moved deeper into the enemy rear it severed communications, cut supply lines and above all 'created data' which confused the enemy and paralysed his response. Enemy commanders could not concentrate their forces to stop the columns – or attack their very vulnerable flanks – because the disruption they caused, and their speed, concealed the direction of their advance. Eventually single columns met (the 'pincer-tactics') at key crossroads and the enemy forces collapsed as their supplies (and coordination) failed. These tactics are still as effective against a numerically superior enemy as they were in 1940, but they require (a) a margin of air superiority, (b) a margin of mobility, and (c) a command structure which allows field officers a real degree of flexibility, to identify and exploit weakly held passage points along a general axis. ◊ *Indirect Approach* for the wider 'parent' concept.

Bloodhound. British surface-to-air   **80**
missile. Bloodhound 2 is an anti-aircraft missile developed from the Mk 1, of which it retains many features. Both models are ramjet-powered with solid-fuel rocket-boosters; both have ◊ *semi-active radar homing*, but the Mk 2 has CW (continuous wave) instead of pulse ◊ *radar*. A radar receiver in the nose-cone 'homes' on signals bounced off the target by a ground-based radar 'illumination' set. In both models, the high-explosive warhead is fitted with a proximity fuse, and flight control is achieved by pivoting the double-wedge wings, while the tail surfaces are fixed.

Bloodhound 2 is used for area defence in several countries, including Britain, Sweden, Singapore and Switzerland.

Specifications: Total length: 27 ft 9 in. Maximum body diameter: 1 ft 9 in. Bloodhound 2s have intercepted supersonic targets at heights below 1,000 and above 40,000 feet.

Blue Steel. British stand-off missile.   **88**
The Blue Steel, a stand-off missile some thirty-five feet long fitted with a thermonuclear warhead in the megaton range, was developed to equip British medium

bombers. It is still available for the ◊ Vulcan 2 and Victor 2 bombers of the RAF, though these no longer have a primary nuclear bombardment mission. The Blue Steel is powered by a liquid-fuel rocket motor and guided by an ◊ inertial system which is linked with that of the carrying aircraft until release. As in the case of other ◊ stand-off missiles, the Blue Steel is intended to deliver the warhead in the final phase of the attack, thus allowing the bomber to avoid enemy air defences. The range depends on the altitude at which the weapon is released: the range envelope of the Blue Steel is classified but a range of 200 miles has been quoted – presumably for a medium-high altitude release. When the V-bombers were converted for low-altitude strikes, the Blue Steel was converted with them and it does have some low-altitude capability, although its range under those conditions is probably quite short. A maximum speed of 'several times the speed of sound' has been claimed but this probably refers to the earliest phase of flight.

Specifications: Power plant: twin-chamber liquid-propellant rocket. Length: 35 ft. Wing span: 13 ft.

BMEWS. US Ballistic Missile Early-warning System. BMEWS is a long-range radar and telecommunications network intended to detect missiles on a ballistic course from the Soviet Union to the United States over the northern or 'direct' route. BMEWS consists of three long-range radar units located in Greenland, Alaska and the United Kingdom linked to the North American air defence system (◊ NORAD) by three communication networks code-named 'Rag-mop', 'Blue-grass' and 'White Alice'. ◊ Radar for a general subject entry. ◊ PAR and MSR for ABM radar units intended for ballistic-missile detection and tracking.

Bofors. 40-mm anti-aircraft gun. Bofors is a Swedish firm which produces a wide range of armaments, including the S-tank (◊ Strv. 103) and the Bantam missile, but is internationally known for its anti-aircraft guns. Its current product, the 40-mm calibre. L.70, is power-operated and mounted on a turntable base which converts into a four-wheel chassis. Tra-

verse and elevation are hydraulically operated and it swings round fast enough to follow the arc of low-flying aircraft. It is not effective against modern fighters manoeuvring at high speeds. Though visual aim is possible, the L.70 is mostly used with a radar-computer system for acquisition and tracking, such as the British F.C.E. A.A. N.7 'Yellow Fever' system. One or more guns are then used with a radar-computer (mounted on a separate lightweight vehicle) and a mobile generator. The cyclic rate of fire is 240 rounds per minute, and as the gun fires, following in parallel the ◊ tracking radar, it takes at least two loaders to keep it firing.

The L.60, a similar but older model, is still widely used.

◊ Artillery.

Bölkow Cobra. German anti-tank missile. A simple and ultra-light wire-guided missile with a ◊ HEAT (shaped charge) warhead. The Cobra is a one-man weapon consisting of the missile itself, control box and wire links. Guidance is visual/manual with the operator keeping the missile on target by joystick corrections.

Specifications: Length: 3 ft 1 in. Body diameter: 4 in. Weight: 22·5 lb. Fibre-paper cylindrical body and plastic cruciform wings; a solid-propellant booster and solid propellant sustainer. A 5·5-lb warhead HEAT, delivered at a top speed of 190 mph over range from 1,310 ft minimum to 5,250 ft maximum.

For the Bölkow HOT and MILAN missiles, ◊ HOT.

Bomarc CIM-10. US surface-to-air missile. The Bomarc CIM-10B (or Super Bomarc, or Bomarc B) is a long-range pilotless interceptor, operational since 1961 with the USAF in· the ◊ NORAD air defence organization. This improved version of the original Bomarc (whose development started in 1949) has a solid-fuel rocket-booster and twin ramjet sustainers and is launched vertically. Guidance is by command direct from the ◊ SAGE Air Defence System supplemented by radar homing for final interception. The CIM-10B has large wings (18 ft 2 in span) and tailplanes which pivot for control. The cone is made of

glass fibre but otherwise it is an all-metal structure. The warhead is nuclear and set just after the nose radome. Bomarc Bs are currently operational with the Air Defence Command; two squadrons also remain with the Canadian forces.
Specifications: Power plant: 2 ramjets. Length: 45 ft 1 in. Diameter: 2 ft 11 in. Weight: 16,032 lb. Cruising speed: Mach 2·8. Range: 440 miles. Operational ceiling: up to 100,000 ft.
◊ *Missile* for general subject entry.

Bomber. A manned aircraft whose primary mission is to deliver weapon loads over medium and long ranges. Short-range or 'tactical' bombers are, by convention, described as 'attack', 'strike' or 'close-support' aircraft, or, more loosely, fighter-bombers. This convention is followed here, and such aircraft are covered in the ◊ *Fighter* entry. Until recently the 'strategic' (large payload, long range) bomber was the key weapon of this class, intended to deliver large weapon-loads from high altitudes. But technological developments in the ◊ *ballistic missile* and air defence fields have on the one hand pre-empted and on the other degraded this role. Modern interceptor aircraft, surface-to-air ◊ *missiles* and their ◊ *radar*-computer control systems would now force the bomber to operate at very low altitudes and high speeds. While this type of flight configuration makes radar detection very difficult, it also makes great demands on the bombers' airframe, fuel consumption and avionics, as shown in the case of the defunct British T S R-2 project and the troubled US ◊ *F-111A* (FB-111A) programme. Automatic flying at supersonic speeds at a few hundred feet above the ground ('contour-flying') requires a radar to detect obstacles ahead, a computer to process the data into flight instructions and control gear to 'fly' the aircraft according to these. It is a very rough ride for both pilot and aircraft, while fuel consumption is high so that range and payload have to be sacrificed. Given the availability of long-range missiles, which can strike, accurately, fixed targets – and do so more cheaply – the cost-effectiveness of strategic low-level bombers is more than doubtful. Another, alternative, response to the effectiveness of modern air defences at cruising heights is the ◊ *stand-off missile*. These weapons can be used to deliver nuclear warheads without exposing the bomber to the inner air defence zone, since the bomber can launch them at the target over long ranges ('several hundred' miles away in the case of the US ◊ *Hound Dog*). The missile is far less vulnerable to air defences than the launching bomber, but with its own avionics, propulsion and warhead the stand-off missile is very costly, and the bomber-plus-missile combination is still less reliable and more expensive than the missile-all-the-way solution.

Since the key factor in efficient air defence is the radar equipment, yet another bomber response to sophisticated air defences is the use of electronic countermeasures (◊ *ECM*). But power-sources and equipment loads are far easier to provide on the ground than in the air, so that ground-based ECCM (electronic counter-countermeasures) have a built-in advantage.

'Medium' bombers and 'light' bombers retain a residual role in 'tactical' situations, but fighter-sized aircraft are more suited for dealing with mobile targets, and tactical ◊ *ballistic missiles* are more efficient against larger and fixed targets – if equipped with nuclear warheads.

◊ *A.M.S.A.* for a current US bomber project.
◊ *Combat Aircraft* for general subject entry.
◊ the US *Stratofortress B-52* strategic bomber, *F-111A* (FB-111A) and *Hustler B-58* medium bombers, and *Vigilante A-5* (developed as a light bomber, though now used for reconnaissance).
◊ the Soviet *Badger Tu-16* and *Blinder Tu-22* medium bombers and the *Bison Mya-4* strategic bomber; also *Bear Tu-20*.
◊ the British *Victor B.2* and *Vulcan B.2* medium bombers; and *Buccaneer* light bomber.
◊ the French *Mirage IV* medium bomber.

Bren. British light machine-gun. Like the American BAR, this is one of the specially designed ◊ *light machine-guns* of the bolt-action rifle age, now being replaced by modified automatic ◊ *rifles*, or modern convertible ◊ *medium machine-guns*. The Bren (which is regarded with affection by all who have used it) is still widely issued in poorer countries. It is, however, no longer in first-line service with the British army (where it was designated M G L.4.) Like many other good small-arm designs, this is a Czech design, though produced in Britain since 1937. The Bren is gas-operated and weighs 20 lb; it is about 45 in long and has a cyclic rate of fire of 450–550 rpm; its effective range is about 600 yds, but it is usually sighted to 2,000 yds. Originally issued in the 0·303 British army calibre, it was later converted to the new NATO 7·62 calibre. It is easy to operate, load and clean, and has variable-length bipods and a characteristic curved magazine.

Brewer. ◊ *Firebar.*

Brigade. An army formation formally subordinate to the ◊ *division*, parallel to the ◊ *regiment* and comprising a number of ◊ *battalions*. In general, a brigade comprises about 3,000–4,000 men, though in some armies this echelon is used only for (smaller) specialized units, while in others, such the Soviet, Warsaw Pact and Chinese armies, it does not exist at all. In the British and Israeli armies the brigade is the focus of decision-making and the effective operational unit.

Brinkmanship. A diplomatic strategy designed to force the opposition to reach an accommodation by the use of a deliberately created risk of nuclear war. This risk must be (a) mutually recognized, and (b) seen as not completely controllable by either party. It implies, on the part of the initiating party, a higher degree of tolerance of uncontrolled risk if the strategy is to be successful. This concept is associated with the supposed policies of John Foster Dulles, sometime US Secretary of Defence (1952–8). Brinkmanship was envisaged as the diplomatic companion of ◊ *massive retaliation*. In the event of any Soviet aggression, even if minor and/or carried out by proxy, the US response would be the threat of a nuclear strike on the USSR. This threat was then supposed to initiate a diplomatic exchange leading to the renunciation of the aggres-

sion, or the beginning of a presumably unacceptable general nuclear war. Both brinkmanship and massive retaliation never represented more than a segment of US policy and were formally abandoned by the end of the 1950s. Current US policy is described as ◊ *Controlled Response*, and its diplomatic companion is normal diplomacy. ◊ also *Assured Destruction*.

Browning. US machine-guns. The noted American arms designer J. Browning produced a large number of small arms, of which the ◊ *BAR* light machine-gun (LMG) and several machine-guns are still widely used. The M.1919 A 4 and M.1919 A 6 are Cal. 0·30 air-cooled medium machine-guns (MMG). Both have the same working parts, but the A 6 has a bipod and shoulder stock to permit operation in the LMG mode. They are recoil-operated and belt-fed and weigh about 45 lb with the tripod; the cyclic rate of fire varies between 400 and 550 rounds per minute; and the length is 41 in without the shoulder stock of the A 6. Related air-cooled weapons were the standard rifle-calibre guns of allied aircraft in the Second World War.

The M.37 tank MMG is a M.1919 adopted for ◊ *armoured fighting vehicles*, and it has different attachment and breech equipment. The M.1917 A 1 is a heavy water-cooled machine-gun originally used in the First World War but still used in some armies. It has a low rate of fire, 400 rounds per minute; and it is fed from fabric belts.

The Browning Cal. 0·5 (M.2 heavy barrel) is a heavy machine-gun, since it fires ammunition of greater than rifle calibre. It has been used as an aircraft weapon, and in multiple anti-aircraft mountings, but only the vehicle version remains widely in use. The M.2 is 65 in long and weighs 84 lb without the mount; the rate of fire is variable, between 450 and 550 rounds per minute.

◊ *Machine-gun*, and *Light*, *Medium* and *Heavy Machine-gun* for competitive types.

BTR (Broneje Transporter). Soviet armoured personnel carrier designation. During the Second World War the Soviet army catered for armour-   **30**

infantry cooperation by having troops
'ride' on the back of tanks. In 1946 the
first Soviet ◊ *armoured personnel carrier*
(APC), the BTR-152, was produced by
adding thin armour to the SIL-151
truck. Like other APCs based on truck
chassis it has an open top, no turret, and
inadequate 13·5-mm armour. The only
fixed weapon is a 12·7-mm machine-gun.
The six wheels are all driven, and their
tyre pressure can be varied en route by
the driver. This and the 95-hp engine
give it adequate mobility. There is a command version with a closed top. The
BTR-152 weighs 8·6 tons.

The BTR-152 is widely used outside
the Soviet Union, but within it has largely
been replaced by the BTR-50(P) in the
armoured divisions and the BTR-60(P)
in the infantry divisions. The BTR-50(P),
which appeared in 1958, was the first
tracked APC produced in the Soviet
Union, and it is based on the chassis of the
◊ *PT-76* light tank. Like the latter, it is a
true amphibian with water-jet propulsion
and a boat-shaped frontal hull. It carries
14 men including the driver. The basic
version has no turret and mounts an external 12·7-mm machine-gun, but various
versions have appeared with command
and weapon turrets. The weight is 14·5
tons, and the wide tracks and the 320-hp
engine give it good all-round mobility.
A new (non-amphibious) APC designated
M-1967, weighing 12 tons, is now in production. This has a 76-mm gun and
attachments for anti-tank missiles and
carries up to 11, crowded, men.

The BTR-60 (P) is a large boat-shaped
vehicle weighing 10 tons with eight driven
wheels and water-jet propulsion, which
first appeared in 1961. This too has no
turret, and it mounts a 12·7-mm machine-
gun in the prow-shaped front; there are
several firing ports on the sides. The
original BTR-60(P) has an open top, but
a variant, 60(PK), has appeared with a
closed top.

The latest version, designated 60(PB),
is fully enclosed and fitted with a turret
for 14·5- and 7·62-mm machine-guns.
Like other BTR-60 versions, this vehicle
has thin (10–13 mm) armour but outstanding mobility in sand, snow or water.
There is also a series of scout cars/weapon
carriers designated BRDM which has

replaced the post-war BTR-40 series.
The BRDM cars weigh 6 or 7 tons
depending on the variant and have four
driven wheels and a boat-like hull. All
BRDM versions are amphibious, and
some are fitted out as missile carriers for
the Soviet series of anti-tank missiles. (◊
Sagger, Snapper, Swatter.)

Buccaneer. British light bomber.
The Buccaneer was designed as a naval
strike aircraft, specifically intended to fly
low over the sea at transonic speeds, but
it is also suitable for low-level ground-
strike missions. It has two engines with
characteristic round air-intakes at the
wing roots; other recognition features are
the long tail-cone and the bulged rear
fuselage. The Buccaneer has a two-man
crew in tandem under the same canopy
and an integrated weapon control and
navigation system. No detailed performance data are available, but it has a maximum weapon-load of 8,000 lb and an un-
refuelled range exceeding 2,000 miles.
No air-to-air missiles or guns are normally
fitted, but standard weapon-loads include
the ◊ *Bullpup* and ◊ *Martel* air-to-ground
missiles. The long tail cone contains a
◊ *MAD* detector.

External dimensions: Overall length:
63 ft 5 in. Height: 16 ft 6 in. Wing span:
42 ft 4 in. Gross wing area: 508·5 sq ft.
Maximum take-off weight: 40,000 lb.
Maximum speed: Mach 0·95. The S.2 is
serving with the British navy and RAF,
and the slightly modified S.50 (with
rocket-boosted take-off) serves with the
South African air force.
◊ *Bomber* for general subject entry.

BUIC. Back-Up Interceptor Control.
An emergency computer and communication system designed to supplement the
main (SAGE) weapon-control system
within the North American anti-bomber
air defence organization, ◊ *NORAD*.
BUIC is intended to operate after a
missile (or even bomber) attack has 'taken
out' the larger SAGE units.

**Bullpup AGM-12 series. US air-to-
surface missiles.** The Bullpup series of
air-to-surface ◊ *missiles* now equips a large
number of US navy, US air force and
allied countries aircraft. It consists of a

large (250-lb or 10,000-lb) conventional or nuclear warhead, a rocket motor and pneumatically actuated control surfaces. The missile is guided by radio commands from the pilot/gunner, who has to see both target and missile with the naked eye. Currently operational types include:

AGM-12B With a pre-packaged liquid-propellant rocket motor and high-explosive warhead.

AGM-12C The Bullpup B, larger version of 12B type, with improved motor and command system. It can attack targets while the parent aircraft is circling around them.

AGM-12D This variant can be fitted with interchangeable nuclear or conventional warheads.

Bulldog. A laser-guided version under development.

BW. ◊ *Biological Warfare.*

BZ. US army designation for an incapacitating chemical agent. BZ has no effect on the body and operates exclusively on the brain. It is not lethal, and (if inhalation stops) after a short period of time its effects are reversed. BZ is a solid of undisclosed composition which is dispensed in aerosol form. Its victims are unaware of its effects, which include: reduced mental and physical activity, hallucinations and giddiness. In experiments on animals and humans BZ appears to cause a failure of resolve. After inhalation of BZ a sentry failed to prevent the passage of strangers, while cats made frantic efforts to escape mice.

◊ *Chemical Warfare* and also *C N, C S,* and *D M.*

Specifications:

	Length	Diameter	Wing span	Weight	Speed	Range
AGM-12B	10 ft 6 in	12 in	3 ft 1 in	571 lb	Mach 1·8	7 miles
AGM-12C	13 ft 7 in	18 in	4 ft 0 in	1,785 lb	Mach 1·8	10 miles

C

C. US designation for transport aircraft.
◊ *Galaxy* for C-5.
◊ *Hercules* for C-130.
◊ *Starlifter* for C-141.

Cactus. ◊ *Crotale.*

CAFDA. The French air defence command. CAFDA is part of the air force and does not control anti-aircraft artillery units. It consists of (a) a chain of radar stations linked by the (as yet uncompleted) STRIDA II ◊ *air defence system*, (b) a fighter force, which includes 3 squadrons of ◊ *Mirage IIIs*, as well as obsolete Vautour IINs, and Super-Mystère B2s. CAFDA's HQ is at Taverny, and it will be linked with ◊ *NATO's* NADGE air defence system, though France no longer participates in joint defence planning with NATO.
For comparable systems ◊ *NORAD, PVO Strany, UK, Strategic Defensive Forces,* and *China, Strategic Defensive Forces.*
For likely opposition, ◊ *USSR, Strategic Offensive Forces* and the same for the US, UK, and China.

Catalytic War. A war brought about by a third party. If for example the Chinese were to simulate a Soviet attack on the United States, and if this were to trigger off an American strike against the Soviet Union, this would be a catalytic war. Catalytic war was one of the nightmares associated with the period in which nuclear powers relied on 'first-generation' delivery systems (which were unprotected and which required longish pre-firing preparation). With such systems there was no time for deliberation and communication in the face of attack or an impending attack, since any hesitation in using retalia-

tory forces would have entailed their destruction. Now that delivery systems are ◊ *hardened* (or otherwise have an ◊ *assured destruction* capability) the possibility of catalytic war between the major powers has receded, since a single or small nuclear strike would not lead to a full and automatic retaliatory response. Instead, there would be communication between the parties, or at least sufficient time to identify and assess the nature of the attack.
◊ *Trigger* for a small nuclear force which may be intended to act in a catalytic role.

CBR. Abbreviation for Chemical, Biological and Radiological agents used as weapons or weapon warheads. ◊ *Chemical Warfare* and *Biological Warfare.* CBR-equipped weapons are sometimes known as 'special weapons'. ABC is synonymous.
The Soviets prefer the more general term 'weapons of mass destruction', which includes nuclear weapons.

CENTAG. A command organization within NATO. ◊ *SHAPE.*

CENTO. Abbreviation for the **Central Treaty Organization** formed under the provisions of the 1955 Baghdad Pact signed by Britain, Iraq, Iran, Pakistan and Turkey. The USA is not a member but is 'associated' with CENTO, and it is linked with its members by various bilateral and multilateral treaties. Iraq has formally withdrawn from the Pact, and Pakistan is no longer an active member.
CENTO does not have a joint command system such as NATO's but works rather loosely through two committees: the Council of Military Deputies and an Economic and Counter-subversion committee. No troops are formally allocated to

CENTO's operational control, but UK and US forces in the area are earmarked to it, as are most local forces in the event of war. CENTO is explicitly directed at 'communist' aggression, and would not operate in the event of member country conflict with a non-communist country.

Central War. War between the USSR and the USA or, more generally, war between the nuclear powers – whether nuclear weapons are used or not; in practice used for a direct confrontation between the USSR and the USA, as opposed to fighting by allies, satellites or otherwise by proxy. Sometimes used as 'nuclear war' between nuclear powers', but this usage is not recommended: 'nuclear war' is adequate as a general term, since, after all, only nuclear powers can wage nuclear war. ◊ *General War* is used for central war that develops into an all-weapon conflict about survival.

19 Centurion. British tank. The first Centurions were already operational in 1945, but the design has been repeatedly modernized into many different configurations and it is currently in service with eleven different armies. In spite of the modifications, its limited mobility, poorly shaped armour (though there is a great deal of it), and petrol engine make it obsolete as compared with the main battle tanks now in service. The layout is conventional with a driver in the hull and a three-man turret crew. In the latest versions (Mk 10 is followed here) the Centurion mounts the outstanding 105-mm gun, which with its ◊ *APDS* rounds is a very effective armour killer. There is a stabilization system for both vertical and lateral oscillations, which keeps the gun on target and allows accurate moving fire. Favourable features include good crew comfort and a large ammunition-carrying capacity. The speed and road range are, however, very limited, and the machine-gun ranging system inferior to the optical systems available on more modern tanks (though not British ones). Apart from the square shapes of all parts of the tank, Centurion can be recognized by the skirting plates which protect the track assembly.

Specifications – Centurion Mk 10 (FV 4015): Length: 25 ft 8 in. Width:

11 ft 1 in. Height: 9 ft 9 in. Weight: 51 tons. Engine: petrol 635 bhp. Road speed: 21 mph. Road range: 115 miles. Ground pressure: 13 lb sq in. Ground clearance: 1 ft 8 in. Main armament: 105-mm high-velocity gun with 63 APDS and HESH rounds carried. Secondary armament: coaxial MG and turret-top MG, smoke dischargers/launchers. Crew: 4.
◊ *Main Battle Tank* for competitive types.
◊ *Armoured Fighting Vehicle* for general subject entry.

CEP. Circular Error Probable. ◊ *Missile* and *Nuclear Weapon*.

CETME G3. Spanish-West German rifle. The CETME, or G3 as it is called in Germany, is the only weapon firing the standard NATO 7·62-mm round which competes with the ◊ *FN* (FAL) rifle adopted by most members of NATO. The CETME was developed in Spain by émigré Germans and re-engineered in West Germany, where it was chosen as the standard rifle after the German army rejected the FN. The G3 uses a different system of operation from that of most current light automatic weapons: delayed blow-back, instead of gas operation. More important, the G3 can be fired automatically as well as semi-automatically, and without the drastic fall in accuracy experienced with fully automatic versions of the FN (FAL). It is also simpler and more reliable than the latter, though neither weapon compares with the Soviet ◊ *AK* in this respect, or with the US ◊ *M.16* in most others.

The CETME is 40·2 in long and weighs 9·9 lb loaded.

Chaparral. US mobile anti-aircraft 72 system. The Chaparral is a new anti-aircraft weapon for the US army. It consists of a modified ◊ *M.113* tracked carrier, mounting an aiming turret fitted out for surface-to-air missiles. In the present configuration the missiles are reliable and economical ◊ *Sidewinder 1C*, modified for ground-to-air work, four being carried ready for launch. These missiles are guided by infra-red homing heads, but they are launched in the general direction of the target by optical fire control from the freely turning and elevating quadruple launcher. Sixteen Chaparral units are to make up one half of the new anti-aircraft battalions of US army divisions, the other half of the battalion consisting of XM 163 20-mm cannon AA vehicles (for which, ◊ *Artillery, US*).

Chemical Warfare (CW). The use of toxic substances as weapons or weapon warheads. The first instance of modern military use of CW occurred in August 1914 when French troops fired tear-gas rifle grenades, but lethal agents were not used until 1915 when pressurized chlorine gas released from cylinders was used by German troops. During the First World War, about 1,000,000 casualties were caused by gas of which about 90,000 were fatal.

Currently available agents which may be used for CW include:

(a) The old lethal agents, such as ◊ *Phosgene* and ◊ *Mustard 'gas'*.

(b) The ◊ *'nerve gas'* group.

(c) Non-lethal harassing agents, such as ◊ *CN*, ◊ *CS*, ◊ *DM*.

(d) Incapacitating agents, such as ◊ *BZ*, and hallucinogenic substances, such as LSD.

(e) Weed-killers and animal poisons which can be used for crop/forest/livestock destruction: these are the agricultural equivalent of non-nuclear bombs used in ◊ *strategic air war* against industrialized countries.

Current CW preparations include the stockage of bombs, shells and missile warheads filled with lethal or harassing agents; the maintenance of CW agent plants on a stand-by condition; and the deployment of associated equipment and personnel. All modern armies have specialized troops for the offensive or defensive role in CW; these tend to be integrated within other units, but in the US army there is a Chemical Corps. The last use of lethal agents in CW was recorded in the Yemen, where Egyptian aerial bombing included the use of gas-filled bombs.

For other ABC/CBR weapons ◊ *Biological Warfare*.

Cheyenne AH.56A. US combat helicopter project. The two-seater Cheyenne was developed as a replacement for the pure helicopters now being used for close-support missions, but the contract was terminated in 1969. Initially designated AAFSS, Advanced Aerial Fire Support System, the Cheyenne had the small wings and 'pusher' propeller of an aircraft as well as the large four-blade rigid rotor and small tail rotor of a helicopter. After take-off, the main rotor was to provide the initial acceleration, with power being switched to the 'pusher' propeller as speed picked up. In high-speed cruising flight almost all the available engine power (3,435 shp) was to be fed to the tail gear-box to turn the propeller and only 300 shp diverted to the rotor in order to reduce drag. In this way, the Cheyenne was intended to fly very fast in level flight (up to 253 mph) with the wings adding to the cruising range, by off-loading the main rotor in cruising flight.

The Cheyenne was developed with 'all-weather' avionics: TFR ◊ *radar* (for very low high-speed flight), doppler ◊ *radar*, and an ◊ *inertial navigation* system. The weaponry intended for the Cheyenne included two rotating turrets (under the fuselage and in the nose) for single 30-mm cannon, a 7·62-mm multi-barrel machine-gun, and a 40-mm grenade-launcher. Four racks under fuselage and wings were to carry up to four ◊ *TOW* anti-tank missiles or other stores.

Specifications: Rotor diameter: 50 ft 5 in. Length: 60 ft. Gross wing area: 260 sq ft. Weight empty: 11,725 lb. Weight at take-off: up to 22,000 lb. Hovering ceiling (no ground effect): 10,600 ft. Design range with maximum standard fuel: 875 miles. Endurance maximum standard fuel: over 5 hours.

Chieftain. British battle tank. The Chieftain which entered service in 1966, marks the British adoption of the main battle tank (MBT) concept in place of the earlier combination of ◊ *Centurion* 'medium' tanks supported by Conqueror 'heavy' tanks. The Chieftain has more armour and a bigger high-velocity gun (120-mm) than any other MBT, and, naturally, it is also the heaviest and least mobile. The general shape and layout are conventional, with the driver in the hull and a 3-man crew in the turret; less con-

20

ventional features include the supine position of the driver, to minimize the height of the hull, and the 0·5-in ranging machine-gun instead of the optical ranging devices used on all other MBTs. This choice is difficult to understand, since the value of the bigger gun is a longer (2,500+ yds) effective range, while range-finding by tracer fire is only effective up to 2,000 yds; but a laser range-finder is to be fitted. Good design features include the gun stabilization system, which allows fairly accurate fire while on the move, and the well-shaped armour, which affords very good ballistic protection. The ammunition is stored in water-jacketed containers to minimize the risk of secondary explosions should the tank be penetrated. Because of its size, each round comes in two separate sections to be joined in the gun breech. As with other elaborate designs, the cost-effectiveness of Chieftain is doubtful, given the availability of cheap and reliable anti-tank missiles.

Specifications: Length: 25 ft 1 in. Width: 11 ft 6 in. Height: 9 ft 6 in. Weight: 51·5 tons. Engine: diesel 700 bhp. Road speed: 25 mph. Road range: 250 miles. Ground pressure: 14 lb sq in. Main armament: 120-mm high-velocity gun firing APDS and HESH rounds; 53 rounds carried. Night-vision aids. Secondary weapons: coaxial 7·6-mm MG and turret-mounted 7·62-mm MG, as well as the 0·5-in ranging gun. Crew: 4.

◊ *Main Battle Tank* for competitive types.

◊ *Armoured Fighting Vehicle* for general entry.

China, Strategic Defensive Forces. The Chinese ◊ *air defence system* includes:

(a) A chain of radar stations with ancillary communication and some computing equipment.

(b) A force of fighter-interceptors, mainly Mig-17s with fewer Mig-19s and yet fewer ◊ *Fishbed* Mig-21s. Their combined number varies from a minimum of 500 operational aircraft to a maximum double that number. Chinese-built Mig-19s are replacing the Mig-17s and the obsolete Mig-15s.

(c) A 'small' number of ◊ *Guideline* surface-to-air missiles.

(d) A large number of anti-aircraft

artillery units equipped with Soviet guns, including: twin 23-mm, single 57-mm and dual-purpose 85-mm and 100-mm guns. There may be some self-propelled twin 57-mm guns with radar control. ◊ *Artillery, Soviet.* At least some of these units are centrally coordinated and integrated within a coherent air defence system.

For comparable systems, ◊ *PVO Strany, NORAD, CAFDA* and *UK, Strategic Defensive Forces.*

For the likely opposition see *USSR, Strategic Offensive Forces,* and the same for *US, UK* and France.

China, Strategic Offensive Forces. The first test of a fission device in China took place in 1964 and a fusion device was tested in 1967. Chinese facilities for the production of weapon-grade nuclear materials are thought to be limited, and no more than a hundred or so warheads were thought to be available in 1971. As is well known, delivery systems are far more difficult to develop than the warheads themselves, and China will not be able to deploy an ICBM force until the mid-seventies. This force is expected to consist of no more than 100 rudimentary ICBMs. Unlike the other nuclear 'also rans', Britain and France, China has not developed a modern medium bomber force, but it may choose to deploy a medium or intermediate range ballistic missile force and a small deployment of MRBMs could take place in the course of 1971–2. The delivery systems which are available consist of a small number of early propeller-driven Tu-4 bombers and some 150 antiquated Il-28 jet bombers. In addition to these, the Chinese navy operates a G-class Soviet submarine with missile tubes, but it is unlikely that China will develop missiles for this vessel.

◊ *US, Strategic Offensive Forces,* and the same for *USSR, UK* and *France.*

◊ also *PVO Strany, NORAD* and *Safeguard* for the defensive systems that Chinese strategic forces would have to penetrate.

Civil Defence. All measures taken to limit the effect of enemy attack upon the general population and civilian society other than active interception. It includes the use of concealment, dispersion, protective construction, duplication of essential services, emergency measures for government, production and distribution and in general preparations for the survival of organized society after an attack. In the present, nuclear, environment the need of civil defence measures is generally felt to be great, but the actual implementation of programmes has been impeded by their very great cost. Except in the case of Sweden, most states have opted for a dual programme of realistic civil defence for key government personnel and facilities, and symbolic civil defence for the general population. Civil defence measures are an integral part of the military posture of a state – especially in nuclear conditions – since they increase the ◊ *credibility* of its threats by degrading the deterrent threat of its opponent. Indeed, an accelerated and realistic civil defence programme on the part of a nuclear power would in itself be read as a threat of an impeding attack on some other nuclear power.

'Passive air defence' is a term used to describe a particular form of civil defence.

CN. A chemical agent developed for riot control but which has also been used in combat: white crystals with an apple-blossom smell which operate by inhalation when dispensed in aerosol form. Its effects are: burning feeling on moist skin, and copious weeping, if 5 to 10 milligrams per cubic metre are present for at least one minute. The effect intensifies with volume until a 50-per-cent lethality is achieved with concentrations of 8·5 grammes per cubic metre per minute. Unless these (exceedingly high) concentrations are present, the normal milder effects last only 'a few' minutes. ◊ *Chemical Warfare*; also *CS* and *DM*.

CO. Commanding officer or commissioned officer. The latter as opposed to NCO or non-commissioned officer.

Cobra. Anti-tank missile. ◊ *Bölkow.*

60 **'Cock', Antonov Antheus An-22. Soviet military transport aircraft.** The An-22 is a long-range heavy transport powered by four turboprops which has been in civil and military service since 1967. The An-22 has a high wing with twin fins and rudders; the landing gear is held in low fuselage fairings. The four turboprops, with a combined rating of 60,000 shp-drive contra-rotating propellers; the maximum level speed is 460 mph.

The An-22 is mainly intended as a freight carrier, but there is a cabin for 28 passengers as well as for the 5–6 man crew. The maximum payload is 176,350 lb to be carried mainly in the separate freight cabin. There is a full-size rear ramp and door unit which allows cabin-sized cargoes to be loaded straight in. The An-22 featured in the Soviet airlift to Egypt in July–August 1967.

Specifications: Length: 189 ft 7 in. Height: 41 ft 1·5 in. Wing span: 211 ft 4 in. Gross wing area: 3,713 sq ft. Main freight cabin: 108·3 by 14·5 by 14·5 ft/in. Equipped empty weight: 251,325 lb. Maximum take-off weight: 551,160 lb. Maximum payload: 176,350 lb. Range with 99,200 lb of payload: 6,800 miles. Range with maximum payload: 3,100 miles. The maximum level speed is about 460 mph.

◊ US counterparts: *Hercules, Star-lifter, Galaxy;* ◊ also the Soviet *Cub.*

Coexistence. A Soviet policy concept which favours the use of covert and peaceful competition with the West as a substitute for overt warfare (◊ *Political Warfare, Economic Warfare, Psychological Warfare*). The concept is quite clearly a renunciation of nuclear war, but the role of the so-called 'Wars of National Liberation' is not clearly spelled out in theory or in Soviet practice. Soviet client states normally receive military assistance as well as the cachet of being 'progressive', but the setting of safe limits to the involvement appears to be a persistent policy problem for the Soviet Union.

Cold War. International conflict in which all means other than overt military force are used; a description of the state of East–West relations in the late-1940s–mid-50s period. The use of a combination of economic warfare, propaganda, subversion and covert operations is now sometimes described as ◊ *confrontation.*

Collective Security. A device of the mind based on the proposition that if only all other states were to oppose any 'aggressor' then peace could be ensured, since the potential aggressor would be deterred by the overwhelming power of all other states. This assumes that all parties are willing to intervene to defend any party to the agreement, regardless of their own interests at that time. The idea found verbal application in the Covenant of the League of Nations and in the Charter of the United Nations, where the Security Council is supposed to deal with threats to peace by organizing collective action. But the five permanent members each have a veto, and these powers are, of course, directly or indirectly responsible for most 'threats to peace'.

Combat Aircraft. Part I: Military Requirements. Part II: Technology.
Part I: Military Requirements
Manned aircraft are the most important class of equipment in modern non-nuclear warfare, with the exception of ◊ *Revolutionary Wars*, where their role is (or ought to be) marginal. The rapid pace of technological development and the evolution of military requirements means that the stock of manned military aircraft operational at any one time includes very diverse aircraft, reflecting both generation and primary mission differences. The two sometimes interact, as in the case of obsolescent fighter-interceptors which are converted into 'close-support' (or 'attack') fighters when their performance becomes insufficient for air-to-air combat while remaining adequate in the less demanding (in the speed sense) air-to-ground role. A basic classification of combat aircraft in terms of primary mission is attempted below:
Air-to-Air Combat
All aircraft whose primary mission consists of fighting other aircraft are sub-classified in the ◊ *Fighter* entry.
Air-to-Surface
(a) Long-range weapon-delivery or 'strategic bomber' role. This class of aircraft is of decreasing importance. The advent of the long-range ◊ *ballistic missile* has pre-empted their main role, while the development of effective surface-to-air missiles has forced them out of the middle-high altitudes where the flight performance of jet aircraft is most efficient. ◊ *Bomber*.

(b) Long-range reconnaissance, or 'strategic reconnaissance', role. Another waning group, some of whose missions have been taken over by satellite systems, though not of course in the case of maritime reconnaissance (◊ *Integrated Satellite System*). Converted (a)-type aircraft are generally used in this role (◊ *Bomber*), but the US ◊ *SR-71* is a special-purpose (b) type.

(c) Long-range troop or logistic transport role. This is partly fulfilled by converted civilian aircraft, but specially developed military types include the US ◊ *Starlifter C-141*, ◊ *Hercules C-130* and ◊ *Galaxy C-5A*, the Soviet ◊ *Cub AN-12* and ◊ *Cock AN-22* and the Franco-German ◊ *Transall C-160*.

(d) Battlefield weapon delivery (or 'close-support' or 'strike' or 'attack') role. This is a key 'mission' area within which various kinds of fixed-wing and rotor aircraft are used for specialized sub-roles (◊ *Fighter*, which includes attack aircraft). Helicopters have acquired great prominence in this role, though attack aircraft are the main weapons. General purpose rotorcraft are adapted for combat by the addition of armour and weaponry, but at least one special-purpose combat helicopter is already operational, the ◊ *Hueycobra AH-1G*. See also the ◊ *Cheyenne* project.

(e) Battlefield reconnaissance. Converted fighters are used in this role, as well as light aircraft and helicopters. (◊ *Fighter*; also for sub-species of electronic counter-measures, ◊ *ECM*, aircraft.)

(f) Battlefield troop or logistic transport and 'assault transport'. Smaller transport aircraft and large helicopters are used in this role. The so-called 'assault transport', which is supposed to be used right in the battlefield, assumes total air superiority (and the suppression of ground fire).

Beyond this general classification there are other kinds of military aircraft designed for special purposes: combat trainers, 'airborne early-warning' aircraft (AEW), and operation-centre/communication-centre aircraft, which provide mobile – and secure – facilities for special purposes, such as ◊ *AWACS*.

Part II: Technology

Four of the five constituents of manned combat aircraft, the airframe, the power plant, the electronics and the weaponry, have been subject to rapid technological development. The fifth element, the crew, has not been similarly affected.

The operational requirements of combat aircraft for different missions lead to incompatible design requirements: a long-range and large-payload combination at subsonic speed requires high by-pass ratio engines, a low wing-loading and low sweep and high aspect ratios for the wings; a high-speed and rapid-acceleration mix requires highly swept wings, low aspect ratios, and high wing-loadings. Other ratios and relationships are also suited to particular speeds at particular altitudes and imply different airstrip requirements. Two new techniques, variable wing-sweep (so-called 'variable-geometry') and VTOL (vertical take-off and landing), resolve some of these design conflicts and permit the design of real multi-role aircraft. For example, 'variable-sweep' as employed for the ◊ *F-111*, the US aircraft already flying, overcomes the following design conflicts:

formance to conventional aircraft when both are up and flying.

STOL (short take-off and landing) appears to be a more promising approach. There are significant advantages in an aircraft which can make do with short landing strips, while at the same time STOL involves far smaller fuel and payload penalties than VTOL. STOL devices include 'boundary-layer control' (where jets of air sweep wing surfaces to increase lift) and trailing-edge devices that retract after take-off. More dramatic is the use of JATO rockets to accelerate for take-off and parachutes to decelerate for landing – but this is an expensive, one-shot device as well as a structurally demanding one.

Airframes are expected in the future to incorporate more and more titanium alloys (since aluminium alloy structures are limited to speeds of about Mach 2·2).

While airframes and engines have shown a steady and gradual progress, navigation, target location, weapon aiming and automatic flight systems have all registered spectacular breakthroughs. The various applications of ◊ *radar* are used in all four areas; ◊ *inertial platforms* and

| | Wing sweep | |
Operational Requirements	High	Low
High speed	suited	unsuited
Smooth transonic flight	suited	unsuited
Short take-off and landing	unsuited	suited
Long endurance at high altitude	unsuited	suited
Long subsonic cruise	unsuited	suited

VTOL aircraft take-off and land by hovering. This can be done by either fold-away rotors or down-pointed turbojets or other thrusters which turn on an axis, as well as by vectoring the thrust of fixed engines, which first lift and then propel the aircraft. This last solution is employed in the only operational VTOL aircraft, the British ◊ *Harrier*. Unfortunately all these techniques involve the penalty of a large fuel consumption and extra equipment, which renders them inferior in all-round per-

automatic astro-tracking have led to a quantum jump in navigation accuracy. The role of the computer is crucial in all four areas, and the steady decrease in their weight and size/per units of data processed makes airborne computers increasingly efficient. The main avionics systems are as follows:

(a) So-called 'all-weather' interception: The ground-based search radar detects the presence of the target (or 'Bandit'), and the tracking unit (radar +

computer) follows and predicts the Bandit's flight path. The ground–air link to the interceptor ('Angel') feeds the tracking-unit data direct into the Angel's flight controls until Angel reaches the Bandit's immediate vicinity. The Angel's airborne nose radar then tracks the Bandit and feeds data direct into the interceptor's flight controls. The Angel's weapons are fired automatically at the Bandit when the nose-radar and computer ('electronic fire-control') have flown Angel within range and into the right attack position.

(b) Terrain-avoidance (i.e. very low altitude) flight system: This is completely airborne with a forward-looking radar to detect obstacles in front of the aircraft, a computer to turn the raw data into flight instructions, and an autopilot to actuate the flight controls. Since land features cannot be seen with the naked eye when flying at high speeds and low altitudes, there is a side-looking radar to keep check points in 'view' so as to allow the pilot to recognize ground features.

(c) 'All-weather' navigation and ground strike: This requires air-to-ground radar mapping to allow the pilot to 'see' in the dark or in bad weather, with doppler radar to measure ground speed and inertial platforms which give accurate position 'fixes' without reference to any external data.

Another branch of avionics is covered in the ⬦ ECM (Electronic Countermeasures) entry.

Aircraft ordnance has also evolved, especially in the area of weapon-avionics. Planes fight other planes with an increasingly diverse family of air-to-air ⬦ missiles (AAM), though cannon of 20-30-mm calibre and with very high rates of fire are retained on most fighters. AAMs are effective but expensive and are rarely suitable against ground targets. The larger family of ordnance intended against surface targets includes the same cannon, unguided rockets (small ones in pods, large ones independently), free-fall bombs, or 'iron' bombs, ⬦ napalm tanks and guided weapons. While airborne targets can be distinguished by electronic means, so that air-to-air missiles can be guided by 'homing' devices, land targets cannot usually be acquired by homing devices. Thus air-to-surface missiles are either

sent on pre-set courses or guided by command from the launching aircraft. Simple and crude missiles such as the ⬦ Nord A S12 use wire links and visual 'joystick' control: ⬦ Bullpup uses radio links in the same way. Both require the aircraft to loiter around the target area while 'flying' the missile to target; television systems such as those used in the ⬦ Walleye and one version of the ⬦ Martel allow the aircraft to remain farther back. Only anti-radar missiles such as ⬦ Shrike can 'home' on surface targets, since ground radar sets provide plenty of recognition data for electronic systems. Anti-shipping missiles can also use homing techniques, given the sharp contrast between target and surface.

Combat Vessels. Those who study military matters in general face special problems when dealing with naval matters. The habits of mind and to some extent even the analytical techniques which are readily applicable to land, tactical air and strategic problems tend to be less reliable when applied to sea warfare. There is, for example, a lack of symmetry between competitive deployments: the US navy maintains a large fleet of aircraft carriers, whereas the Soviet Union has none; the Egyptian navy deploys 13 submarines against a single Israeli destroyer. Such asymmetries can be explained largely in terms of geographic considerations, though not by simplistic geopolitics based on the brief perusal of small-scale maps. The geography that matters is that of narrow sea-lanes, access routes, trade flows, trade dependence and the quantity and quality of ports. In other words, detail. It is perhaps this mass of detail which has enabled most naval establishments to resist the application of mathematical-logical analysis to force requirements, basing arrangements and tactical planning.

The US navy, for example, finds it necessary to maintain aircraft carriers in the Mediterranean although the US disposes of large and well-equipped air-bases in south-east Turkey, south and north Italy and southern Spain; the Soviet Union, on the other hand, is engaging a fleet of destroyers in traditional 'gunboat diplomacy' without any air cover at all. These two apparent paradoxes illustrate the complexity of air-power/sea-power interaction and the role of visual symbolism in naval matters. A squadron of fighters flying at 30,000 ft may have a greater effective 'punch' than a fleet of destroyers, but it is the latter which have a powerful political impact. And while airbases can provide more cost-effective air power than aircraft carriers, in any one location, vulnerability considerations, the combat radii of preferred aircraft and the political costs of such bases may well justify even permanently stationed aircraft carriers in an area already 'covered' by land-based aircraft.

In order to find a logical path through the maze of naval 'hardware' and tactical planning, certain modes of naval action – and interaction – are selected for appraisal. First, it is assumed that a certain kind of naval power is required and this involves the deployment of certain kinds of naval forces which we will call 'primary'. The opposition can be expected to deploy forces intended to degrade the survivability and effectiveness of these 'primary' forces. We call such forces the 'threat'. This 'threat' has to be negated in turn in order to maintain the value of the 'primary' forces, and this third category of forces we will call 'support'. On this basis the following clusters of interactions readily emerge:

(a) *Sea-based air-power* (*essentially tactical*)
Primary: Aircraft carriers.
Threat: Aircraft carriers. (A parallel alignment cancels out.)
Submarines.
Support: Anti-submarine carriers.
Anti-submarine destroyers.
Hunter-killer submarines.
Secondary threat: Guided-missile ships versus destroyers.
More submarines, including hunter-killer types.
Secondary support: More air power, including anti-submarine aircraft.
(b) *Sea-based strategic missile forces* (*submarine-launched*)
Primary: 'Polaris' type submarines.
Threat: Hunter-killer submarines.
Support: Anti-submarine destroyers.
Anti-submarine air-power.
Secondary threat: Guided-missile ships (destroyers).

Secondary support: Aircraft carriers, guided-missile destroyers.
(c) *Amphibious capability*
Primary: Landing ships, tank-landing craft, assault landing vehicles.
Gun and rocket cruisers and other surface vessels for fire support.
Helicopter carriers.
Threat: Land-based air power.
Support: Aircraft carriers.
(Take up chain (a) above.)
Anti-aircraft (guided-missile) cruisers/destroyers.
Secondary threat: Submarines.
Secondary support: Anti-submarine destroyers, aircraft and hunter-killers.
(d) *Interdiction of sea lanes.* (by submarines; otherwise see (a))
Primary: Submarines.
Threat: Anti-submarine surface vessels and aircraft.
Support: Air power and submarines.
Secondary threat: Hunter-killer submarines and air power.
Secondary support: Hunter-killer submarines.

As is clear from these chains of action/reaction, the pivotal combat vessels are ◊ *aircraft carriers* and ◊ *submarines*; these are discussed in both general and country-by-country entries. The Soviet fleet of cruisers and destroyers is discussed in some detail because, in the absence of aircraft carriers, these vessels perform the primary surface-to-air and surface-to-surface functions: ◊ *Cruisers, Soviet,* and *Destroyers, Soviet.* The US navy's fleet of cruisers, largely equipped with surface-to-air missiles, is also given a detailed entry (◊ *Cruisers, US*) but the fleet of frigates, destroyers and escorts is given summary treatment in the ◊ *Destroyer, US* entry. British, West German, Japanese, Italian and French destroyers and similar vessels are also given country-by-country entries. Mine warfare ships, patrol ships, command and communication ships, amphibious warfare ships and the whole panoply of auxiliary supply and service craft are not mentioned, for reasons of space (and technical complexity).

The small Soviet fast patrol boats armed with ◊ *Styx* cruise missiles (whose capabilities are much overrated) are entered under ◊ *Osa,* the main class name.

Commando. A term originally used to describe irregular cavalry units formed by South African Boers. Now used to describe (a) units of the British Royal Marines (Naval Infantry), (b) members of elite units. Since the word has connotations of bravery, efficiency, ruthlessness, strength, etc., it is widely used to flatter assorted inferior soldiery. Thus, it is used by the Fatah (Organization of Palestinian irregulars) to describe their untrained recruits and assorted desperadoes.

Commando M.706. US armoured car. The first post-war US military armoured car, not as yet issued to the US Army but which is being made available to certain foreign countries. It is a lightly armoured 4 × 4 vehicle which can swim (slowly) in calm waters and carry twelve men; it can mount a variety of turrets, but a two-machine-gun one appears to be standard. The poor suspension limits cross-country use, but otherwise it has very good mobility.

Specifications: Length: 18 ft 8 in. Width: 7 ft 5 in. Height: 7 ft 2 in. Weight: 6·35 tons. Engine: 190 bhp. Road speed: 65 mph. Range: 550 miles. Water speed: maximum 4 mph. Armament: 0·5-in and 0·3-in MGs mounted in small turret.

▷ *Armoured Car* for competitive types; also *Armoured Personnel Carrier*.

Company. An army formation formally subordinate to the ▷ *battalion* and comprising two or more ▷ *platoons*. The size of a company varies around a median figure of 100 for infantry units and perhaps 60 for armoured and other units, though US army infantry companies number about 200 men, and Soviet tank companies only 40.

Comprehensive Test Ban Treaty. ▷ *CTB*.

Confrontation. A continuing conflict in which all means other than overt warfare are used; a description of the Indonesia–Malaysia conflict (1963–6). The term can be used to describe a conflict in which a combination of ▷ *economic warfare*, ▷ *political warfare*, ▷ *subversion*, and covert and clandestine operations is used. The ▷ *Cold War* was in effect a confrontation;

the Arab–Israeli conflict has also taken this form at various times.

Conscientious Objection. A denial of the state's right to compel the participation of a given individual in the activities of its armed force organizations, this denial being based on that particular individual's adherence to stated values and consequent rules of behaviour which are claimed to transcend the individual's obligations to the state with respect to military service. The objection can occur at several levels of generality, and the state's recognition of the objection and the consequential grant of exemption from military service is often related to the generality of the objection.

1. Objection to any participation in any armed force activity in any contingency. Such objection is usually part of the individual's adherence to a religious faith or ideology such as pacifism or Buddhism. If comprehensive adherence can be documented, exemption is often granted in those communities where the prerogative rights of individuals over certain matters are generally recognized; activities totally outside the armed force organization or direct war production may then be substituted. It should be pointed out that in states with developed economies which are waging general war *any* economic activity is ultimately related to the war effort. Thus total objection logically implies abstention from any economic activity other than consumption.

2. Objection, *in any contingency,* to activities which imply the obligation, or eventual necessity, of killing those designated as 'enemies'. Such objection, being also general with respect to the contingency, but less general with respect to the exemption claimed, is more easily granted in those communities where the individual's prerogative rights are accepted.

3. Objection to activities within the armed force organization in one stated contingency. Such objection is *sui generis* 'political' rather than 'religious' in that the individual claims the right to formulate policy at variance with the policies of the state's leadership. The overt acceptance by the competent authorities of such an objection therefore implies the right of

any individual to formulate his own policies with respect to military activities actual or potential. In view of this, no coherent political entity overtly grants exemption on these grounds. If the individual can claim that his objection is *not* political, though particular to a stated contingency, the right is sometimes granted. Such non-religious and non-political objection is generally related to membership in a particular ethnic group related in some way to the ethnic composition of the prospective enemies. Thus Jewish troops were removed on request from the British invasion armada which sailed to the Middle East in the Suez war of 1956 (when the target of the attack was officially not yet determined); similarly, the state of Israel grants exemption from military service to all its citizens of Arab descent.

Exemption from military service on conscientious grounds, the criteria of exemption and the nature of alternative services imposed, if any, vary widely. First, the authorities of those states where *no* prerogative right of the individual is recognized (i.e. 'totalitarian' states) do not grant *de jure* exemption. Such states include the Soviet Union and the communist states of Eastern Europe, and the Far East as well as Spain, Portugal, Greece and certain Latin American states. In many economically backward countries it is generally easy to evade military service – especially if the conscientious nature of the evasion is concealed. If the nature of the objection is advertised in order to gain recognition for the right to avoid, rather than evade, military service, then compulsion is usually enforced.

Objection type (2) is generally granted, subject to more or less exacting documentation requirements, in those states where certain prerogative rights are recognized, essentially the industrialized countries of Western Europe, most English-speaking communities elsewhere, and certain non-European non-industrialized states such as India, Ceylon and Israel. Most African and Asian states have no compulsory military service. Almost all countries recognize the legal existence of religious organizations in some form or another, and the full-time clergy are generally granted exemption from all compulsory service.

Conscription. The process whereby members of a political entity or other population are selected and compulsorily inducted into an armed force organization, or otherwise obliged to render specified services.

Contour-flying (Terrain-avoidance). A mode of flying intended to take advantage of the poor performance of air defence radars at low altitudes. Since nearly all radar systems are ineffective against targets below the horizon, an aircraft can avoid radar detection by flying so low that it is shielded by folds in the ground. But in order to do this at high speeds, a radar-computer-autopilot system is required to 'fly' the aircraft, since human pilots cannot see obstacles or correct the flight path within the time available. ◊ *Radar* (TFR) for the technique employed.

Controlled Response. A policy in which the response to enemy attack is deliberately kept within certain definable limits, on the assumption that the enemy recognizes the existence of these limits and also the element of deliberation in keeping within them. This policy, associated with the former U S Defence Secretary McNamara, is supposed to apply even where nuclear weapons are used by the parties to the conflict. The policy implies (a) the existence of mutually recognized degrees of conflict (◊ *Threshold*), (b) a comparable range of weaponry, and (c) a comparable range of targets. In the absence of such symmetry, the stabilizing effect of the policy will be impaired. If for example A destroys a 'medium' target but lacks medium targets itself, B will be forced either to show weakness by attacking a 'small' target ot to escalate by attacking a 'large' target.

The adoption of a policy of 'controlled response' implies the development of a full range of military capabilities, as opposed to the earlier policy of massive retaliation for which (in theory) only nuclear delivery systems were needed. Because of this requirement, a state going from a massive retaliation to a 'controlled response' posture may be seen by its opponents as merely engaging in an arms race with respect to conventional weapons.

'Controlled response', the present US policy, is sometimes described as 'Flexible Response'. ◊ *Massive Retaliation* for what was presumed to be earlier US policy in this field. ◊ *Escalation.*

Conventional Warfare. An inadequate term for any non-nuclear armed conflict. ◊ *War* for a list of concepts and interpretations. ◊ *Threshold* for gradations of conflict.

38 **Corsair II A-7. US attack fighter.** The Corsair II has been in service since 1967 with the US navy as a supplement to and replacement for the ◊ *Skyhawk A-4* and the old Skyraider A-1; following the precedent set by the ◊ *Phantom F-4*, the USAF is also using a variant for land-based use, designated A-7D. The Corsair is a relatively cheap subsonic aircraft based on an existing airframe design, that of the ◊ *Crusader F-8*, and 'state of the art' avionics. Like the Crusader it has shoulder-mounted wings and a large air-intake below the radar nose-cone. It is powered by a single turbofan (originally a US engine rated at 11,350 lb of thrust, later a Rolls-Royce engine of 14,250 or 15,000 lb) without afterburning. It has two 20-mm guns in the nose and racks for up to 15,000 lb of external stores, though normal armament loads are expected to be around 6,000 lb. The weapon control avionics are very comprehensive, so that ◊ *Bullpup*, ◊ *Walleye*, ◊ *Shrike* and other missiles can be carried; there is also an elaborate navigation system with radar and a computer all-altitude reference system.

The first naval version, the A-7A (to which the details above refer) is already in service; the USAF version, A-7D, has an uprated 14,250-lb turbofan, a multi-barrel 20-mm gun, and more avionics. A developed naval version, the A-7B, is now in production with a slightly uprated 12,000-lb engine, while a later naval version (A-7E) will also have the 20-mm Vulcan cannon and more avionics.

Specifications (A-7A): Length overall: 46 ft 1·5 in. Height: 16 ft 2 in. Wing span: 38 ft 9 in. Gross wing area: 375 sq ft. Weight empty: 14,875 lb. Maximum take-off weight: 32,500 lb. Maximum

level speed: 578 mph (clean). Ferry range: 3,800 nautical miles.
◊ *Fighter* for general subject entry.

Corvette. ◊ *Destroyer.*

Counterforce Capability. A quantitative and qualitative level of armaments which is considered sufficient to destroy or absorb enough of an enemy's nuclear capability to make the latter's retaliation acceptable. The possession of such a capability therefore depends on (a) an adequate number of sufficiently accurate missiles for strikes against the enemy's missiles and air or naval weaponry to deal with other systems, (b) an anti-missile defence, (c) civil defence measures, (d) a tolerance of damage which compensates shortfalls in (a) + (b) + (c). If (d), the level of damage regarded as acceptable, is high enough, it can be said that a counterforce capability can be achieved without any 'hardware'. (If Mao Tse-tung's reported willingness to engage in a general nuclear war with the United States ever materializes, the famous Chinese slogan that the US is merely a 'paper tiger' may be said to represent a counterforce capability on its own.)

In concrete terms, a counterforce capability implies both large numbers of missiles accurate enough to centre hardened targets, and intelligence about the location of enemy installations. Two developments – the introduction of development ◊ *MIRV* warheads, and the satellite observation systems (◊ *Surveillance Satellite*) – have made it considerably easier for either super-power to have a counterforce capability against the other.
◊ *Assured Destruction* and *First Strike.*

Counterforce Strategy. A strategy in which the enemy's retaliatory forces, and general military forces, are the targets of a nuclear attack, rather than its population centres. The concept was unveiled at the Ann Arbor speech of Secretary McNamara (16 June 1962): ' . . . The principal military objectives, in the event of a nuclear war stemming from a major attack on the Alliance, should be the destruction of the enemy's military forces, not of his civilian population.' Such a strategy implies the possession of nuclear delivery systems in sufficiently large numbers and accurate enough to destroy hardened and dispersed missile sites and bomber bases, much more difficult targets than large and exposed cities. If a power has this capability, this in turn implies that it can also mount a successful ◊ *pre-emptive* first strike. Thus, though the speech – and the policy – were intended to relax tension by removing the threat of nuclear strikes on Soviet cities, the Soviet leadership may have read it as a threat, since the U S could now mount a pre-emptive strike aimed at depriving the Soviet Union of its deterrent. Support for this interpretation can be found in a later McNamara speech: ' . . . In planning our second strike force we have provided a capability to destroy virtually all the "soft" and "semi-hard" military targets in the Soviet Union and a large number of their fully hardened missile sites . . . ' (Senate Armed Services Committee, 1964). ◊ *Counterforce Capability*. But as the Soviet Strategic Offensive forces developed in 1964–6, U S policy drifted towards ◊ *assured destruction*.

Counter-insurgency (Warfare). Action taken by (or on behalf of) a constituted government against a party or parties waging ◊ *Revolutionary War* or conducting a localized armed rebellion (insurgency). As a result of the Vietnam war and various colonial conflicts, counter-insurgency warfare has taken a definite form; methods and techniques have been developed, but not so far with very satisfactory results of general applicability.

These usually take the form of 'civil action' at the village level: technical assistance, economic aid and propaganda plus the use of small independent infantry forces in small-scale action against guerrilla forces; at the national level, economic aid and technical assistance plus the use of conventional forces in large-scale operations. There is some reason to believe that village-level operations are more likely to be cost-effective (where 'cost' includes political as well as military costs). But since the main arm of revolutionary war is ◊ *subversion*, it seems that the present balance of intelligence versus active military operations should be

altered in favour of the former. ◊ *Guerrilla Warfare*, for the secondary arm of revolutionary war.

Counter-value Capability. A nuclear strike capability so limited by the number and/or accuracy of the delivery systems that only large cities are feasible targets. The present French, British and Chinese nuclear delivery systems are so limited, in so far as they have any effective capability at all *vis-à-vis* super-power defences. The Soviet Union and the United States have vastly greater capabilities, including ◊ *counterforce capabilities*. For other (qualitative) assessments of nuclear offensive forces, ◊ *First-Strike Capability* and *Second-Strike Capability*. For actual weapons and systems ◊ *U S, Strategic Offensive Forces*, and the same for U S S R, France, U K, and China. ◊ also *Deterrence*.

Counter-value Strategy. A 'strategy' (actually a tactic) in which available nuclear strike forces are targeted on the enemy's cities. Such a strategy may be imposed by the lack of any but a ◊ *counter-value capability* – where delivery systems are limited by quantity and accuracy to this role; or it may be a deliberate ◊ *second-strike strategy*. Such a strategy may also result from deliberate choice rather than from poverty of means, where a policy of minimum deterrence is adopted (◊ *Deterrence*).

Credibility. The expectation on the part of others that a threat or promise related to a specific contingency will in fact be carried out. Credibility is therefore another's estimate of one's resolve to act as announced in a particular way in a particular setting. In the nuclear context it is another's estimate of our own view of a situation which includes his strike capability. Credibility is therefore related to one's opponent's military capability rather than to one's own. Since the magnitude of a threat is capability plus credibility, and since the latter can be increased by, say, improving the ◊ *civil defence* system, it is obvious that even 'purely defensive' measures can increase the credibility and therefore the magnitude of a threat in the eyes of others. The credibility of a threat depends on the degree of implied automatic behaviour; thus a computer-operated, purely independent system is most credible; a very vulnerable strike capability is also very credible, since hesitation or deliberation are seen as unlikely. A completely invulnerable weapon system with a recall mechanism is least credible, since it allows for many more than two ('strike'/'no strike now') options, and is therefore subject to post-attack negotiations or simple blackmail. It should be clearly realized that the main factor in credibility is the assumed importance of the national interests which are being protected by the 'threat' or 'promise'.

73   Crotale. A French low-level air-defence system (not yet in production). It consists of small solid-fuel surface-to-air missiles, a search radar for detection, a computer unit for tracking, and subsidiary manual control. For general mode of operation, ◊ *Rapier*, the British equivalent. The Crotale system's missiles have a top speed of Mach 2·5+ and this is reached within 2·5 seconds of launch. A high kill-probability is claimed for aircraft flying at or below Mach 1·2 at low altitudes. The whole system can be housed in a variety of marine mountings as well as in two small jeep-sized vehicles or towable units. This equipment is being supplied to South Africa under the name Cactus.

◊ *Missile* for general entry and competitive types.

Cruisers, French. The French navy operates two vessels described as 'anti-aircraft cruisers'. These large and very vulnerable ships have undergone repeated changes of function, nomenclature and ancillary equipment, in an effort to justify their continued deployment.

The *Colbert* was completed in 1959 after six years of construction during which its configuration was recast several times. Its high castellated superstructure mounts 18 gun turrets: 8 twin 5-in guns and 10 twin 57-mm AA guns. The ship is partially armoured. At 11,100 tons of full-load displacement this is the nearest thing to a battleship in any European navy. There are facilities for a helicopter, room for 2,400 equipped soldiers, and extensive radar and communication equipment intended for the air-control and flagship roles. There is no anti-submarine weaponry.

The *De Grasse* was laid down in 1938 and completed in 1956 after extensive design modifications. There are 6 twin 5-in gun turrets, though some have been removed when a further extra-high mast for long-range communications was fitted. There is no anti-submarine weaponry.

Cruisers, Italian. The Italian navy   **12** deploys four cruisers, all equipped with surface-to-air missiles (SAMs), one of which also has four tubes for ballistic missiles. Three of the vessels have been completed since 1964, and one is a fully modernized pre-war gun-cruiser. The

French Cruisers – Specifications:

	Colbert	*De Grasse*
Displacement, full-load (tons)	11,100	12,350
Length (feet)	593	618
Guns	8 twin 5-in turrets	6 twin 5-in turrets
	10 twin 57-mm turrets	—
Engine power (shp)	86,000	105,000
Maximum speed (knots)	32·4	33
Radius (miles)	4,000 at 25 kts	5,200 at 18 kts
Complement	964	952
Year laid down	1953	1938
Year completed	1959	1956

modern vessels are very effective and very original designs with ample provisions for anti-submarine helicopters, very clean lines and outstanding handling qualities.

The *Vittorio Veneto* was completed in 1969, and a second, improved version, *Trieste* is under construction. At 8,850 tons of full-load displacement its size is comparable with that of the latest US destroyer leaders, but good design has fully exploited the tonnage. There are provisions for 9 helicopters for AS work, as well as two triple AS torpedo tubes. Anti-aircraft defences include a twin launcher for ◊ *Terrier* SAMs. There are also eight 3-in AA guns.

The two vessels of the *Andrea Doria* class have the same armament as the *Vittorio Veneto* but carry only 4 helicopters. Like the latter they are very wide in relation to their length, though somewhat smaller at 6,500 tons of full-load displacement.

The *Giuseppe Garibaldi* was laid down in 1933 but fully reconstructed and com-pleted in late 1962 as a guided-missile light cruiser. Apart from the ballistic missile tube, for which no missiles are available, the armament is the same as that of the modern cruisers except for two twin 5·3-in gun turrets forward. There are no hangar provisions for helicopters. At 11,335 tons of full-load displacement this is the largest ship in the Italian navy.

Cruisers, Soviet. The Soviet navy has been building a fleet of modern guided-missile ◊ *destroyers* with anti-aircraft and anti-submarine weaponry, and, like other navies, it has no real missions which its gun cruiser fleet can fulfil. Since guided missiles are not being fitted to these vessels, it is thought that they are not in-cluded in the Soviet navy's combat plans.

The 14 *Sverdlov* class cruisers were launched between 1951 and 1958 and are of post-war design. Though their draught is 24·5 ft, one of their intended missions was fire-support from inland river or canal waters. These ships are fitted with mine-

Italian Cruisers – Specifications:

	Vittorio Veneto	*Andrea Doria* (2 vessels)	*Giuseppe Garibaldi*
Displacement, full-load (tons)	8,850	6,500	11,335
Length (feet)	558	490	614
Missiles	1 twin launcher for Terrier, forward	1 twin launcher for Terrier, forward	1 twin launcher for Terrier, aft
Helicopters	9 AS type	4 AS type	none
Guns	8 single 3-in automatic	8 single 3-in automatic	2 twin 5·3-in forward and 8 single 3-in turrets
Torpedo tubes 12-in AS	2 triple	2 triple	none
Engine power (shp)	73,000	60,000	100,000
Maximum speed (knots)	32	31/32	30
Radius at 20 kts (miles)	6,000	6,000	4,000
Complement	550	478	694
Year completed	1969	1964	First completed 1937, fully reconstructed 1957–65

laying equipment, torpedo tubes, surface and anti-aircraft guns, and a great deal of armour. The anti-aircraft bridge is very large and rises well above the forward funnel. The *Admiral Lazarev, Admiral Nakhimov, Admiral Senjavin, Admiral Ushakov, Alexsandr Nevskii, Alexsandr Suvorov, Dmitri Donskoi, Dmitri Pozharskiy, Dzerzhinski, Kosma Minin, Mikhail Kutusov, Murmansk, Oktyabrskaya Revolutsiya* and the *Sverdlov* are still in commission, and various modifications have been made to the superstructure and weaponry. Only the *Dzerzhinski* has been fitted with a twin missile-launcher aft. The *Ordzhonikidze* was transferred to the Indonesian navy which renamed it *Irian.*

Three *Chapaev* class cruisers (*Komsomolets, Kuibyshev* and *Zheleznyakov*), laid down before the war and completed in 1948–50, are being used as training ships.

Specifications (*Sverdlov* class): Displacement: standard 15,450 tons, full-load 19,200. Length: 689 ft. Hull beam: 70 ft. Armour on decks, sides, turrets: from 1 to 5·9 in. Guns, surface: 4 triple 5·9-in. Guns, dual-purpose: 6 twin 3·9-in. Guns, anti-aircraft: 16 twin 37-mm. Torpedoes: ten 21-in tubes. Mines: 140–250. Engine power: 130,000 shp. Maximum speed: 34 knots. Radius: 5,000 miles at 20 knots. Crew: 1,050.

13 **Cruisers, US, includes Guided-Missile Cruisers.** The US navy deploys two classic gun cruisers, nine guided-missile cruisers (CG), and the *Long Beach*, a nuclear-powered guided-missile cruiser (CGN). In addition to these, there are some seventeen other gun cruisers in the reserve fleet, but these ships are to be converted or otherwise disposed of, since heavy gun vessels no longer have a significant role in American naval planning. The primary mission of the gun cruiser is to fight other surface ships, and this has been taken over by carrier-borne strike aircraft. The naval gun duel is considered obsolete and, with it, the conventional gun cruiser, except for the minor task of shore bombardment. But the mid-1950s trend to do away with guns altogether has been reversed, and pairs of 5-in guns have been retrofitted to the *Long Beach*, which originally appeared without any guns at all. Such guns are dual-purpose, i.e. they

can deal with slow aircraft and with those surface targets which do not warrant an air or missile strike.

The *Long Beach* (CGN-9) was completed in 1961 as the first nuclear-powered surface fighting ship. Its external appearance is unusual, with a large square box-like castle, whose outer surfaces are the antennas of its ◊ *phased-array* radar. The ship is equipped with twin launchers for the ◊ *Talos* missile, with a computerized loading and elevation system, two ◊ *Terrier* twin launchers and a single launcher for the ◊ *Asroc* anti-submarine missile. The *Long Beach* is also fitted with a variety of experimental radars, and eight vertical tube silo/launchers for ◊ *Polaris* missiles (though these are not used) and various other devices including the NTDS, tactical data communication and display system, and advanced ◊ *sonar*. The total cost of the ship, excluding later modifications, was 332,500,000 dollars.

The fleet of conventionally powered guided-missile cruisers currently includes three main types:

Three completely modernized heavy gun cruisers of wartime design, the *Albany* CG-10, the *Chicago* CG-11, and the *Columbus* CG-12. The entire gun armament has been removed and replaced with twin Talos launchers forward and aft and twin ◊ *Tartar* launchers on either side. Two 5-in guns were subsequently retrofitted for dealing with targets which do not warrant an expensive missile. The reconstruction has included almost the entire superstructure, and there are a high castle and twin combination stack-masts ('smacks') which support the search radar antennas. Apart from the torpedo tubes, these ships carry Asroc anti-submarine missiles.

The six former *Cleveland* class (CLGs) are partial conversions, with the forward half retaining the original gun armament while the aft section is rebuilt and fitted with a twin launcher for the Talos or the Terrier missiles. The Talos-armed *Galveston* CLG-4 and *Oklahoma City* CLG-5 have 46 missiles, while the Terrier-armed *Providence* CLG-6, *Springfield* CLG-7 and *Topeka* CLG-8 carry 120 of the smaller Advanced Terrier missiles. CLG-3 and CLG-8 have retained two triple 6-in turrets, while the others have a

single turret. These two ships also have three twin 5-in dual-purpose turrets, while all the others have a single turret.
The two oldest guided-missile cruisers

It has a shoulder-mounted wing of the 'variable incidence' type and a characteristic air-intake under the fuselage nose for the single turbojet. Since the first version,

US Cruisers – Specifications:

	Long Beach	CG–10/11/12	CLG–3 to 8	CAG–1 to 4
Displacement (tons)				
full-load	15,000	17,500	14,600	17,500
Length (feet)	721	674	610	674
Missile launchers				
Talos	1 twin	2 twin	⎫	none
Terrier	2 twin	none	⎬ variable	2 twin
Tartar	none	2 twin	⎭	none
Asroc	1 8-tube	1 twin	none	none
Guns, surface	none	none	3 6-in	2 triple 8-in
dual-purpose	2 5-in	2 5-in	2 5-in	10 5-in
anti-aircraft	none	none	none	12 3-in
Torpedo tubes, 12-in				
anti-submarine	2 triple	2 triple	none	none
Armour	none	Side 6 in	Side 5 in	Side 6 in
		Deck 3 in	Deck 3 in	Deck 3 in
Engine power (shp)	80,000	120,000	100,000	120,000
Maximum speed (knots)	31	34	33	34
Radius (miles)	100,000	not available	7,500 at 15 kts	not available
Crew	985	1,010	1,012–1,077	1,273

◇ *Destroyers, US,* which includes all other surface vessels except aircraft carriers.
◇ *Combat Vessels* for general subject entry.

the *Boston* CAG-1 and the *Canberra* CAG-2, retain their six 8-in guns forward (two triple turrets), five twin 5-in gun turrets, and twelve 3-in guns, but have a pure missile aft with two twin launchers for the Terrier missile. The missiles are fed to the launchers by a computerized storage system which can 'reload' every 30 seconds. As in the other ships, the firing installation has been converted for the longer 'extended range' version of Advanced Terrier.

Crusade. ◇ *Holy War.*

35 **Crusader F-8. US naval fighter.** The Crusader is a naval fighter developed in the United States which has been in service since 1957. It now equips the *Oriskany* class of US carriers and the two French aircraft carriers, the *Foch* and the *Clemenceau.*

the F-8A, reached operational status, development versions designated F-8B, C, D, E, H and J have introduced progressive improvements in avionics, power plant and armament, as well as minor structural modifications. All the versions have provision for air-to-air armament, with four 20-mm guns in the fuselage nose and 2 or 4 ◇ *Sidewinder* missiles, but the E has also been modified to carry some air-to-ground armament. The French version, F-8E(FN), has been modified to fit the smaller decks of the French carriers and to carry the ◇ *Matra R.530* missile in addition to the Sidewinders. With a top speed just below Mach 2, and with a modern search and fire-control radar system, the later Crusader versions are adequate as air defence aircraft, but no further versions are planned. A training version, the TF-8A, and two reconnaissance versions, the RF-8A and G, have also been produced; the basic airframe has also been used as a development basis for the ◊ *Corsair II* attack fighter.

Specifications (E): Power plant: one turbojet rated at 16,000 lb st with after-burning. Length: 54 ft 6 in. Height: 15 ft 9 in. Wing span: 35 ft 2 in. Gross wing area: 350 sq ft. Maximum take-off weight: 34,000 lb. Maximum level speed: 'nearly' Mach 2.

◊ *Fighter* for general subject entry.

CS. A non-lethal chemical agent developed for riot control but which has also been used in combat. It is a white crystalline substance dispensed in aerosol form; it has a peppery smell. Doses of 1 to 5 milligrammes per cubic metre of air per minute cause: stinging, burning feeling on skin, coughing, tears, chest tightness and nausea. The effect is immediate, but CS does not kill unless exceedingly large concentrations are absorbed. First developed in the UK, it is now exported to a large number of countries.

◊ *Chemical Warfare*; also *CN* and *DM*.

CTB. Comprehensive Test Ban Treaty. A treaty proposal intended to ban all nuclear detonations. The subject of US–Soviet negotiations within the framework of the permanent Eighteen-Nation Disarmament Conference at Geneva. As long as the development of new warheads for ◊ *ABMs* remains a necessity for the USA and the Soviet Union, the signature of a CTB remains unlikely. At the time of the signature of the ◊ *Test Ban Treaty*, underground test-

ing was excluded because the limitation could not be inspected on the ground (Soviet refusal), while remote inspection was ineffectual.

'Cub' Antonov An-12. Soviet military transport aircraft. The An-12 is serving as a paratroop and freight transport in the Soviet, Indian, Egyptian, Algerian, Iraqi and Indonesian air forces. It is a four-turboprop high-wing design derived from the An-10 airliner, Cat. The rear tail is sharply elevated and contains a tail gunner's position, and there is a ramp for straight-in loading as well as for airdrops. The An-12 is powered by four turboprops rated at 4,000 ehp each and it has a top speed of 373 mph, though the normal cruising speed is 342 mph at 25,000 ft. The maximum payload is 44,090 lb, and there is room for up to 95 passengers in addition to the 5-man crew.

Specifications: Length: 44 ft 3·5 in. Height: 32 ft 3 in. Wing span: 124 ft 8 in. Cargo hold: 44·3 by 9·10 by 7·10 ft/in. Maximum take-off weight: 134,480 lb. Take-off run: 2,790 ft. Landing run: 2,820 ft. Range with 22,050 lb of cargo and one hour reserve: 2,110 miles. Service ceiling: 33,500 ft.

◊ US counterparts *Hercules, Starlifter, Galaxy* and the new Soviet *Cock*.

CW. ◊ *Chemical Warfare.*

CW Radar. ◊ *Radar.*

Cyclonite (RDX). ◊ *TNT.*

D

Davy Crockett. US short-range nuclear weapon. An obsolescent lightweight unguided rocket fired from a tube launcher and provided with nuclear or other warheads. The Davy Crockett, now withdrawn from service, was the smallest nuclear delivery package ever deployed: it could be fired, bazooka-fashion, from a jeep or used as a mortar on the ground. With a maximum range of 8,500 yards, it was intended as a front-line weapon, under the control of correspondingly junior commanders. This, and the lack of feasible fail-safe controls, has led to its withdrawal.

Deep Sea Treaty. At the present time, most of the world's seabed area remains entirely unexplored, but up to 95 per cent of it is already reachable for certain military and civilian purposes. While the 1958 Geneva Convention on the Continental Shelf provides a jurisdiction for the seabed less than 200 metres deep (or indeed beyond this limit, where natural resources can be exploited), there was, until 1970, no agreement upon the military uses of the seabed.

The seabed is now being used for the deployment of passive sonar chains, which form a grid across much of the world's ocean areas. These grids are intended to detect and locate submarine movements (◊ *Anti-submarine Warfare*). Another way in which the seabed could be used for military purposes (but is not) is for the deployment of nuclear delivery systems. With current ICBM accuracies, land-based long-range missiles in their steel and concrete silos have become vulnerable to ICBM strikes. Underwater tractors crawling on the seabed could, therefore, provide a more secure form of deployment. Both the US and the Soviet Union were interested in some form of ◊ *arms limitation* agreement,

but when both submitted draft treaties to the Eighteen Nation Disarmament Committee (◊ *ENDC*) in March 1969 the two drafts were conflicting. The Soviet draft covered the seabed outside a twelve-mile limit, and all military installations, and would therefore have rendered illegal the passive sonar grids deployed by the US. The American draft cut down the 'coastal' strip of seabed to just three miles, but was limited to actual weapons – nuclear, biological or chemical – and did not cover such things as sonar grids. But by October 1969 a compromise was reached. The US accepted the Soviet twelve-mile limit, while the Russians accepted the American interpretation of what should be limited. This will probably remain as an agreed settlement, though inspection may present some problems, since missile silos can no doubt be disguised as plausible military installations of the non-forbidden variety.

Defence in Depth. A system of ground defence based on successive lines or independent perimeters, as opposed to linear defence where all available forces are deployed in a single frontal system. 'Defence in depth' was developed as a countermeasure to ◊ *Blitzkrieg* or, more generally, ◊ *indirect approach* tactics of deep penetration. Instead of the conventional 'thick' front with 'soft' supply lines behind it, defensive forces are deployed in mutually supporting lines or perimeters. These are intended to resist direct assault by the concentrated deployment of anti-tank (and anti-aircraft) weapons within them, while the artillery they contain is used to interdict the passage of the enemy's 'soft' troops towards or between the perimeters or system of lines. While a linear defence is disrupted once enemy armour penetrates it, since supply and communication lines are cut, independent perimeters are unaffected by envelopment since they are self-sufficient (at least for a while) with respect to supplies and ancillary services. Instead, it is the enemy attack which is disrupted since the hostile armour – which can pass between the defended zones – is separated from its unarmoured infantry and supply train, which is stopped by artillery fire. Once the attack is weakened, the mobile forces of the defence emerge from the rear and counter-attack the by now unbalanced enemy forces.

The Soviet application of this tactic relies on the use of successive lines rather than perimeters; their flanks are supposed to rest on natural obstacles and the artillery is intended to destroy forces trapped between the lines rather than between the perimeters. There is not much in this difference since where there are no suitable 'natural obstacles' the Soviet application requires a complete perimeter too.

If the available armoured, or more generally, mobile, forces are not held in reserve in the rear but are instead deployed within the defended points, the defence loses its 'sword' and becomes all 'shield'. This means in effect that the concept is degraded into a linear defence which consists of a series of points rather than an actual line, and there is very little in the distinction. In other words, what is valuable about the tactic is the optimization of different types of forces, and if mobile and less mobile forces are put together then its effectiveness is lost.

Delivery System. A vehicle or a group of vehicles which transport a warhead to target. It includes associated guidance, communication and basic maintenance equipment. The term is currently fashionable and is sometimes used for simple vehicles, but it is more generally used for missile systems which consist of ground control, launching mechanisms, missile motor, missile guidance or homing device(s) and warhead.

A Nuclear-Capable Delivery System is one which can be used to carry nuclear warheads to their target. When a nuclear warhead is actually fitted to a vehicle, it becomes a Nuclear Delivery System. The

number and quality of these is the commonly used arithmetical measure of nuclear (strategic) offensive capability, since delivery systems rather than the warheads are the constraining factor. The production of nuclear devices appears to be less demanding in terms of skills and resources than the vehicles needed to carry them – unless these are simple aircraft which are vulnerable to modern air defences; and the missiles themselves are relatively cheap compared to the ground or mobile facilities needed in ◊ hardened and properly controlled sites. Thus ◊ Minuteman 2 missiles had original capital costs of a million US dollars, their silos about two millions, and the multi-missile ground control facility about the same per missile. The actual warhead – included in the missile cost – is thus a fraction of the total cost. In more expensive missiles – and Minuteman is the cheapest land-based missile system – warhead cost makes up an even smaller proportion of the total cost. ◊ Missile, Bomber and F B M S.

Delta Dart F-106. US fighter. The F-106 has been the 'first-line' air defence fighter of the US air force since 1959 serving with the Air Defense Command within the ◊ N O R A D air defence organization. It is a single seat delta-wing single-jet based on an even earlier interceptor, the Delta Dagger F-102, of which it retains the basic structural features. The F-106 has remained an effective weapon because its mission, the interception of long-range bombers over the continental United States, requires speed, endurance and a large weapon/avionics payload but does not require the combat agility which is one of the features of more recent Mach 2 designs. The F-106 is fitted with advanced interceptor radar and fire-control systems which give it a so-called 'all-weather' capability: interception at stratospheric altitudes and/or in poor visibility. The F-106 is equipped with the ◊ Falcon missile and the ◊ Genie unguided nuclear-warhead rocket; both these and the ◊ E C M and weapon-control avionics have been repeatedly up-dated. In 1968 it was announced that the F-106 would be adapted for the period 1969–73 by further weapon/avionics improvements, including 'downward-looking' radar to deal with contour-flying bombers ($\diamond$ *A.M.S.A.*). This new version, called F-106X, is to be controlled by the new $\diamond$ *AWACS* airborne air-defence system instead of the present ground-based $\diamond$ *SAGE* and $\diamond$ *BUIC*.

Specifications (F-106): Length: 70 ft 8 in. Height: 20 ft 3 in at the tail. Wing span: 38 ft 3 in. Gross weight: 30,000 lb plus. Maximum level speed at optimum height: exceeds Mach 2; in 1959 an F-106 set a (then) speed record at 1,529 mph. Operational ceiling: over 60,000 ft.

$\diamond$ *Fighter* and *Radar* for general subject entries.

Demilitarized Zone (DMZ). A defined area within which it is prohibited to station military forces or to retain or establish military installations of certain kinds. The nature of prohibited installations and the arrangements for internal security forces, if any, are defined by the particular agreement under which the DMZ was established. Since demilitarized zones are established between neighbours who are likely to come to blows they tend to be nasty places to take a walk in. Both clandestine and sharp overt conflicts are common in the three main DMZs: Israel–Syria (now defunct), North–South Vietnam, and North–South Korea. A $\diamond$ *nuclear free zone* is a DMZ in terms of nuclear weapons only and liable to be advocated by a party with 'conventional' superiority in the area.

Destroyer (includes Frigate, Destroyer Escort, Corvette and Escort). Destroyers are conventionally defined as 'speedy combat vessels with light or no armour and heavy armament'. Originally developed as platforms for torpedo weapons and for operation against torpedo boats, destroyers are now multi-purpose ships variously armed with missiles, guns, torpedoes and depth charges.

The largest variety of vessel in this group is variously known as destroyer-leader, or frigate in American usage. In British usage a frigate is smaller than a standard destroyer, whose typical size is (full-load displacement) over 3,000 tons. All three types of vessels have speeds of the order of 30 knots and crews of about

300 men. Smaller and somewhat slower destroyers are known as destroyer escorts, or simply as escorts. As is shown in the country-by-country entries below, there are many sub-classifications within these groups, and the US navy's coding system (in approximate descending order of size) gives some idea of these:

DLGN Nuclear-powered guided-missile destroyer-leader
DLG Conventionally powered guided-missile destroyer-leader
DL Destroyer-leader
DDN Nuclear-powered destroyer
DDG Guided-missile destroyer
DDR Radar picket destroyer
DD Destroyer
DER Radar picket destroyer escort
DEG Guided-missile destroyer escort
DE Destroyer escort

In addition to these, smaller but quite fast vessels of 1,000–2,000 tons full-load displacement are sometimes described as corvettes (though this description is used in France for quite large destroyers).

More important than confused nomenclature are the configuration and mission profile of these vessels. Three primary roles tend to be emphasized:

1. Anti-aircraft. Both as radar pickets for early warning and as carriers of surface-to-air missiles (SAMs) destroyers provide protection for amphibious operations, convoys and aircraft carrier task forces. In the latter case, their role is sacrificial in the context of a major engagement: their early warning helps to alert the carriers' air defence fighters, but the latter are unlikely to provide a local umbrella for the picket ship in the event of a major attack.

Apart from SAMs such as $\diamond$ *Terrier* and $\diamond$ *Tartar*, $\diamond$ *Seaslug*, $\diamond$ *Sea Dart*, and $\diamond$ *Seacat*, destroyer-type vessels also mount dual-purpose guns of 4–5-in calibre as well as specifically AA cannon in the 57-mm to 30-mm calibre range.

2. Anti-submarine. This is the primary role of destroyers in most Western navies, and a whole group of detection devices and weapons is used:

VDS, variable depth sonar ⎫
Long-range fixed sonar in hull ⎬ $\diamond$ *Sonar*
Dip sonar suspended by shipborne helicopters ⎭

DC (depth-charge) racks. Simple release

slides for depth-controlled explosive charges DC mortars. These project DCs by explosion (◊ *Hedgehog*) DC rocket projectors.

This group of weapons is gradually being supplanted by: Sonar-guided homing torpedoes (typically with a 13-minute life at 30 knots) DC and homing torpedo-carrying 'hunter-killer' helicopters Missiles such as ◊ *Ikara*, ◊ *Asroc* and the French Malafon (◊ *Anti-submarine Warfare*).

3. Surface. Though Soviet destroyers (see below) are equipped with surface-to-surface cruise missiles, Western navies, oriented around aircraft carrier task forces, do not deploy such weapons, though most larger SAMs can be adapted for use against surface targets.

The primary role of carrier-based and shore-based aircraft is the interdiction of surface vessels and gun equipment for this role has become correspondingly less important. Thus instead of mounting triple 5 or 6-in turrets for a heavy first salvo, present gun armament tends to consist of single turret automatically loaded guns in the 4–5-in calibre range. The early-1960s' trend of almost gunless destroyers appears to have been reversed, and it is interesting to note that the US navy's most recent vessels tend to have more gun power than their predecessors of the early 1960s.

Destroyers, British (includes Fri-   **14**
gates). The Royal Navy maintains some 60 vessels of the guided-missile, gun-destroyer, anti-submarine frigate and general purpose escort categories. These are mostly modern or fully modernized vessels with advanced electronics, good accommodation and modern weaponry.

Type 82. A new class of guided-missile destroyers now in the advanced development stage. At 6,750 long tons of full-load displacement they will approach light cruiser size, but the crew required is to be kept down to 433 by the use of automated steering, engine control and store handling. Their combined gas turbine (for emergency start and 'peak') and steam turbine

Specifications:

	Type 82	*County*	*Daring*
Displacement (tons)			
full-load	6,750	6,200	3,600
Length (feet)	490	505	366
Missiles	1 twin launcher for Sea Dart 1 single-Ikara	1 twin launcher for Seaslug, 2 × 2 quadruple for Seacat	none
Guns, 4·5-in	1	4	6
40-mm	2	0	6/2
20-mm	0	2	0
Torpedo tubes	0	0	5/0
Helicopter	1	1	none
Depth-charge, mortar	1 3-barrelled	none	1 3-barrelled
Engine power (shp)	74,600/30,000	60/30,000	54,000
Maximum speed (knots)	32	32·5	34·75
Radius (miles)	5,000 at 18 kts	n.a.*	4,400 at 20 kts
Crew	433	440	297
Year first completed	—	1962	1952

* not available.

(for economy cruise) propulsion system is thought to be highly efficient in terms of output/space/weight. These vessels will be equipped with the ⬦ *Sea Dart* SAM and the ⬦ *Ikara* AS missiles, as well as a single rapid-fire 4·5-in dual-purpose and two 40-mm AA guns. There will also be a depth-charge mortar and a light helicopter with some AS capability.

The eight *County* class of large destroyers are the backbone of the fleet. Their armament includes ⬦ *Seaslug* 2 and ⬦ *Seacat* SAMs, two twin rapid-fire 4·5-in gun turrets and two single 20-mm cannon. This rather varied equipment configuration is a result of the many different mission requirements these ships have to fulfil: coastal bombardment for amphibious support, fleet air defence, and 'police', escort and 'showing the flag' duties. Anti-submarine equipment is limited to a 'hunter-killer' helicopter fitted out with dip sonar and homing torpedoes. There are no torpedo tubes, depth-charge mortars or provisions for the Ikara AS missile.

Daring class. The four vessels of this class, completed between 1952 and 1953,

represent the immediate post-war design wisdom: too many guns, no missile-launchers and ten torpedo tubes (later reduced to five and now fully suppressed). At 3,600 tons full-load displacement they are large by Second World War standards but just over half the size of the *County* class. They are armed with two twin 4·5-in guns forward and a twin turret aft; there is also a three-barrelled depth charge mortar. They have the sharp depression in the aft hull characteristic of classic destroyers, and a rather cluttered aft superstructure.

There are also some *Battle* and *Ca* class destroyers, some of which have been rearmed with Seacat SAMs. The *Battle* class vessels are conversions to the early-warning picket role, and have a very large search radar mounted on the centre mast.

Leander class. 26 vessels of this type, described as frigates, have been or are to be built. At 2,800 tons of full-load displacement they are smaller than modern destroyers (in line with British nomenclature, which reverses the US usage of 'frigate' for the larger ships) but larger

Leander F-AS-GP	Tribal F-AS	Rothesay F-AS	Whitby F-AS	Blackwood F-Utility
2,800	2,700	2,600	2,560	1,456
360	350	360		300
1 quadruple launcher for Seacat	n.a.*	none		none
2 twin	2 single	2 twin		none
2/0	2 single	1	2	2
0/2	0	none		none
0	0	n.a.*		none
1 light	1 light	none		none
1 3-barrelled	1 3-barrelled	2 3-barrelled		2 3-barrelled
30,000	20,000	30,000	30,430	15,000
30	28	30	31	27·8
n.a.*	n.a.*	n.a.*	n.a.*	4,000 at 12 kts
263	253	200	221	140
1963	1961	1960	1956	1955

than many wartime destroyers. These vessels are equipped with an AS helicopter, depth-charge mortars (◊ *Hedgehog*), and advanced sonar. For air defence these vessels originally had two 40-mm AA guns, but in later ships these were replaced with Seacat SAMs and two 20-mm guns. All these vessels are equipped with a twin forward 4·5-in dual-purpose gun. The superstructure is very clean with a large air defence search radar on a separate mast and a central mast for surface radar and radio antennas.

Tribal class. The seven vessels of this class were completed over the years 1961 to 1964 and were mainly intended for 'police' or 'colonial' duties. This means that their armament and fittings are generally inadequate for specialized fleet duties, while their endurance, accommodation and communication facilities are suitable for sustained independent operations. As well as two 4·5-in guns, in single turrets forward and aft, there are also two single 40-mm AA guns. Seacat SAMs may be fitted to this class, but with the exception of one vessel AA protection is now limited to the rather inadequate guns. A light helicopter and a depth-charge mortar (◊ *Hedgehog*) are provided for anti-submarine work.

The *Whitby* and *Rothesay* classes are essentially similar, the nine *Rothesay* class vessels completed in 1960–1 being based on the six *Whitby* class vessels completed in 1956–8. The full-load displacement of the *Rothesays* is slightly larger (2,600 tons as against 2,560 tons), while their crew is smaller (200 against 221), reflecting the use of more labour-saving machinery. Both classes are equipped with twin 4·5-in guns forward, two depth-charge mortars, and 40-mm AA guns, a twin turret in the *Whitby* class and a single turret in the *Rothesay* class. Some of these vessels are being equipped with Seacat missiles. The *Leopard* class of four vessels is similar in tonnage and general appearance to this group though it has diesel propulsion instead of steam turbines, and there are two twin 4·5-in gun turrets.

Blackwood class. Twelve vessels of this class were completed between 1956 and 1958. At 1,456 tons of full-load displacement they are quite small, and their cost

was further reduced by prefabricated welded construction. They are essentially old-style AS vessels with depth-charge mortars only, and no provisions for a search helicopter and no launchers for AS homing torpedoes or AS missiles. There are also two 40-mm AA guns but no heavier dual-purpose guns.

Apart from these classes, the British navy also deploys a number of wartime destroyers fully modernized and converted to AS work, four modern aircraft-direction frigates intended to assist carrier task forces, and a few modernized wartime frigates.

Destroyers, French (includes Frigates and Corvettes). The French navy deploys a number of modern French-built vessels of original design and armament configuration, though there are also some older vessels of US origin which are now being retired. One feature of this fleet is the lack of standardization in armaments, a problem which seems to affect the French armed forces in general. (◊ also *Cruisers, French*.)

Suffren class. These two vessels are guided-missile frigates (in the US meaning of the term, i.e. large destroyers). They are equipped with Masurca SAM and Malafon ASW missiles. A very large ball-shaped radar antenna is prominent on the forecastle, and these ships are supposed to have advanced sonars and other electronics. Gun armament is limited to 3·9-in and 30-mm AA weapons.

Aconit. Five ships of this class are to be built. They are officially described as corvettes, though their tonnage (3,560 full-load) puts them in the destroyer class. They are primarily intended for ASW and will carry Malafon AS missiles.

La Galissonnière. This ship was completed in 1962 as a destroyer/command ship. A helicopter is provided as well as extensive sonar and communication equipment. The full-load displacement is 3,910 tons.

Surcouf class. Seventeen of these ships were completed between 1955 and 1958, five are fitted out as radar pickets, four are equipped with ◊ *Tartar* SAMs, and some are to be converted to a more specialized ASW role (with fewer guns, the Malafon AS missile and depth-charge projectors).

With their sharply depressed aft hull and cluttered superstructure these ships do not have a very modern appearance; but their design does make effective use of their 3,850 full-load tonnage.

Frigates:

Commandant Rivière class. Nine of these general-purpose frigates were completed between 1962 and 1968, and are used for a variety of missions including training and fishery protection.

Le Normand class. Fourteen of these were built between 1956 and 1960 as primarily anti-submarine vessels with specialized armament.

but the fleet is increasingly equipped with modern Italian-built ships of excellent design.

Impavido class. The two vessels of this class are guided-missile destroyers completed since 1963. With a tonnage just below 4,000 tons full-load they are medium-sized by present standards, but their armament is well balanced and comprehensive: a twin 5-in gun turret forward, four 3-in AA guns in single turrets, a ◊ *Tartar* SAM launcher aft, and torpedo tubes and a helicopter for anti-submarine work. Their radar equipment and interior fittings are advanced.

French Destroyers – Specifications:

	Suffren	Aconit	Surcouf	Commandant Rivière	Le Normand
Displacement, full-load (tons)	5,700	3,560	3,850	1,950	1,795
Length (feet)	518	417	422	338	326
Missile launchers					
surface-to-air	Masurca 1 twin	none	Tartar single	none	none
anti-submarine	Malafon single	Malafon single	none	none	none
Guns, 20-mm	none	none	6	none	2
30-mm	2 single	2 single	none	2 single	none
57-mm	none	none	6	none	3 twin
3·9-in	2 single	2 single	none	3 single	none
5-in	none	none	6 twin	none	none
Torpedo tubes					
anti-submarine	4	4	6	6	4 triple
anti-shipping	none	none	6	none	none
Depth-charge					
mortar	none	1 quadruple	none	1 quadruple	1 quadruple
rocket	none	none	none	none	2
rack	none	none	none	none	1
Engine power (shp)	70,000	27,200	63,000	16,000	20,000
Maximum speed (knots)	34	27	35	25	29
Range (miles)	5,000	5,000	5,000	4,500	4,500
at knots	18	18	18	15	12
Complement	446	252	336	214	200
Year first completed	1967	1970	1955	1962	1956

Destroyers, Italian (includes Frigates). The Italian navy operates a number of older vessels of US origin as well as several pre-war Italian vessels,

Impetuoso class. These two ships are the immediate predecessors of the *Impavido* class and embody early-1950s design concepts: too many AA guns, no SAM missiles, no helicopter, and a larger crew for somewhat smaller vessels.

Apart from these, there are also some older US and Italian destroyers. Frigates (i.e. destroyer escorts): *Alpino* class. Two ships in this class are being completed. At 2,700 tons full-load

trend in anti-submarine weaponry: a single-barrel depth-charge mortar instead of a ◊ *Hedgehog*; anti-submarine torpedo tubes instead of depth-charge racks. The anti-aircraft armament has also changed, with two 3-in guns in lieu of two twin 40-mm guns.

Italian Destroyers – Specifications:

	Impavido	Impetuoso	Alpino	Bergamini	De Cristoforo
Displacement, full-load (tons)	3,941	3,800	2,700	1,650	940
Length (feet)	430	419	352	308	263
Missile launchers	1 Tartar	none	none	none	none
Helicopters	1 light	none	2 light	1 light	none
Guns, 40-mm	none	16 single	none	none	none
3-in	4 single	none	6 single	3 single	2 single
5-in	2 twin	2 twin	none	none	none
Torpedo tubes, 12-in anti-submarine type	2 triple	2 triple	2 triple	2 triple	2 triple
Depth-charge, mortars	none	1 3-barrel	1 single	1 single	1 single
Engine power (shp)	70,000	65,000	31,800	15,000	8,400
Maximum speed (knots)	34	35	28	26	24
Radius (miles)	3,300 at 20 kts	3,400 at 20 kts	4,200 at 18 kts	4,000 at 10 kts	4,000 at 18 kts
Year first of class completed	1963	1958	1969	1961	1965

displacement they have provisions for two helicopters of the AS variety and mount 6 single 3-in AA guns. There are also AS torpedo tubes and a depth-charge mortar. These are somewhat slower than destroyers but have good endurance.

Centauro class. Four ships in this class were completed between 1957 and 1958. Their armament is very light, and four 40-mm guns are being retrofitted.

Bergamini class. The four vessels in this class have a full-load displacement of 1,650 tons and are rated as 'light frigates'. They carry a light helicopter and have the usual AA and AS armament.

De Cristoforo class. Four of these vessels were completed in 1965–6 and a fifth is under construction. At 940 tons of full-load displacement these are in fact small escorts, and their design is based on the earlier *Albatros* class of the mid-fifties. The difference between the two illustrates the

Destroyers, Japanese (includes Frigates). The Japanese navy still operates a number of modernized wartime vessels of both Japanese and American origin, but most of the fleet consists of very efficient modern vessels built in Japan. Their armament is, however, largely American in origin.

Improved *Moon* class. Two of these are already operational and two more are being built. They are intended primarily for anti-submarine work, for which they have a very comprehensive armament. There are neither AA guns nor SAM missiles, and this is in line with the Japanese expectations that these vessels will operate under a land-based air umbrella.

Cloud class. Three of these vessels are operational and four more are being built. At 2,050 short tons of standard displacement they are 1,000 tons smaller than the

Improved *Moon* class and have no provision for an AS helicopter, though again they are primarily intended for ASW.

Amatsukaze. At 4,000 tons full-load displacement this is the largest ship in the Japanese navy, and it is closer to US design trends than the two classes above. There is provision for a helicopter, and, like the *Cloud* class, it mounts 3-in guns, though the single ◊ *Tartar* launcher is the main AA weapon.

Moon class. Two of these ships were completed in 1960 as flotilla leaders and have duplicated radar and sonar equipment. Though their full-load displacement is only 2,860 tons, they carry a whole range of armaments for both surface and ASW work. On the other hand, there are no helicopter provisions or SAM fittings.

Wave class. The seven ships of this class were completed during the 1958–60 period and are primarily intended for ASW. Like other of the smaller Japanese vessels their armament includes a wide range of ASW equipment and 3-in AA guns.

Rain class. The three ships in this class were completed in 1959, and their design was based on the two *Wind* class vessels completed in 1956, which are somewhat smaller but quite similar (*Rain* = 2,500 full-load tons; *Wind* = 2,340 full-load tons). Both classes mount three 5-in guns in single mountings, though the more recent *Rain* class has better (54-cal.) guns. AA gun armament consists of four 3-in in the *Rain* class and 2 quadruple 40-mm in the *Wind* class; this configuration is generally regarded as quite inefficient as compared to the former. Both have a comprehensive array of AS weaponry but no SAM or helicopter fittings.

In addition to these there are a number of wartime US destroyers which are to be retired.

Frigates:

River class. The four vessels of this class were completed during the 1961–4 period, and they fit the DE (destroyer escort) designation, with their 1,700 full-load tonnage and the relatively slow 25 knots speed. Their primary mission is ASW, and gun armament is limited to 2 twin 3-in AA guns.

'Type B' class of destroyer escorts. Two

of these have diesel-power plants, while a third one has a steam turbine. All three were completed in 1956 with similar gun armament. This has since been modified, and the two diesel types now have two 3-in and two 40-mm AA guns, while the steam turbine *Akebono* has only two 3-in AA guns. Unlike all other ships of the DE and D groups these vessels have no provisions for homing torpedoes.

There are in addition a number of older DEs not here entered. Also of importance are the *Mizutori* (10 vessels; see Specifications), *Umitaka* and Diesel SC fast patrol vessels of 330 to 450 tons (standard). These are equipped with 40-mm AA guns as well as depth-charge mortars, homing torpedoes and depth-charge racks.

Specifications are overleaf.

Destroyers, Soviet. The navy has   10 9 8
traditionally played a marginal role within the Soviet armed forces as a result of the Soviet preoccupation with land conflicts and later with strategic air conflicts. Now it seems that the Soviet Union is attempting to make the transition from the status of a 'super-power' to that of a 'world' power and, naturally, the surface fleet is to be one of the instruments of its 'presence'. The Soviet Union has no aircraft carriers (though there is the ◊ *Moskva* class of helicopter carrier/assault ships), so that apart from its old *Sverdlov* class cruisers this fleet consists of modern (or modernized) missile-platform destroyers, supported by older gun destroyers, land-based aircraft and small missile boats (◊ *Osa*).

The NATO-designation *Kresta* covers a group of five of the more modern vessels deployed by the Soviet Union. They are equipped with ◊ *Goa* surface-to-air missiles and ◊ *Shaddock* anti-shipping missiles, as well as torpedo tubes and depth-charge rocket-launchers for anti-submarine missions, and there are facilities for helicopters. The only guns mounted are two twin 57-mm anti-aircraft guns which are radar-controlled. The *Kresta* class is thought to be an efficient combination of the best qualities of two earlier types, the *Kashin* and the *Kynda*. The *Kashin* is a class of AA/ASW vessels which carry no anti-shipping missile mountings. There are two high open-trellis tower-masts and two smaller radar

Japanese Destroyers – Specifications:

	Improved Moon	Cloud	Amatsukaze
Displacement,			
full-load (tons)	3,050 std	2,050 std	4,000
Length (feet)	446	374	430
Missile launchers			
Asroc	1 octuple	1 octuple	none
Tartar	none	none	1 single
Guns, 40-mm	none	none	none
3-in	none	2 twin	2 twin
5-in	2 single	none	none
Torpedo tubes			
12-in anti-submarine	2 triple	2 triple	2 single
21-in anti-shipping	none	none	none
Hedgehogs	none	none	2
Depth-charge racks	none	none	none
rockets	1 4-barrel	1 4-barrel	none
projectors	none	none	none
Engine power	60,000 shp	26,500 bhp	60,000 shp
Maximum speed (knots)	32	27	33
Complement	270	210	290
Year first of class completed	1967	1966	1965

towers. These vessels are equipped with a pair of twin SAM launchers and multi-barrel projectors for anti-submarine rockets. 15 *Kashins* are thought to be operational.

The four *Kynda* class vessels are primarily equipped for anti-shipping missions with two very large quadruple mounts for Shaddock cruise missiles, multi-barrel ASW rocket-projectors and a single twin launcher for Goa anti-aircraft missiles. The *Kynda* class, the first of which was laid down in 1960, is distinguished by the large conical masts which are forward of the two funnels. Four *Kyndas* are currently operational.

The *Krupny* class is somewhat older (though this class is also modern by naval standards, having been laid down in 1958–9) and is also known as *Kanin*. The 8 *Krupny* vessels are equipped with Strela anti-shipping missiles and torpedo tubes. The *Krupnys* originally mounted 16 57-mm AA guns in quadruple mounts, but Goa SAMs have now replaced these. Apart from these ships, the Soviet navy also operates the *Kildin* group and other

missile conversions based on the *Kotlin* class of gun-destroyers. These have either a single Strela anti-shipping missile-launcher (*Kildin*) or twin anti-aircraft missile-launchers.

The Soviet fleet also includes about 50 older *Skory* gun destroyers.
▷*Combat Vessels* for general subject entry.
▷ *Cruisers, Soviet.*

Destroyers, US (includes Frigates, Escorts). The US navy maintains a fleet of about 60 guided-missile destroyers and frigates (destroyer leaders), 6 smaller guided-missile vessels classified as escorts, and about 200 gun-equipped vessels classed as frigates, destroyers and escorts. (There are in addition large numbers of ships in reserve.) This fleet consists of more than thirty different classes and there are significant differences between the sub-classes within these groupings; only a few of the more representative classes are selected for the comments below and tables of specifications. The designation initials are those of the US navy. ▷ *Destroyer.*

Moon	Wave	Rain	River	Mizutori
2,890	2,500	2,500	1,700	420–450
387	358	354	309	197
none	none	none	none	none
none	none	none	none	none
none	none	none	none	1 twin
4 twin	3 twin	4	4	none
3 single	none	3 single	none	none
2 single	4 single	8 single	2 triple	2 single
1 quadruple	1 quadruple	none	1 quadruple	none
2	2	1	none	1
1	none	1	1	1
1	none	none	1 4-barrel	none
2	2	1	1	none
45,000 shp	35,000 shp	30,000 shp	16,000 shp	3,800 hp
32	32	30	25	25
330	230	250	180	70
1960	1958	1959	1958	1960

Soviet Destroyers – Specifications:

	Kresta	Kashin	Kynda	Krupny	Kildin
Displacement (tons)					
standard	6,000	4,800	4,300	3,650	3,000
full-load	n.a.*	6,000	5,200	4,650	4,000
Length (feet)	508·5	492	475	453	426·5
Beam (feet)	55·8	51	53	44	42·7
Missiles, surface	2 × 2	none	2 × 4	2 × 1	1 × 1
AA	2 × 2	2 × 2	1 × 2	none	none
ASW rocket-launchers	2 × 12	2 × 12	2 × 12	none	2 × 16
	2 × 6	2 × 6			
Torpedo tubes	2 × 2	1 × 5	2 × 3	2 × 3	n.a.*
Guns	2 twin	2 twin	2 twin	16 57-mm	16 45-mm
	57-mm	85-mm	85-mm		
Engine power (shp)	100,000	100,000	85,000	80,000	80,000
Maximum speed (knots)	35	35	35	34	35
Crew	400	n.a.*	390	360	300

* not available.

(a) Guided-Missile Frigates, 5,800–9,200 tons (full-load displacement) *Truxtun* (D L G N). Four of these ultra-modern nuclear-powered guided-missile frigates are to be built, while the *Truxtun* itself was completed in 1967. At 9,200 tons of full-load displacement and with a crew of 479 men this ship approaches cruiser size. In line with the latest fashions in weapon 'mix' it has a single twin missile-launcher unit aft for both ◊ *Terrier* (S A M) and ◊ *Asroc* (A S) missiles; tubes for standard 21-in torpedoes and also for 12-in anti-submarine homing torpedoes. An anti-submarine helicopter is carried and provided with hangar, maintenance and flight-deck facilities. The standard dual-purpose 5-in gun is mounted forward, and there are two 3-in A A guns for targets which do not rate a missile. The *Truxtun* class is equipped with the latest thing in radar, sonar long-distance communication and a slot in the N T D S system of data-handling and transmission. It is noteworthy that the somewhat smaller nuclear-powered *Bainbridge* D L G N-25, and the *Leahy* class completed in 1962–4 have 2 twin launchers for Terrier missiles and a separate Asroc launcher.

Belknap (D L G-26 to 34). Nine vessels of this class were completed between 1964 and 1967. Their power plant is conventional and as in other ships of modern design their masts and stacks are combined into conical 'smacks'. The armament and ancillary equipment is almost identical with that of the *Truxtun* class though at 8,150 full-load tons they are rather smaller.

Leahy class (D L G-16 to 24). These nine vessels were completed between 1962 and 1964 and embody the earlier armament configuration which overemphasized missiles at the expense of guns. As well as two twin Terrier launchers fore and aft, the *Leahy* has an 8-tube *Asroc* A S missile-launcher. The only guns carried are four 3-in A A guns in two twin turrets. The anti-shipping 21-in torpedo tubes reinstated in the subsequent classes are also missing, and there are only two triple 12-in tubes for A S homing torpedoes. Another 'advanced' feature found to be of limited value is the very high top speed of 34 knots. This was reduced to 31 knots in the more recent *Belknap* class and further

rated at 30 knots in the nuclear-powered *Truxtun* class.

Coontz class (D L G-6 to 15). These ten vessels were the immediate predecessors of the *Leahy* class and inaugurated the D L or frigate concept of large near-cruiser size (D L G vessels equipped with missiles). There is a twin Terrier launcher aft, an 8-tube Asroc launcher, and the usual 5-in gun turret. The four 3-in A A guns, two twin, and A S (small, 12-in) torpedo tubes complete the armament. These vessels are being retrofitted with N T D S and similar advanced gear.

(b) Guided-Missile Destroyers, 4,500–4,730 tons (full-load displacement)

Charles F. Adams class (D D G-2 to 24). Twenty-three vessels of this class were completed between 1960 and 1964. Their full-load displacement of 4,500 tons is about double the size of destroyers of wartime vintage. The first batch, D D G-2 to D D G-14, were equipped with twin ◊ *Tartar* S A M missile-launchers, but in D D G-15 to D D G-24 this was replaced with a single launcher. All carry A S torpedoes (2 triple 12-in tubes) and the A S Asroc missile, as well as two dual-purpose 5-in guns, in single turrets fore and aft. At 35 knots they are very fast indeed, and their living accommodation is suitable for sustained operations (including general air-conditioning). The super-structure is made of aluminium. This class was based on the *Mitscher* class of four vessels laid down in 1949–50 and repeatedly converted to various roles and weapons including, in the final version, Tartar and Asroc missiles.

Forrest Sherman group. The eighteen vessels of this group comprise two classes (*Hull* and *Decatur*) of which about ten have been or will be converted to D D Gs (guided-missile destroyers) while the others remain D D s. At 4,050 tons of full-load displacement these ships are thought to be remarkably successful in terms of the basic structure, handling characteristics and crew comfort. They embody the design wisdom of the early 1950s (still valid now), with more gun power aft than forward, and aluminium superstructures to enhance stability. In the original form they are armed with two single 5-in guns aft and one forward. In the D D Gs only the forward gun is retained. There are also ◊ *Hedgehog* AS mortars, six 12-in tubes for homing AS torpedoes, and four 3-in AA guns, or two in the DDGs. The DDGs are equipped with Asroc AS and Tartar SAM missiles and are also provided with a drone AS helicopter.

Gearing group (DD). More than eighty vessels of this group were completed in 1945–6 and have been variously modified and modernized under different class names. At about 3,500 tons of full-load displacement they represent the apogee of the wartime destroyer, with a rated speed of 35 knots, a radius of action of 5,800 miles at 15 knots, and crews of about 270 men. The basic *Gearing* group have been modernized by the addition of Asroc launchers and variable depth sonar and the removal of the AA guns. Two twin 5-in guns are retained, as are AS torpedo-launchers. A drone AS helicopter is provided with an aft flight deck.

The basic *Gearing* hull and propulsion have been used to develop a number of

Hedgehog, and 2 triple 12-in tubes. Asroc AS missiles and drone helicopters are also carried. The *Kenneth B. Bailey* (DDR) class are radar pickets derived from the *Gearing* hull by the addition of long-range search radars and the removal of the aft 5-in twin guns. Their mission consists of providing early warning for fleet air defence.

Fletcher group. More than 100 vessels of this group were built during the war. They are now classified into 'Converted *Fletcher*', 'Later *Fletcher*' and '*Fletcher*' *tout court*. Another fifty destroyers, of the *Allen M. Sumner* and *English* classes, were also based on the basic *Fletcher* design. The basic propulsion units are retained in all these ships – geared turbines rated at 60,000 shp. The performance characteristics are essentially the same for all these vessels. A top speed of 34 knots and a radius of 6,000 miles at 15 knots. It is interesting to compare the evolution of the armament fitted on these vessels:

Armament Trends in US Destroyers

	Fletcher	Later *Fletcher*	Converted *Fletcher*
Guns, 5-in	5	4	2
3-in	none	6	4/0
40-mm	6	10	0
Hedgehogs	2 fixed	2 fixed	trainable Hedgehog
Torpedo tubes, 21-in	5	0	0
AS 12-in	6	6	6
Drone helicopter	none	none	yes

This of course ignores intra-class differences and modifications, but is indicative of the trends in destroyer armament during the 1940s and 50s.

variations for various purposes. The *Lloyd Thomas* class have two drone helicopters, four 5-in guns, and a trainable Hedgehog. The *Carpenter* class is similar but has only two 5-in guns, no Hedgehog and an Asroc launcher. Six tubes for 12-in homing torpedoes are mounted in both classes, but the *Lloyd Thomas* retains two 21-in tubes while the *Carpenter* class does not. The *Basilone* class, also based on the *Gearing* hull, are also primarily AS types with two twin 5-in guns (basic *Gearing* equipment), a fixed

(c) Guided-Missile Escorts (DEG) and Escorts (DE)

Brooke class. Six vessels of this class were laid down between 1962 and 1965. At 3,426 full-load tons displacement they are of destroyer size, and what makes them DEs is the slightly lower speed, 27 knots. They are fitted with AS drone and manned helicopter facilities, a single Tartar SAM launcher and an 8-tube Asroc AS missile-launcher. There is a single 5-in gun; forward there are two tubes for wire-guided torpedoes and six

US Destroyers – Specifications:

US navy code	Truxtun DLGN	Belknap DLG	Leahy DLG	Coontz DLG
Displacement (tons)				
full-load	9,200	8,150	7,800	5,800
Length (feet)	564	547	533	512
Launchers for:				
Terrier	1 twin	1 twin	2 twin	1 twin
Tartar	0	0	0	0
Asroc	1 twin	1 twin	1 8-tube	1 8-tube
Guns, 5-in	1	1	0	1
3-in	2	2	4	4
Torpedo tubes, 21-in	2	2	0	0
12-in	6	6	6	6
Hedgehogs	none	none	none	none
Engine power (shp)	60,000	85,000	85,000	85,000
Maximum speed (knots)	30	31	34	34
Radius at 15 kts	Nuclear, unlimited	—	—	—
Complement	479	395	372	355
Year completed (first of class)	1967	1964	1962	1959

Each set of specifications refers to a representative class or vessel within each group or class.

launchers for 12-in AS torpedoes of the homing variety. Their basic hull and propulsion unit is also used in the ten *Garcia* class vessels which are classified as DEs since they mount a second 5-in gun aft instead of the *Tartar* launcher, but which are otherwise quite similar.

The forty-odd *Knox* class DEs are similar to the *Brooke* class, though slightly larger at 4,100 tons of full-load displacement, but have a single 5-in gun turret forward (though they do not carry the SAM launcher). All three classes have integral sonar built into the bow for long-range search as well as variable-depth sonar.

John C. Butler class. More than eighty vessels of this class were completed in the course of 1944 as destroyer escorts, though they have since been re-rated as escorts. At 2,100 tons of full-load displacement they are representative of the later wartime anti-submarine escort designs, with a relatively slow speed, 24 knots, a radius of action of 5,000 miles (at 15 knots), and 190-man crews. They are armed with single 5-in guns fore and aft, two 40-mm AA guns, Hedgehogs, and depth-charge dispensers. The cluttered superstructure, the separate antenna masts and smoke stacks, and the lack of integral sonar fittings are wartime design features shared by the *Rudderow*, *Buckley* and *Edsall* classes, which are not separately discussed.

Destroyers, West German (includes Frigates, Escorts, Corvettes). The West German fleet still includes some wartime US and British vessels, but new ships built in West Germany since 1960 constitute the main part of the fleet.

For the German-built *Charles F. Adams* class of guided-missile destroyers, ⟡ *Destroyers, US.*

Hamburg class. Four of these large gun destroyers were completed between 1964 and 1966. They are equipped with 4 single 100-mm guns, the standard dual-purpose (AA and surface) gun of the West German navy, 8 40-mm AA guns in twin mounts, as well as depth-charge rocket mortars, anti-shipping torpedoes (5 21-in tubes) and

C. F. Adams DDG	*Forrest Sherman* DD/DDG	*Gearing* DD	*Fletcher* DD	*Brooke* DEG	*John C. Butler* DEG
4,500	4,050	3,479	3,050	3,426	2,100
431	419	391	377	415	306
0	0	0	0	0	0
2/or 1	1 (DDG)	0	0	single	0
8-tube	1 8-tube	2 triple	0	1 8-tube	0
2	3 (DD)	4	5	1	2
0	4 (DD)	0	0	0	[2 40-mm]
0	0	0	5	0	0
6	6	6	6	6	0
none	2 (DD)	none	2	none	2
70,000	70,000	60,000	60,000	35,000	12,000
35	33	35	34	27	24
—	—	5,800	6,000	—	5,000
333	276	274	249	241	190
1960	1955	1945	1943	1966	1943

West German Destroyers – Specifications:

	Hamburg	*Köln*	*Rhein*	*Thetis*
Displacement (tons) full-load	4,330	2,550	2,540	680
Length (feet)	440	361	324	230
Guns, 100-mm	4	2	2	0
Guns, 40-mm	8	6	4	2
Torpedo tubes, 21-in	5	0	0	0
A S	2	2	0	0
Depth-charge mortars (4-barrel)	2	2	0	1
Engine power	68,000 shp	38,000 shp	11,400 bhp	6,800 bhp
Maximum speed (knots)	35·8	30	22	24
Radius (miles)	920 at 35 kts	920 at 30 kts	1,625 at 15 kts	not available
Complement	282	210	110	48
Year first completed	1964	1961	1961	1961

Specifications refer to main class type; there are individual variations.

two A S torpedo tubes. These are essentially short-range coastal vessels with a very high maximum speed (35·8 knots) and extensive radar equipment.

Köln class. These are classified as escorts or frigates, though their tonnage (2,550 full-load) and speed (30 knots maximum) lie in the destroyer bracket. Their armament consists of two single 100-mm guns, six 40-mm A A guns, two depth-charge mortars and two A S torpedo tubes. Their very large, low funnel is a recognition feature.

Rhein class. The thirteen ships of this class are officially classified as 'tenders' though they are in fact fighting ships and their 2,540 tons full-load displacement would allow the addition of extra weaponry beyond the two 100-mm dual-purpose and the four 40-mm A A guns.

There is also the *Deutschland*, a light cruiser of 5,500 tons (full-load) designated as a training vessel. Completed in 1963, it has four 100-mm guns, depth-charge mortars, six 40-mm A A guns, a helicopter and A S torpedo tubes.

Corvettes, or coastal-patrol-ships, are important in the narrow sea configuration of the West German navy. Ten very large 2,000-tonners are to be built, but a medium 1,100 tonner has been operational since 1963.

The *Thetis* class of 680 tons full-load displacement are classic corvettes with a squarish superstructure. These vessels are equipped with two 40-mm A A guns and a depth-charge mortar/rocket projector.

Deterrence. A measure or a set of measures designed to narrow an opponent's freedom of choice among possible policies by raising the cost of some of them to levels thought to be unacceptable. The term is usually used in the more specialized sense of discouraging a nuclear attack by arousing the fear of retaliation. Deterrence has always been central to relations between states: it is based on the assumption that an opponent's intentions will go as far as his capabilities unless his behaviour is modified by deterrence. In order to believe that deterrence is an invalid concept it must be explicitly assumed that states would not use – or threaten to use – what weaponry they have except for self-defence.

It is clear that in order to deter a nuclear attack retaliation must be possible even after such an attack has taken place. If an attacker can expect to destroy the means of retaliation, as well as other targets, he will not be deterred. Effective deterrence therefore depends on the survival of the means of retaliation as much as their capabilities once unleashed. ◊ *Assured Destruction* for an explicit formulation of the requirements of a valid deterrent force – from the U S point of view.

Active deterrence: A strategic threat which is specifically intended to prevent a particular move on the part of an opponent (other than direct attack on the owner of the deterrent).

Extended deterrence: A declared extension of the intention to retaliate if an attack is made on some specified third party territory: i.e. a statement that from now on an attack on X-land will be considered an attack on, say, the Soviet Union is a case of extended deterrence. The limits of extended deterrence are set by the ◊ *credibility* of the threat: will the Soviet Union really risk nuclear war over X-land?

Minimum deterrence: A plausible but technically incorrect strategy based on 'a few' weapons which are assumed to be (a) sufficiently destructive to inflict unacceptable damage – in this context defined as the destruction of one or two large cities – and (b) invulnerable. The idea is to obtain a secure strategic defence at low cost, and it is historically associated with the development of the first protected nuclear delivery system, the ◊ *Polaris* submarines. Its enthusiasts argued that a small Polaris force would achieve stability, security and economy, so that the deployment of other systems (land-based I C B Ms, bombers, etc.) would be wasteful. This ignores the basic requirement of deterrence: a secure retaliatory force. The deployment of 'excess' weapons, beyond the megatonnage required to inflict high levels of damage, is necessary in order to ensure that enough gets through *after* an attack and after discounting for technical failures and defensive barriers. Reliance on a single type of system is vulnerable to a single technological breakthrough on the other side. And there would be every incentive for the other side to invest in research and development to achieve such a breakthrough in order to obtain a disarming ◊ *counterforce capability*. Finite deterrence is a variant in which a small number of weapons is deployed against a small and stated number of targets. The term has been introduced by advocates of minimum deterrence to escape the connotation of 'minimum' with respect to the deterrence of aggression.

Dew Line (Distant Early-Warning Line). A radar, computer and communications network which is intended to provide tactical warning of impending penetrations of the North American airspace from the Soviet Union. The perimeter of the DEW line forms an arc from the Aleutian Islands to Iceland, through Alaska, Canada and Greenland. Since Soviet aircraft could reach North American airspace by longer routes which avoid the DEW perimeter, this is supplemented by a network of Airborne Early Warning aircraft and radar-picket ships which give 'lateral' cover. The DEW line is primarily intended against manned aircraft and non-ballistic missiles. Warning of ballistic-missile attack (though only over the polar route) is provided by the ◊ *BMEWS* network. Tactical intelligence provided by both DEW and BMEWS is fed into the ◊ *NORAD* air defence organization which controls the active defences through the SAGE and BUIC systems. The ◊ *Safeguard* system of ballistic-missile defence will probably replace all these systems in association with ◊ *AWACS*, an airborne defence against manned aircraft. The DEW line, which became operational in 1957, is maintained under contract by a private US corporation.

'Dirty Bomb'. Popular name for a ◊ *nuclear warhead* which has a fission trigger, a fusion (thermonuclear) core and an outer jacket made of U-238, which undergoes fission when triggered by the fusion core. This fission–fusion–fission bomb is 'dirty' because it produces a great deal of radioactive debris out of the U-238 casing, which is not fissionable unless fusion temperatures are present.

Disarmament. A reduction in the personnel and/or equipment of armed force

organizations, whether unilateral or international, total or partial, controlled or uncontrolled. Whenever what is perceived as an ◊ *arms race* takes place, the effect is usually much talk about disarmament together with accelerated development and deployment of weaponry. When third parties can be profitably affected, some measure of ◊ *arms limitation* may be agreed upon by the principal arms racers. In recent times many ◊ *arms control* measures have been effectively blocked by interested parties who have called for 'general and complete disarmament'. Such calls have usually been accompanied by accusations that any inspection procedures suggested by 'the other side' as part of a disarmament agreement are 'covers' for espionage. Plans for general and complete disarmament usually provide for the retention of small forces for internal security purposes and a UN-controlled multi-national force. Contrary to what is widely believed, historical evidence shows that there is a low correlation between arms races and war, and a high correlation between uninspected or unilateral disarmament (whether formal or not) and war.

Division. An army formation comprising two or more ◊ *brigades* and subordinate to a multi-divisional corps. In some cases, for example the US Army and Marine Corps, the division is the viable operational unit and the focus of decisionmaking. In the Soviet and the Warsaw Pact armies, however, the division is a mere component of larger multi-divisional forces, and the support and ancillary units are attached to higher echelons such as the corps, army or even army group. Other armies are organized at the brigade level and the operational unit is the 'brigade group', with its own permanently attached support units. Thus in the British and Israeli armies the 'division' has been little more than an administrative unit. The Chinese (People's Republic) army is organized into 'Armies', one for each of the 13 military regions. Artillery, armour and engineers are assigned to the 'Army' since the division is not intended for use as an independent unit.

The US 'Re-organized Objective Army Division' (ROAD) is a large force of some

16,000 men with its own air wing, artillery, logistic and support units, which are assigned to a headquarters echelon which usually controls about 10 *battalions*. The ROAD division is in turn supported by an 'Initial Support Increment' force, also of 16,000 men, which includes specialized units such as armoured cavalry regiments and logistic facilities. The combined ROAD + ISI force is in turn backed up by a 'Sustaining Support Increment' – also of about 16,000 men – which provides full logistic support, theatre-level ancillary facilities, and rear area security. Bearing in mind these differences, the table below shows the numerical strength of combat-ready full-strength divisions:

connaissance and ground-attack versions, performing very well in all these roles. The Draken has a double-delta wing (which appears to be a peculiarly Swedish design concept) and oval air-intakes at the wing-roots. The following versions are currently operational:

J.35B: An early version for the interceptor role. Power plant gives 14,400 lb thrust with afterburning. Nose radar, automatic fire-control and other 'all-weather' features.

J.35C: Two-seat training version without nose radar.

J.35D: Basic multi-purpose version. Upgraded power plant giving 17,650 lb of thrust with afterburning; nose scanner radar, fire-control equipment and auto-

Divisional Establishments, Combat Echelon Only

	United States	Soviet Union	China	West Germany	France
Infantry	16,000	10,500	12,000	15,500	14,000
Armoured	15,500	9,000	10,000	14,500	16,000
Airborne	13,500	7,000	6,000	12,000	14,000

DM. A non-lethal chemical agent, also known as Adamsite. Yellow to green crystals, almost odourless, and dispensable in aerosol form. Doses of 2–5 milligrammes per cubic metre per minute cause: headache, sneezing, coughing, chest pains, nausea, and vomiting. After inhalation, effects last for about thirty minutes. It has been used in combat. ⟡ *Chemical Warfare* and also *CS* and *CN*.

DMZ. ⟡ *Demilitarized zone*.

Doppler Effect. ⟡ *Radar*.

48   Draken (J/S/Sk.35). Swedish Mach 2 fighter. The Draken, when first flown in October 1955, was the first European fighter with a Mach 2 potential. Originally designed as a high-altitude interceptor, it has since been developed into training, re-

pilot. The weapon-load is two 30-mm cannon and up to 6,000 lb of external ordnance or fuel tanks.

J.35E: Reconnaissance version with seven cameras. There is a zero height and zero-speed ejection seat.

J.35F: Main production version, an improved J.35D with collision-course fire-control equipment and ⟡ *Falcon* air-to-air missiles; one 30-mm cannon retained.

J.35X: Long-range 'intruder' version of the J.35F. Increased range and/or store-carrying capacity. Two 30-mm cannon. This has been described as a version intended for export and has been purchased by Denmark.

Specifications (J.35D): Length: 50 ft 4 in. Height: 12 ft 9 in. Wing span: 30 ft 10 in. Gross wing area: 529·6 sq ft. Take-off weight with no external stores: 22,530 lb. Take-off combat weight: 27,050 lb. Maximum level speed: Mach 2 (clean), and Mach 1·4 at normal load. Time to 36,100 ft: 2·5 minutes (clean), and 3·5 minutes, with a normal combat load. Maximum combat radius: 447 miles.

▷ *Fighter* for general subject entry.

Dunkirk Treaty. A 1947 Treaty between France and Britain which provided for consultation and action in the event of German aggression against either party. It also affirmed both countries' commitment to constant consultation about economic matters.

E

Early-warning Systems. One of the elements of an ◊ *air defence system*. Usually refers to ◊ *radar* and communication units intended to detect the presence of aircraft in the relevant airspace.

The B M E W S (Ballistic Missile Early-Warning System) is an early-warning system intended to detect ballistic missiles, but ◊ *A B M* for the early-warning requirements of ballistic-missile defences. A E W (Airborne Early-Warning) consists of search radar units mounted on aircraft and used to fill gaps in land-based radar systems or to extend their reach. Shipborne or submarine E W systems are used for similar purposes.

◊ *P A R* for the E W component of the 'Safeguard' A B M.

Eastern European Mutual Assistance Treaty. ◊ *Warsaw Pact*.

EBR. French heavy armoured car. The E B R (*Engin Blindé de Reconnaissance*) is a large and complex 8-wheel vehicle used by the French army for various reconnaissance roles. The E B R has two outer wheels on each side with pneumatic tyres as well as four centre wheels with solid and deeply spudded metal tyres. There are two driving stations so that the vehicle has no real 'front' or 'back'; in the centre there is (in the latest model) an oscillating turret with a smooth-bore 90-mm gun which fires fin-stabilized ammunition. There are two gearboxes and a centre differential all driven by the long and low 12-cylinder air-cooled engine. In other words, the E B R is an extremely complex, expensive and delicate vehicle (the French had to purchase British ◊ *Ferrets* for use in Algeria) not suitable for sustained use in combat.

Specifications: Length: 18 ft 2 in.

Width: 7 ft 11 in. Height: 7 ft 4 in. Weight: 13 tons. Engine: air-cooled petrol 200 bhp. Road speed: 62 mph. Road range: 400 miles. Armament: various 75-mm guns, but now a 90-mm Mecar gun is being standardized.

◊ *Armoured Car* for general entry and competitive types.

ECM. Electronic Countermeasures (includes Electronic Counter-Countermeasures, E C C M). E C M techniques are mainly used to degrade the performance of enemy *radars*, especially air defence surveillance, tracking and homing guidance units. E C M techniques are conventionally divided into confusion measures, intended to hide real targets by cluttering the enemy radars, and deception measures, which produce false echoes in order to simulate real targets. Bomber aircraft such as the ◊ *Stratofortress B-52s* rely heavily on E C M to penetrate sophisticated air defences; indeed without E C M such bombers would have little chance of penetrating such defences, since surveillance and tracking radars can fix their position at long ranges with a degree of accuracy sufficient to ensure interception and destruction by fighters or missiles. Simple confusion jamming can be achieved by high-power continuous-wave transmissions (C W) on the enemy radar frequency; these completely obliterate enemy radar displays but require a great deal of transmitter power, which is difficult to provide on board an aircraft. Further, the enemy frequencies may not be known, in which case the entire bandwidth must be jammed. When the enemy frequencies are known, all available energy can be concentrated on them, this being known as 'spot jamming'. A counter to this (an instance of E C C M) is the rapid shift from one frequency to the other, so as to leave the 'spot jammer' behind. 'Frequency agility' is a typical and common form of ECCM, and some sets can change frequency with each pulse. Another confusion ECM technique is 'sweepthrough jamming', where a whole range of frequencies are swept at rapid intervals so that many different radars are jammed at the same time (though very briefly) until the next (random) sweep comes through. Deception ECM require less power than confusion measures but need more complex equipment, and often more information. A 'repeater jammer', for example, picks up the enemy signals and sends out its own, and these simulate the 'echoes' which provide data for the enemy. By introducing a slight delay, the false echoes deceive the enemy radar as to the range or elevation of the target. ECM 'transponders' are an application of the secondary radar system used in IFF (Identification Friend or Foe) units (◊ Radar). The transponder is automatically switched on when the beams of the enemy radar set sweep over the aircraft, and transmits a stored replica of the enemy signal, which again displaces the radar 'image' of the target. Other, more complex, deception ECM are used to break the 'lock' of enemy tracking radars on their target. Tracking radars keep their pencil beam on target by oscillating it around its expected course, so that any deviations are registered and the tracking course is suitably adjusted. A 'range-gate stealer' operates by sending a signal which synchronizes with the real echo and which initially reinforces the 'lock' by strengthening the real echo. The signal is then slowly shifted in timing so that the apparent position of the target is made to deviate from its real position. If the false 'echo' is stronger than the real echo, the tracking circuits will follow the false signal from the jammer and ignore the real (but weaker) echo. This goes on until the target escapes the 'range-gates' or oscillation limits of the antenna, and then the simulator can be turned off, leaving the tracking radar without a target.

The standard ECCM response to these techniques is to use a signal with complex frequency modulation (a given pattern of frequency changes) which is difficult to mimic, so that false 'echoes' can be identified.

'Passive' ECM techniques rely on 'chaff', various kinds of decoys or radar cross-section reduction. 'Chaff' (also known under the old British code-name, 'Window') consists of a large number of foil strips whose length is intended to 'fit' the frequency of enemy radars. 'Spot chaff' is used to deceive by simulating targets; 'corridor chaff' is the 'confusion' counterpart where aircraft continuously release bundles of chaff that create a radar 'cloud' through which other aircraft can pass. But chaff is quickly left behind by high-speed aircraft, and enemy radars can then discriminate between chaff and targets by comparing their velocities. One simple remedy is to dispense the chaff by means of rockets fired ahead of the aircraft; but this is ineffectual in a sophisticated radar environment. Decoys, designed to simulate aircraft or missile warheads, can sustain the deception for much longer and against more sophisticated radar systems. The US ◊ Quail for example, is a turbojet-powered mini-aircraft equipped with radar enhancement devices that simulate the larger carrying aircraft; it is also equipped with small ECM devices which simulate those of the mother plane. In order to be effective such decoys must have a reasonable range and a speed similar to that of the aircraft they protect.

The radar cross-section (or image) of an aircraft (or other potential target) can be reduced at the design stage by using materials or shapes which have poor reflecting properties. In particular, double-curvature surfaces reduce the cross-section very considerably. Special paints (electromagnetic absorbents) have been developed which can be very effective if the frequency range of the enemy radar is known. If broadband protection is required – which is usually the case since enemy frequencies can be variable – much thicker layers are required, and these are usually unacceptable on high-speed aircraft. Neither type of paint is effective against certain types of radar, including those with widely separated antennas for reception and transmission.

The ECM versus ECCM battle, like similar confrontations, continues, but the advantage is usually with the ground-based system, whose power supplies can always exceed those of an aircraft system. This is obviously not true when the technological sophistication of the two sides is very different, and a single EA-6B (▷ *Intruder A-6*), a US carrier-based ECM aircraft, can probably neutralize the radar-ECCM resources of most countries. ▷ *Radar.*

Economic Warfare. The manipulation of an enemy's foreign trade by means of boycotts. embargoes and financial controls. More direct measures such as a blockade or a 'quarantine' are acts of war and therefore usually occur in association with armed conflict.

▷ *Strategic Air War* for a more direct attack on the war potential of nations.

EDC. ▷ *European Defence Community.*

Egyptian Missile Programme. The existence of an Egyptian missile programme was revealed publicly in 1962. Two types of rockets were then displayed in a parade, and their successful firing was announced. At the next year's parade, in July 1963, three vehicles purporting to be usable missiles were displayed. These were described as the Al-Zafir, with a 1,000-lb warhead and a claimed range of 235 miles, the Al-Kahir, with a 1,500-lb warhead and a claimed range of 375 miles, and the Al-Ared with a claimed range of 440 miles. These missiles are controlled by the UAR Missile Command, intended as the local counterpart of the US Strategic Air Command Missile Command and the USSR Strategic Rocket Forces. It appears that guidance systems are not yet perfected, so that the 4,000 technicians, scientists and soldiers who belong to the command cannot predict even approximately the impact area of the rockets.

▷ *Missile* for general subject entry.

Electronic Countermeasures (and Electronic, Counter-Countermeasures). ▷ *ECM.*

ENDC. Eighteen-Nation Disarmament Conference. Sits permanently in Geneva. For its recent activities ▷ *NPT* and *GCD.*

ENTAC. ▷ *NORD.*

Escalation. An increase in the scope or intensity of an armed conflict between states. Escalation is generally seen as being a deliberate attempt at 'winning' a war by bringing it to some higher threshold where the escalating party believes that it has military superiority. What is seen as escalation in this sense may simply reflect the consequences of structural differences between the parties to a conflict. If, for example, a sophisticated military power is engaged in a conflict with guerilla + subversion forces (▷ *Revolutionary War*) what is in fact a tit-for-tat matching response may be seen as escalation, for example defoliation or aerial bombing in response to the assassination of cadres or sabotage attacks in cities. The absence of symmetry in the weapons available to the parties in a conflict will inevitably result in qualitatively different moves and counter-moves, and, regardless of intent, these can be described as escalation. In other words, the concept is meaningful only if there are clearly recognized ▷ *thresholds* between various kinds of military actions, and if these thresholds are available to all the parties to the conflict.

Escort. ▷ *Destroyer.*

Espionage. Trade name for classic operative-based intelligence activities, i.e. where the primary source is an individual sent to collect the data, as opposed to the use of sensory devices. Espionage as so defined provides a very small fraction of the data available to the major powers (▷ *Strategic Intelligence*).

European Defence Community (EDC). An international armed force organization which was to be established by treaty between Belgium, France, West Germany, Italy, Luxembourg and the Netherlands. The treaty was duly signed in May 1952, but the French parliament refused ratification under communist-Gaullist pressure. The EDC would have involved a fully integrated defence with mix-manned forces, a single budget and a single general staff; the project was replaced by the milk-and-water ▷ *WEU* (West European Union), whose members are the proposed EDC partners plus the UK.

European Security, the Problem of.

The government of the Federal Republic of Germany and other European NATO countries regard Russia as a potential aggressor which has to be continuously deterred by adequate defence measures. European governments disagree about the intensity of the Russian threat, but they all agree that their security cannot be assured without a balancing American presence. Only the ex-leader of France believed or affected to believe that (a) the Russian threat is largely unreal, (b) there are other potential threats, and (c) these can be countered by France (or more credibly by Europe) on its own if adequate nuclear forces were developed. The present NATO structure is built on a complete reversal of these three propositions; it dates from the 1949 North Atlantic Treaty and embodies 1949 assumptions; (a) that the Russian threat is very real (still valid); (b) that it can be countered only by a large American presence in the area since Europe cannot provide its own defence; (c) that only the US can provide effective nuclear deterrent forces. Propositions (b) and (c) imply further that major strategic decisions can be made only in Washington, and NATO's main military commanders must be American. The divergence between current reality and these two 1949 assumptions is the problem of European security. The US government is still willing to accept the burden of defending Europe but wants to do it by a combination of an American nuclear guarantee and European non-nuclear forces. This scheme would have three advantages for the United States: (a) it would reduce the foreign currency and manpower burden by withdrawing most of its naval, air and ground forces from the area; (b) it would retain strategic control over the situation because of its effective nuclear monopoly; (c) it would arrest further proliferation of independent nuclear forces. This is a key policy objective of the United States, for which it can count on the support of the Soviet government.

The European members of NATO resist this policy since:

(a) any reduction of the American presence 'on the ground' impiles a greater burden on their own defence budgets;

(b) they believe, or affect to believe, that an American nuclear guarantee is of limited value in the absence of the physical presence of American troops, given that Russia too has an effective deterrent:

(c) there are pressures within Italy and Germany for the development of independent nuclear deterrents.

France has responded to these pressures by withdrawing from its participation in NATO and expelling NATO personnel from its national territory. Its defence forces are now supposed to provide it with a self-sufficient and all-round defence, though, as usual with France, its armed forces are in reality very weak indeed. Britain and some other NATO members are in sympathy with the USA but fear that the American position could lead to the development of independent nuclear forces by those allies which felt that an American nuclear guarantee from across the Atlantic would be insufficient; in other words, they fear a 'nuclear Germany'. Various plans have been mooted to solve the impasse: a small ◊ tripwire deployment of American forces plus an American nuclear guarantee; a multi-national NATO sea-borne force armed with ballistic missiles under 'joint' (= US) control which would provide a 'European' nuclear guarantee (◊ MLF). Meanwhile the disagreements within Europe and with the USA are unresolved and NATO continues with its picturesque command structures and American commanders: SACLANT and ACLANT, SACEUR and SHAPE, ACCHAN, ACE, as well as STANAVFORLANT, AFCENT, AFNORTH and AFSOUTH. ◊ NATO for a guide to the maze.

In the meantime the introduction of tactical nuclear weapons, needed to balance Russian and ◊ Warsaw Pact ground superiority, has been achieved without ◊ nuclear proliferation by means of 'joint' control, where American personnel have one key of a two-key system which activates the warheads. This kind of arrangement would be useful to plug gaps in a sound system but in this context it only highlights the basic and as yet unresolved problem: the nature of the American contribution to European defence.

F

F. US designation for fighter aircraft.
◊ *Phantom* for F-4.
◊ *Freedom Fighter* for F-5.
◊ *Crusader* for F-8.
◊ *Voodoo* for F-101.
◊ *Starfighter* for F-104.
◊ *Thunderchief* for F-105.
◊ *Delta Dart* for F-106.

F-111A. US multi-purpose aircraft.
A twin-jet two-seat multi-purpose fighter-
bomber with variable sweep on the outer
wing portions. This 'variable geometry'
wing solves a basic design conflict by
achieving optimum lift/drag throughout
the speed range. The aluminium, steel and
titanium fuselage is built around a T-
section keel under the arms of which the
two engines are hung. Each of the two
turbofans is rated at close to 20,000 lb with
afterburning in the basic F-111A version,
but more powerful engines were scheduled
for later versions. The F-111's avionics
were intended for ◊ *contour flying* at high
speeds. This required a strong fuselage as
well as TFR radar, a flight computer and
an ◊ *inertial platform*. In addition to this,
the ◊ *Phoenix* and ◊ *SRAM* missiles were
to be controlled by a sophisticated fire-
control computer.
But this very advanced aircraft turned
out to be a costly failure. Its initial
engineering problems were quite normal,
and in fact the aircraft is very satisfactory
in technical terms. But the real problem is
its over-ambitious mission profile. The
F-111 was to be both fighter and bomber,
but it lacked the agility needed for air
combat while its payload was inadequate
for strategic bombing.
Six versions were initially prospected.
F-111A: USAF strike fighter version.
Detailed specifications below.
F-111B: Naval 'air-superiority' carrier-

based interceptor. Various modifications
have been tried but this variant has
been rejected by the US navy.
F-111C: A fighter-bomber for the Australian air force. Basically similar to the
F-111A with modified avionics.
F-111K: A strike-reconnaissance aircraft based on the A with advanced –
and partly British – electronics.
FB-111A: A 'strategic' bomber for the
US strategic Air Command. With a
payload which was to be up to 50 750-lb
bombs, longer wings, and fittings for
the SRAM missile.
RF-111A: A reconnaissance version with
cameras and other sensory equipment
in the weapons bay.
Specifications (F-111A): Length: 73 ft
6 in. Height: 17 ft 1 in. Wing span: 70 ft to
33 ft 11 in. Wing sweep: from 16° to
72° 30'. Maximum take-off weight:
70,000 lb. Armament: missiles and bombs
externally; two bombs internally. Electronics: TFR radar plus computer system
for supersonic low-level and contour
flying. Nose radar for electronic fire-
control. Maximum level speed at height:
Mach 2·5. Maximum speed at sea level:
Mach 1·2. Ceiling: 60,000 ft. Take-off
run: 3,000 ft. Landing run: under
3,000 ft. Maximum range (internal fuel
only): 3,800 miles, with no other loads
('clean').
◊ *Fighter* for general entry.

Fail-safe. A procedure whereby US
strategic bombers with nuclear weapons
proceed to target only if specific orders to
do so are received at a certain point on their
intended path. These communication
arrangements are designed to prevent a
nuclear strike resulting from technical
failure. The alternative procedure is
'recall', where the delivery system pro-

51

ceeds as programmed unless actively re-
called. Fail-safe was introduced by the US
Strategic Air Command as a replacement
for the traditional recall procedure used in
non-nuclear bombing. In SAC it is known
as 'positive control'.

**Falcon (includes Super and Nuclear
Falcons). Series of US air-to-air
missiles.** The Falcon series of air-to-air
missiles are standard equipment on all
American fighter-interceptors. The initial
model, GAR-1 was the first air-to-air
guided weapon adopted by the US air
force; since then a whole family of air-to-
air missiles has been developed from it.
Currently operational types of the Falcon
series include:
AIM-4D. This equips the ⟡ *Phantom
F-4s,* ⟡ *Voodoo F-101B* and Delta
Dagger F-102A fighters. It has a
cylindrical body with small cruciform
nose fins and long-chord cruciform
wings; solid-fuel rocket motor.
Guidance is by ⟡ *infra-red homing.*
AIM-4E Super Falcon. A somewhat
larger missile with better performance
all round. Guidance by ⟡ *semi-active
radar homing.* It equips the ⟡ *Delta
Dart F-106,* the USAF's first-line
interceptor.

Fallout. Popular name for those radioactive particles produced by the explosion
of a ⟡ *nuclear weapon* which reach the
lower atmosphere or the earth's surface.

**FBMS. Fleet Ballistic Missile Sub-
marine (or System).** ⟡ *Polaris* for the
original US submarine missile system; ⟡
Redoutable for the French copy; ⟡ *Resolution* for the British one. ⟡ also *Submarines,
Soviet* (Y class).

Federation of Arab Emirates. An
embryo common market and defence
organization formed by a treaty between
Bahrain, Qatar, and the seven 'Trucial'
states Abu Dhabi, Ajman, Dubai,
Fujairah, Ras al-Khaimah, Sharjah and
Umm-al-Qaiwain. The *raison d'être* of the
Federation is common defence after the
British military withdrawal from the
Persian Gulf which is to take place in (or
by) 1971. The treaty, which was signed in
March 1968, leaves some crucial matters
undecided, such as the nature of the proposed common budget (Abu Dhabi may
soon have the highest per capita income in
the world; Sharjah has one of the lowest).
The main security problems of the Federation arise from Persian offshore claims,
and Saudi claims which could become a

Specifications:

	Length	Diameter	Weight	Speed	Range	Warhead
AIM-4D	6 ft 7 in	6 in	122 lb	Mach 2+	5 miles	high explosive
AIM-4E	7 ft 2 in	6·6 in	150 lb	Mach 2·5	7 miles	high explosive
AIM-4F	6 ft 9 in	6·6 in	145 lb	Mach 2·5	7 miles	40-lb HE warhead
AIM-26A	7 ft 0 in	11 in	203 lb	Mach 2	5 miles	nuclear

AIM-4F Super Falcon. As above, but
with infra-red homing guidance.
AIM-26A/X Nuclear Falcon. First air-
to-air missile with a nuclear warhead.
Similar to Super Falcon, and has semi-
active radar homing guidance. AIM-
26B is the same as the A/X but with
a non-nuclear warhead. A variant of
AIM-26B equips Swedish and Swiss
interceptors.
⟡ *Missile* for general entry.

hot issue if there is a change of government in Saudia.

Ferret FV 701. British armoured car.
General name of a series of four-wheel
drive light armoured cars in service since
1954. These car-sized vehicles have petrol
engines, two-man crews, very light
armour and fairly good mobility. Several
versions have been developed on the basis
of the original chassis:

Mk 1: A turretless liaison and light reconnaissance vehicle.

Mk 2: With a small machine-gun turret.

Mk 2/6: With a makeshift installation for two ◊ *Vigilant* anti-tank missiles.

A new chassis was developed from the Mk 2 which has larger-diameter wheels, and larger dimensions all round. This, the Mk 4, has a simple machine-gun turret, but Mk 5 mounts a ◊ *Swingfire* missile installation. The Mk 4 and 5 are useful in the reconnaissance and light-weapon carrier role; they are also useful in security operations, but they can be used in assault roles only against very unsophisticated opposition. The Ferret has been widely exported.

The Ferret series is to be replaced by the new ◊ '*Fox*' armoured cars.

Specifications (Mk 4 and 5): 4 × 4 drive. Length: 13 ft. Width: 7 ft. Height: 6 ft 8 in. Weight: 5·35 tons. Engine: petrol, 129 bhp. Road speed: 50 mph. Range: 190 miles. Weapons: 0·3-in Browning MG. In Mk 5 Swingfire anti-tank missile.

◊ *Armoured Car* for competitive types.

◊ *Armoured Fighting Vehicle* for general subject entry.

45   Fiat G 91. Italian fighter-bomber.
The Fiat G 91 was produced as a joint NATO light ground-attack plane but it was eventually bought only by the West German and Italian air forces. It is a subsonic 'daylight' attack aircraft requiring little servicing and with limited capabilities: the combat load consists of two 30-mm cannon and about 1,500 lb of bombs or rockets.

An improved version, the G 91 Y, is being produced. This has a greater combat load and radius of action, but is still subsonic.

Some G 91s have been transferred from West Germany to Portugal.

◊ *Fighter* for general subject entry.

'Fiddler' Tu-28 (YAK-42). Soviet fighter. The Fiddler is a new Tupolev design though it was initially thought to be a Yakovlev design and given the YAK-42 designation. The Fiddler, a heavy twin-jet long-endurance fighter with a long-range strike capability, has sharply swept wings, side-by-side jet pipes and shoulder air-

intakes. It is thought to have a maximum take-off weight of about 100,000 lb – almost twice that of the ◊ *Phantom F-4* – and this must translate in very long range/ endurance or – in a 'strike' configuration – in a large weapon-load (or much greater electronic gear than is externally visible). What *is* visible includes a large nose radar, the four ◊ *Ash* air-to-air missiles under the wings, and the two-seat canopy. The Fiddler has been credited with a top speed of Mach 1·7 and a ceiling of 60,000 ft, and if this is correct it must have been designed as an air defence (anti-bomber) rather than an 'air-superiority' (anti-fighter) plane. The Fiddler is about 65 ft long.

◊ *Fighter* for general subject entry.

Fighter (includes Attack Aircraft).
The jet fighter is the key weapon of modern non-nuclear warfare, other than ◊ *Revolutionary War*, where its role is marginal – or ought to be. The aircraft consists of three elements: the airframe, the power plant, and the sensory and control ◊ *avionics*. What makes it a 'fighter' is the payload; the crew, the weaponry and – increasingly important – the weapon-control avionics. Different types of mission require different kinds of aircraft and payload; although these requirements can sometimes be combined there are also some incompatible requirements, which has led to the development of specialized aircraft intended for a specific primary mission. The two basic combat configurations are air-to-air and air-to-ground, but within them there are some further 'mission' classifications:

1. *Air-to-Air*

(a) Interception. In the purest form an aircraft which can climb rapidly to high altitudes in order to identify and/or destroy other aircraft. This requires a high wing loading and a highly swept and low aspect–ratio wing. In addition, it requires a nose-cone air-intercept (AI) radar, as well as communications and weapon-control gear for a so-called 'all-weather' capability. This does not mean that the plane can land in all weathers but refers to the fact that the plane can make a 'blind' intercept where ground control guides the aircraft to the target area, until the airborne radar can take over and home in for the final intercept. An interceptor therefore needs only a limited range, a one-man crew and a low weapon payload. This usually consists of air-to-air missiles as well as guns with very high rates of fire; these are fired automatically by the ranging radar or 'electronic fire-control' unit, which releases the weapons when the target is in range and within the 'kill' area. The pressure for increased speed performance has largely derived from the speed superiority requirement of gun-armed fighter-interceptors which could attack targets only from astern. Now that there are air-to-air missiles, such as the ◊ *Red Top* and the ◊ *Sparrow*, which can be fired on collision courses against targets of superior altitude and speed, the pressure for ever increased speed has abated and the speed requirement is stabilized at around Mach 2·5 – about the maximum feasible with aluminium airframes. The British ◊ *Lightning* and the Soviet ◊ *Fishbed Mig-21* are close to the pure interceptor class, with high rates of climb, short range, and a limited payload. (b) The Long Endurance Fighter (or 'Air Superiority' Fighter). This too is an air-to-air mission, but instead of a one-shot intercept the requirement is for a fairly long patrol time followed by target engagement at supersonic speeds and high altitudes. The design conflict between a good low-speed and medium-altitude fuel-economy on the one hand and a capability for rapid acceleration and climb on the other, leads to expensive and complex aircraft, typically twin-jets. The fact that the strategic bomber threat has waned has resulted in the eclipse of the pure interceptor, while the persisting requirement for battlefield 'air superiority' has maintained the importance of this type of aircraft. This is so because efficient attack (or close-support) aircraft (see (e) below) need air-to-air protection and also because the long-endurance fighter is also useful for strike and reconnaissance (i.e. air-to-ground) missions. Since these are large aircraft they have a large payload option and can also carry additional (air-to-ground) avionics. The US ◊ *Phantom F-4* and the Soviet ◊ *Foxbat* are in this class, with long-endurance, Mach 2 + performance, a large weapon/avionics load-option, and a two-man crew.

2. *Air-to-Ground Missions*

(c) Strike and Reconnaissance. This type of mission requires a large payload/long-range option, high speeds at low altitudes, and a good air-to-air combat capability. In practice, this role is performed by the (b)-type aircraft (as above). The Phantom F-4C and F-4E are (c) conversions of (b)-type aircraft, but the Soviet ◊ *Fitter Su-7* approximates the pure (c)-type, though in this case the 'mission' category clashes with the 'generation' category, since this is an older plane.

(d) Close-support or Attack. Since most battlefield targets can still be recognized only visually, the high speeds of (a), (b) and (c)-type aircraft, and their concomitant poor performance at low speeds, disqualify them for this role. The sophistication of modern (a), (b) and (c)-type aircraft also means that they are an uneconomic way of delivering weapon-loads on the battlefield. This requires only limited avionics, and simple airframes. Thus a new breed of attack aircraft has emerged, the ◊ *Skyhawk A-4*, the ◊ *Corsair II* and the Anglo-French ◊ *Jaguar* being typical in their unambitious speed, relatively simple avionics, moderate cost and very high weapon payloads. These aircraft are tactically efficient but cannot protect themselves against Mach 2 fighters, so that (b) and (c)-type aircraft must provide them with 'cover'. The US Corsair II is the latest aircraft of this type: its avionics are a great deal more sophisticated than those of the Skyhawk A-4, but it still lacks a real air-to-air capability. The ◊ *Mirage 5* is a (d)-type aircraft obtained by removing the expensive avionics of a (b)-type aircraft (the ◊ *Mirage III-E*) and using the additional payload for rockets and bombs. The ◊ *Intruder A-6* is another variant of this theme and close to the light-bomber category: it is subsonic but its avionics are exceedingly sophisticated and intended for air-to-ground strikes in all weathers and at night. An 'all-weather' capability in this context means a doppler ◊ *radar* and an electronic display system which enables the pilot to 'see' targets and land features at night and in bad weather. While strategic targets are usually fixed and can be attacked with ballistic missiles, and airborne targets can be attacked with homing missiles (since their background is clear), ground targets cannot yet be distinguished by electronic means. The only alternative to visual target-acquisition is remote-control television, as employed in the ◊ *Martel* missile and the ◊ *Walleye* glide bomb. Thus such targets can be attacked only with visually aimed rockets or bombs or by command-guided ◊ *missiles* such as the ◊ *Bullpup*. Until land targets can be distinguished from their background by automatic means, (d)-type aircraft will retain their importance – and their sub-sonic speeds.

(e) Tactical Reconnaissance. Land targets and features still have to be found by visual means but (e)-type aircraft can do a great deal more than that. (c)-type aircraft and (d)-type aircraft are adapted for this role by the removal of fixed weaponry and the fitting of cameras (including oblique ones), infra-red sensors to spot enemy movements at night, side-looking radar for mapping in bad weather, and radio receivers for communication-interception. A variant of this type is the newest breed of electronic countermeasures aircraft, such as the Intruder *EA-6A* (◊ *ECM*).

◊ *Combat Aircraft* for general subject entry, including VTOL, STOL and 'variable-geometry' developments.

◊ *Radar* and *Inertial Guidance* for general entries.

◊ *Missile* for fighter aircraft weapons.

'Fire and Movement.' Infantry tactics evolved in response to First World War trench fighting conditions when the availability of high volumes of defensive fire made straightforward frontal assaults impossible or very costly. From the initial idea that in order to advance enemy artillery had to be silenced, and that direct fire from machine-guns and indirect fire from mortars were needed to force the enemy to keep his head down, various patterns of assault have emerged. These all consist of dividing forces into two or more teams. While one fires (the 'fire-team' with machine-guns and heavier weapons), the other dashes forward. Then *their* fire keeps enemy heads (now presumed to be fewer) down, while the 'fire-team' advances. Sometimes a third, smaller team was added, intended to act as a reconnaissance and leadership unit, as in the

now discarded American Able-Baker-Charlie squad system.

For competitive tactics, ◊ *'Marching Fire'*.

'Firebar' Yak-28 (includes Brewer light bomber). Soviet fighter. A two-seat, twin-jet supersonic development of the Yak-25 ◊ *Flashlight*. Like other heavy-weight fighters, the Firebar has been developed into a secondary bomber version:

Firebar Yak-28P is a long-endurance fighter, in service since 1959, with nose radar and fire-control electronics (i.e. 'all-weather' capability) used for the 'air superiority' role within the Soviet air defence system (◊ *PVO Strany*). Its maximum speed is estimated at Mach 1·1. It is usually seen with ◊ *Anab* air-to-air missiles.

Brewer is the NATO code name for the fighter-bomber version. This has a glazed nose, an air-to-ground radar, and two guns. There is also a version with a bomb-carrying blister but no data are available on the weapon load/fuel options. Estimated maximum speed Mach 1·1. The Brewer is about 59 ft long, and its maximum take-off weight is estimated at 35,000 lb.

◊ *Fighter* for general subject entry.

Fire Power. Theoretically quantitative but actually qualitative measure of the power of weapons or groups of weapons used within a military unit. Not a measure of the 'power' of the unit itself. Rarely expressed in numbers, but, if so: either the weight of shells fired per shooting cycle, or total rounds of stated calibre ammunition per time period. Potential effectiveness of a military unit can be thought of as fire power + mobility. (Actual effectiveness = fire power + mobility + quality of decision-making + data + motivation and skill of personnel.)

Firestreak. British air-to-air missile. A 'first-generation' air-to-air missile with ◊ *infra-red homing* guidance. Like ◊ *Sidewinder*, its US counterpart, the Firestreak is a fair-weather pursuit-course weapon, but it is larger and more complex. Apart from the nose-cone infra-red sensor, there are two further sensor sets set in rings farther down the missile; in operation, the nose-cone sensor achieves initial homing with the other two sets locking on the target and controlling the detonation of the warhead. Firestreak has been succeeded by the ◊ Red Top a more advanced British air-to-air missile. Specifications: Length: 10 ft 5 in. Diameter: 8 in. Weight: 300 lb. Warhead weight: 50 lb. Cruising speed: Mach 2 +. Range: 5 miles.

First Strike (Capability, Strategy). A 'first strike' is the first use of nuclear weapons in armed conflict. 'First-strike capability' refers to the possession of a delivery system for nuclear weapons which is presumed to be unable to survive a (nuclear) attack made upon it. This vulnerability really depends on the capabilities of the enemy, but it is often thought of in relation to the number and nature of the delivery systems (i.e. unprotected missiles requiring lengthy prefiring preparations, or bomber forces whose range requires basing near the hostile perimeter).

A first-strike-capability-only implies therefore first use (in the expectation of enemy attack), since the force cannot survive to deliver a retaliatory blow after the attack has taken place. This is described as a 'first strike strategy'.

Very confusingly the phrase has a second, completely different meaning as follows: a 'first-strike capability' is also one which is sufficiently effective to destroy the enemy's retaliatory forces. In other words, a disarming ◊ counterforce capability. Thus while one meaning implies the possession of few and/or low-quality delivery systems, the other implies the opposite. This second version of the phrase was used by the new US administration's statements in 1969, during the debate of the Safeguard system of ballistic-missile defence, when it was asserted that the Soviet Union had the capability of depriving the US of its ◊ assured destruction forces by a counterforce surprise attack.

49   **'Fishbed' Mig-21. Soviet fighter.** The Mig-21 is a short-range delta-wing fighter used as an interceptor in Soviet service and which has been supplied in large numbers to satellite and foreign countries. Four versions of the Mig-21 have been identified, but the F and PF (Fishbed C

and Fishbed D) are the two operational combat aircraft, while a third version, Mongol, is a two-seat trainer, and the fourth has not reached operational status.

The Mig-21F has a combat radius of only 375 miles (clean) and a top speed of Mach 2 in ideal conditions (at 36,000 ft) – again with no under-wing weapons or fuel (i.e. clean). The Mig-21F is a relatively light plane, and among other limitations it has a correspondingly limited weapon load: two 30-mm cannon and two ◊ Atoll air-to-air missiles (or perhaps two bombs of the 500-lb class). The radar and control electronics are also limited, and the plane is usually described as a 'clear-weather' only fighter. A finned external fuel tank of about 150 imperial gallons can be carried under the fuselage.

The Mig-21 PF or Fishbed D is an improved version with a more powerful, 13,120 lb, thrust turbojet, a larger nose radar and other additions. There are several further improvements on progressive versions of the PF.

As it forms the main opposition to Western aircraft in the Middle and Far East the Mig-21's capabilities have been subject to a great deal of debate. It is now generally accepted that the PF is a very good interceptor, a poor 'air-superiority' and a very inadequate 'strike' plane – a performance which roughly fulfils its design requirements.

Specifications Mig-21 F (Fishbed C): Power plant: one turbojet rated at 9,500 lb static thrust dry and 12,500 lb with afterburning; two jato rockets can be fitted. Overall length: 55 ft. Wing span: 25 ft. Take-off weight with external tank: 16,700 lb. Armament: 2 30-mm cannon 600 rpm; 2 Atoll AAMs. Top level speed at 36,000 ft: with no external stores, Mach 2; with external fuel tank and 2 Atolls, Mach 1·5. Rate of climb at sea level: 30,000 ft per minute. Combat radius with no external fuel or payload: 375 miles.

◊ *Fighter* for general subject entry.

'Fishpot' Sukhoi-9 (Su-9). Soviet fighter. The Fishpot is an 'all-weather' fighter similar in appearance to the well-known Mig-21 ◊ *Fishbed*. It is, however, a larger and heavier aircraft with more avionics, longer range and, generally, an 'air superiority' rather than a limited 'interceptor' type of performance (⟡ *Fighter*). Fishpot was first seen in a prototype form in 1956 but has since undergone various changes, though the current types retain a bulged cylindrical fuselage, delta wings, a pitot boom above the fuselage air-intake, and a nose radar forming a centre-body within it. The main recognition features, which distinguish it from the Mig-21, are the cleaner airframe, the absence of a ventral fin and fuselage fairings, and the rearward-sliding canopy. The Fishpot has a 26-ft wing span, an overall length of 56 ft and an estimated top speed at optimum height (36,000 ft) of about Mach 1·8. Fishpot was initially seen with ⟡ *Alkali* air-to-air missiles (and no guns), but more recently ⟡ *ANAB* missiles have been seen to be carried by it – and still no guns. The Fishpot appears to be suitable for patrol and interception *vis-à-vis* a strategic-bomber threat; it is too slow for air-superiority missions against modern fighters. The gross weight is estimated at 29,000 lb.

'Fitter' Sukhoi-7B (Su-7B). Soviet fighter-bomber. The Fitter is in use in the Soviet, satellite and Egyptian air forces as a strike fighter. It is a swept-wing aircraft of conventional appearance with a large fuselage air-intake partly masked by a small nose radar. The precise weapon-load versus fuel options are not known, but the Fitter can carry at least two rocket pods and two fuel tanks or bombs in the 750-lb class, as well as two 30-mm cannon in the wing roots.

Fitter is 56 ft long and has a 30-ft wing span. The gross weight is estimated at 30,500 lb.

The top speed in ideal conditions (36,000 ft altitude and no external stores) is estimated at Mach 1·6, and the combat radius – at optimum height – is of the order of 500–600 miles. In the more demanding low-level mission configuration its range is estimated at little more than 250 miles.

⟡ *Fighter* for general subject entry.

'Flagon' Su-11. Soviet fighter. This delta-wing heavyweight fighter, attributed to the Sukhoi design bureau and first seen publicly in 1965, is deployed with ⟡ *PVO Strany*, the Soviet air defence organization. The Flagon has a delta-wing configuration and two engines side-by-side in the fuselage. The maximum airframe Mach number has been estimated at 2·5, and it is thought to have a large weapon payload (or alternatively a large fuel load). The gross weight is estimated at 50,000 lb. The Flagon is about 68 ft long.

⟡ *Fighter*, for general subject entry.
⟡ *Foxbat* for the other first-line Soviet heavyweight fighter.

'Flashlight' Yak-25. Soviet fighter. An obsolescent two-seater, twin-jet fighter of which three main versions are known:

Flashlight A: fighter version with nose radar and other 'all-weather' electronics. There are two 37-mm guns in the forward fuselage and provision for under-wing unguided air-to-air rocket pods. This version is probably obsolete.

Flashlight B: a fighter-bomber version with a glazed nose for the navigator/aimer and a single-seat cockpit for the pilot; air-to-ground radar. It is still used in the tactical air force, and it retains the gun armament of Flashlight A.

Flashlight D: a tactical reconnaissance version with a single 30-mm gun and slightly modified wings.

Specifications: Overall length: 62 ft. Height: 14 ft 6 in. Wing span: 38 ft 6 in. Gross wing area: 340 sq ft. Top speed in ideal conditions: Mach 0·95.

⟡ *Fighter* for general subject entry.

FN. Belgian-designed small arms. The Belgian Fabrique Nationale d'Armes de Guerre, usually known as FN, develops, produces and exports a wide range of small arms of which the 7·62-mm NATO rifle and the MAG 7·62-mm medium machine-gun are used in very many armies. FN also produces various vehicles and aero engines under licence. The rifle (known as Fal in Belgium and elsewhere, SLR L1A1 in Britain, C1 in Canada) is a gas-operated 10·4-lb weapon, 41·4 in long, and capable of either full or semi-automatic fire. It has a 20-round magazine, a pistol grip and very good general finish. There are several optional attachments, including bayonet, grenade-launcher, various flash hiders and sight.

As in the case of other modern rifles,

there is a conversion model with a simple bipod, better sights and a 30-round magazine (called Falo in Belgium) which is used in the ◊ *light machine-gun* role. ◊ CETME for a competitive weapon; ◊ *Rifle* for general entry.

The MAG 7·62-mm NATO medium machine-gun has also been adopted by many armies. (It is known as GPMG in the British army.) The MAG is gas-operated and fed from a metallic link belt; it fires full automatic – at a cyclic rate of 700 to 1,000 rounds per minute. The MAG has a built-in bipod but is converted to the medium machine-gun role with an additional 15·9-lb tripod. It is 49·2 in long and weighs 23·9 lb with butt and bipod; without these two the MAG/GPMG is used as a tank and vehicle weapon. The air-cooling and quick barrel change allow sustained fire accurate to about 2,000 yds.

◊ *Medium Machine-gun* for competitive types.

◊ *Machine-gun* for general subject entry.

FOBS. Fractional Orbital Bombardment System. Under the terms of the 1966 Outer Space Treaty, nuclear warheads cannot be sent into orbit; but anything less than a full circle around the Earth does not violate the treaty. The FOBS method of nuclear delivery takes advantage of the orbital attack configuration while staying within the terms of the treaty. While a normal ICBM follows a very high parabolic path to target, highly 'visible' to defending radars, a weapon in low orbit (say, 100 miles) can make a sharp descent to Earth, thus cutting radar warning time very substantially, to about 3 minutes. A FOBS path therefore consists of a blast-off movement into low orbit, a partial circle to the target earth zone and a rapid descent; this would seriously prejudice ballistic missile defence systems as now conceived (◊ ABM). There is, however, a loss of accuracy and payload which makes FOBS weapons unsuitable for a ◊ *counterforce* role against ◊ *hardened* missile silos.

Forward Defence. A term applied to the official NATO strategy for the defence of Western Europe. This involves the defence of Germany as far forward as possible, as

opposed to the earlier strategy of a defence on the Rhine with West Germany as a nuclear no man's land. This policy was reportedly decided in 1952 (as an inevitable result of the German rearmament decision), but announced only in 1961.

For another strategic concept proposed for NATO, ◊ *Tripwire*.

Fox. British armoured car. The Fox, which is now replacing the ◊ *Ferret* series, is a small and lightly armoured vehicle with four driven wheels. Its chassis is very similar to that of the Ferret, but it mounts a large two-man turret fitted with a gun and a coaxial machine-gun. Its Rarden 30-mm single-shot cannon is intended as an anti-armour weapon, but, given its small calibre, the Rarden could penetrate only lightly armoured targets such as ◊ *armoured personnel carriers*. A vehicle of this size could not, of course, mount the large-calibre guns required to penetrate the much thicker armour of battle tanks, but a low-velocity gun with a large ◊ *HEAT* shell could have done this. This is the solution used in the ◊ AML French armoured car, which is no heavier than the Fox.

The Fox has a three-man crew, and it is powered by a petrol engine which is a militarized version of a cheap civilian engine. Such armour as it has consists of welded light alloy plates, and it is fitted with a flotation screen. The Fox may turn out to be a useful vehicle, but the choice of armament is perplexing.

Specifications: Length: 16·7 ft. Width: 6·9 ft. Height: 7 ft. Battle weight: 12,500 lb. Engine: 195 bhp. Road speed: 45 mph. Range: 200 miles.

'Foxbat' Mig-23. Soviet fighter. The Mig-23 is thought to be the main Soviet design for the latest generation of air-superiority and strike fighters. It has a cropped-delta wing, twin fins and slanted rectangular air-intake trunks. The thin wings mounted high on the fuselage indicate a high-speed/low-altitude capability. This is a large plane and apart from the large nose radar is reported to have a large avionics payload. A Mach 3 top speed has been claimed for this plane, and from performance data derived from the related competition aircraft, the E-266, it is thought to have a ceiling of 90,000 ft plus, and a speed of Mach 2+ with a 4,000-lb payload. The Mig-23 is a multi-mission fighter in the same category as the US ◇ *Phantom* F-4 (though more advanced), and the US F-111A, but the Phantom has been operational since 1962, while the F-111A is an altogether more advanced concept.

Dimensions: Length: 69 ft. Wing span: 40 ft. Gross weight (estimated): 80,000 lb. ◇ *Fighter* for general subject entry.

France, Strategic Offensive Forces· French nuclear-capable systems include:

(a) 45 ◇ *Mirage IVA* medium bombers supported by US-built C-135F tankers for in-flight refuelling.

(b) 18 IRBMS in armoured silos to be operational in 1971. Both the planes and the missiles are under a separate Strategic Air Command (CFAS). For details of equipment and tactics, ◇ *Mirage IV*.

(c) A force of 3 ballistic-missile Polaris-type submarines (◇ *Redoutable*) to be operational in 1970–2.

When the first Mirage IVAs became operational in 1964, France became the second country in the Western camp to acquire an 'independent deterrent', that is 'independent' of the United States. Because of the limited capabilities of this force a counter-value targeting policy has been adopted; this was supplemented by the policy of 'Défense tous azimuts', whereby the Soviet Union ceased to be the only conceivable target. The events of May 1968, De Gaulle's resignation in 1969, and their financial repercussions will probably affect the programme for second and third generation delivery systems. And the French thermonuclear programme has been delayed for purely technical reasons, so that only 60 kiloton fission bombs are available as yet. Communications, computation and guidance mechanisms for the IRBMs are also meeting development problems. Apart from these considerations the programme has been criticized on the grounds that (a) its cost weakens the French armed force organization as a whole, and (b) each system will be obsolete *vis-à-vis* potential defence by the time of completion (◇ *PVO Strany*,

NORAD, ABM). This force is on the other hand an effective deterrent *vis-à-vis* the UK and smaller powers. For comparison purposes, ◇ *US, Strategic Offensive Forces* and the same for USSR, UK and China.

Freedom Fighter F-5. US lightweight Mach 1 fighter. The F-5, a single seat twin-jet, is a combat development of the T-38 trainer, which like the British ◇ *Gnat* is marketed as a cheap 'lightweight fighter' to countries which do not need – or cannot afford – a 'real' Mach 2 fighter. Like the T-38, it is powered by two small turbojets built into the rear fuselage, with very small air-intakes above and forward of the low-mounted wings. The basic F-5A version has twin 20-mm guns in the fuselage nose and racks for up to 6,200 lb bombs, missiles (◇ *Sidewinder* and ◇ *Bullpup*) rocket pods or extra fuel tanks. The ease of maintenance and the cheapness of the F-5A are largely due to the very limited avionics: there is little more than communications equipment, though there are provisions for fitting IFF transponders, missile command systems and fire-control electronics. The USAF has bought only a small number of F-5As for combat evaluation in Vietnam, but the aircraft has been made available to 'friendly and allied' nations under various military assistance programmes. Iran, South Korea, Norway, Spain, Formosa, Greece, Turkey, the Philippines, Canada and the Netherlands have all received or bought the F-5A or the two-seater training version, the F-5B. The aircraft is also manufactured in Canada/Netherlands (jointly) and Spain. The first operational unit equipped with these aircraft was formed in Iran, in 1965.

Specifications (F-5A): Power plant: two turbojets each rated at 4,000 lb st with afterburning. Length overall: 47 ft 2 in. Height: 13 ft 2 in. Wing span: 25 ft 3 in. Gross wing area: 170 sq ft. Weight empty: 7,800 lb. Maximum take-off weight: 20,040 lb, of which the weapon-load is 6,200 lb. Maximum level speed at 36,000 ft: Mach 1·4. Cruising speed: Mach 0·96–0·85. Range with maximum payload: 368 miles. Range with maximum fuel: 1,750 miles.

◇ *Fighter*.

Frigate. ◊ *Destroyer.*

'Frog'. Soviet surface-to-surface un-guided rocket series. Frog (Free Rocket Over Ground) is the NATO designation for a series of Soviet unguided rockets similar in concept to the US Honest John (MGR-1). These spin-stabilized rockets are sufficiently accurate for nuclear bombardment of rear areas, and are seen as the modern counterpart of classic heavy artillery.

Frog 1 First of the series and in service since 1957. Six tail fins, solid-propellant spin-stabilized sustainer, no booster. It is 31 ft long, weighs about 6,000 lb and has an estimated range of 40 miles. It is carried on, and launched from, a converted heavy tank chassis.

Frog 2 A single-stage, solid-propellant rocket, spin-stabilized and about 29 ft 6 in long. Carried and launched from a converted ◊ *PT-76* amphibious tank chassis. The range is about 15 miles.

Frog 3 A two-stage, in tandem, rocket. Large-cylindrical bulbous warhead. PT-76 transporter/launcher. It has a range of 20–30 miles.

Frog 4 Long slim two-stage rocket. Solid-propellant booster and sustainer in tandem. It is about 33 ft long and weighs about 4,400 lb. Estimated range about 30 miles.

Frog 5 Similar to Frog 4, but with different (chemical?) warhead.

Frog 7 First seen in 1965. A single-stage spin-stabilized rocket. Wheeled trans-porter/launcher.

◊ *Scud, Shaddock,* and *Ballistic Missile* for Soviet missiles.

FUG 65/66. A light armoured car, Hungarian-assembled on the basis of various Soviet components. This un-distinguished 4-wheel truck-chassis type vehicle has well shaped frontal armour, but elsewhere appears to be very poorly protected. The FUG 65 is turretless, while the 66 mounts a small turret with a (Soviet) 23-mm light automatic cannon and a 7·62-mm machine-gun. It has a three-man crew and weighs about 6·5 tons.

FV. British common designation for ◊ *armoured fighting vehicles,* as in FV 433 ◊ *Abbot.* For FV 432, ◊ *Trojan.*

G

g. The acceleration due to gravitational force, equivalent to about 32 ft per second. Acceleration tolerance for airborne equipment, including pilots, is measured in gs.

57 **Galaxy C-5A. US long-range transport aircraft.** The C-5A which is now in production is a four-jet transport with two decks and shoulder-mounted wings. Its payload is about three times as large as that of the ◊ *Starlifter C-141*, the current long-range transport of the US air force. The four turbofans have a combined thrust rated at 164,000 lb and the aircraft is equipped with a wide range of standard military communication gear as well as extensive radar, computer and control equipment for navigation. In 1969 the C-5 developed 'cost escalation' problems, which may slow its deployment. Though so much larger than the Starlifter (and previous transport) the C-5 can operate from 'short' (5,000 ft) unpaved airstrips. The C-5A can accommodate up to 345 fully equipped troops (as well as 20 crew, including relief crew, loading assistants, etc.), but it is mainly intended as a logistic transport, for freight. The C-5A is the first aircraft which can accommodate the standard US battle tank, the M.60 – of which two can be carried – but the 220,000 lb-plus payload will no doubt be used for more efficient cargoes. Alternative loads include 10 Pershing missile systems or five M113 armoured personnel carriers and two trucks.

Specifications: Length: 245 ft 11 in. Height: 65 ft 1·5 in. Wing span: 222 ft 8·5 in. Gross wing area: 6,200 sq ft. Internal dimensions: lower deck 41,480 cu ft, upper deck 4,190 cu ft. Lower deck floor area: 2,373·6 sq ft. Weight empty: 323,904 lb. Maximum payload: 220,000 lb. Maximum level speed: 571 mph. Econo-

mic cruising speed: about 540 mph. Range with 90,596 lb payload and fuel reserves: 7,150 miles. Range with maximum payload: 3,580 miles.
◊ also the Soviet *Cub* and *Cock*.

'Galosh'. Soviet surface-to-air missile for ballistic missile defence. A large (65 by 8 ft) multi-stage solid-fuel surface-to-air missile with a range of several hundred miles and a 1–2 megaton warhead. The Galosh is similar to the US ◊ *Spartan* and, like the latter, is intended to intercept incoming nuclear warheads while these are still outside the atmosphere. The limited deployment of Galosh missiles around Moscow provides a thin area defence.
◊ *ABM*.

Gamma Rays. Gamma rays are very short electromagnetic waves, part of the radiation produced by nuclear explosions. Gamma rays have wavelengths of 0·01 to 0·001 Angstrom or 1/100,000th of a millimetre; this ensures very good penetration. Their effect on humans arises from the ionization of body cell atoms: ionization consists of the liberation of electrons from parent atoms, a process which is invisible but lethal. The amount of gamma radiation 'liberated' by a nuclear explosion varies with the intensity of the heat produced, so that fusion weapons ordinarily produce much more that fission ones, and as megaton power increases gamma-ray emission increases more than proportionately.
◊ *Nuclear Weapon*.

'Ganef'. Soviet surface-to-air missile. 7 A ramjet-powered, rocket-launched missile first seen in 1964. It is about 30 ft long and has a maximum diameter of 2 ft 8 in at the annular air-intake. There are four solid-fuel rocket boosters but sustained flight is by ramjet. The Ganef weighs about 2,000 lb, and its range has been estimated at about 40 miles. It is probably intended for mobile battlefield defence. Guidance is by radio command on the basis of radar tracking.

GCD. General and Complete Disarmament. Plans and proposals for general and complete disarmament have been around since the 1950s and were often used to camouflage the rejection of less ambitious – but more feasible – arms control agreements. But since the enunciation of 'agreed principles' by the United States and the Soviet Union, the more frivolous use of GCD proposals has been replaced by sustained and detailed study of its problems and requirements. This proceeds within the framework of the Eighteen-Nation Disarmament Committee (ENDC) which works in Geneva under a UN mandate. Recent US and Soviet proposals have benefited from extensive research into the problems of GCD: the stages along the way, inspection requirements, the post-disarmament international police force, etc. Despite this intellectual progress GCD is as unlikely as ever, but the meetings of the ENDC have been put to better use in working over various arms control measures, such as the ◊ *Partial Test Ban Treaty* of 1963, the ◊ *Outer Space Treaty* of 1967 and the ◊ *Non-Proliferation Treaty* of 1968.
◊ *Disarmament.*

General War. Armed conflict between the super-powers (plus allies and satellites) in which all available weapons are used, including nuclear weapons, and where national survival appears to be threatened. Professional term for total war, now in literary use. For an armed conflict between the super-powers below this threshold, central war is the recommended term. ◊ *War* for a list of other concepts and interpretations.

Geneva Convention. In common usage, one of a series of codifications of international humanitarian law, of which the latest is that of 12 August 1949. The conventions define approved behaviour with respect to non-combatants, prisoners of war and populations under military occupation; they also outlaw the use of certain weapons such as barbed arrows, dum-dum bullets, as well as chemical and biological agents. They do not outlaw the use of nuclear weapons. The extent to which this pseudo-legal system is adhered to depends on power relationships rather than legal ones. Thus Anglo-American P.O.W.s were given largely 'Convention' treatment in Germany during the Second World War (even though Allied strategic bombing violated the Convention) because of the possibility of retaliation on German P.O.W.s in Allied hands. Where retaliation could not be direct, as in the case of 'occupied' populations, the latter did not in fact receive 'Convention' treatment.

Genie AIR-2A. US air-to-air unguided rocket. The Genie is an unguided rocket with a nuclear warhead which currently equips F-106A ◊ *Delta Dart* and F-101B ◊ *Voodoo* interceptor fighters of the USAF. It is a simple and reliable system, which was first tested in 1957, and a large number are operational. The missile is fired by an electronic fire-control device which also detonates the nuclear warhead at interception; this is armed a few seconds before firing. The ◊ *fallout* which would be produced by the Genie's detonation at medium to high altitudes would be negligible.
Specifications: Length: 9 ft 7 in. Body diameter: 1 ft 5 in. Weight: 820 lb. Speed at burn-out: Mach 2+. Range: 6 miles.

Glide Bomb. An unpropelled warhead which descends to its target in wing-supported flight. Unguided weapons of this kind would be very inaccurate. The US ◊ *Walleye* is a guided glide bomb.

Gnat Mk I. British lightweight fighter. This small swept-wing jet, light, simple and cheap, was a 1955 version of the 'poor man's fighter' concept roughly equivalent to the US ◊ *Freedom Fighter F-5* currently being marketed. With a maximum take-off weight of 8,885 lb, the Gnat weighs less than half as much as 'normal' jet fighters and this translates into a correspondingly limited armament and avionics payload. The maximum speed is only transonic (Mach 0·98), but the Gnat has a reasonable range and endurance. The Gnat Mk 1 is serving with the Finnish and Indian air forces.

Specifications: Armament: 2 30-mm cannon with 115 rounds each and up to 1,000 lb of under-wing stores. Climb to 45,000 ft: 5·25 minutes. Ceiling: 50,000 ft plus. Radius of action: 500 miles. Endurance with under-wing tanks: 2 hours 30 minutes. There is a two-seat trainer version, used in the British Air Force and designated T.1.

'Goa' (SAM-3). Soviet surface-to-air missile. A compact two-stage SAM, used as a ship-borne air defence and also as a low altitude 'point' defence missile. Its diameter is 2 ft 3 in, and its overall length 20 ft. The Goa is solid-fuelled, and roughly similar to the US ◊ Hawk in terms of its configuration. Guidance is by ◊ semi-active radar homing, and the effective range is about 10 miles. The Goa is carried in pairs on transporter-launchers.

GPMG. ◊ FN.

Grenade. A small explosive or chemical missile thrown by hand or shot from an adapted rifle or a special-purpose grenade-launcher. The most common type are the pineapple-shaped defensive or 'fragmentation' grenades. These, which also come in smooth ball shapes, rely for their effect on splinters produced by the explosion of a small cartridge within the thick metal container. Since the usual hand-thrown range is about 35 yds, and since the splinter radius of such grenades is at least 10 yds, offensive grenades have been developed for assault purposes: these contain more explosive and have thinner walls which do not produce splinters, so that they can be used to hit particular targets without the user having to stop and take shelter as the grenade explodes. Chemical grenades include incendiary, illumination and smoke types; the first are sometimes known as 'phosphorus' and the second as 'magnesium' grenades.

All these grenades can be shot from rifles, where both have been adapted for the purpose by the fitting of grips and extensions, though many modern rifles have built-in grenade-launchers. Specialized rifle grenades are usually anti-tank types, such as the Energa AT, but these are effective against modern tanks only in special circumstances. Grenade-launchers are of course more efficient projectors than adapted or modified rifles. The US M.79, for example, fires 40-mm grenades at ranges of up to 300 yds, as opposed to the 100–150 yds of the rifle grenade and the 30 yds or so of the hand-thrown one.

'Griffon'. Soviet surface-to-air missile. A large two-stage rocket missile which first appeared in 1963, it may have a small third stage in the warhead section. Described as an 'anti-missile-missile' it is thought to be in reality a long-range anti-aircraft missile with some ballistic or stand-off missile intercept capability. It is 54 ft long and has a maximum diameter of 3 ft 6 in. There appears to be a radar homing nose-cone.

Guerilla Warfare. Military operations conducted by informal forces which operate within territory nominally controlled by the enemy and which have no permanent bases or other permanently defended territory. Usually thought of as small-scale and independent operations, as in ◊ insurgency or ◊ revolutionary war situations, guerila warfare can also be a secondary adjunct to ordinary military operations, as in the case of the Palestinian 'commandos'.

Guerilla warfare is based on temporary concentrations on the offensive, followed by dispersion on the defensive; this may involve going clandestine (using concealment by natural cover, e.g. forests) or covert (disguise by assuming a civilian 'cover'). The actual tactics are conventional rule-book ones for any inferior force: avoidance of battle, ambuscades, and the slow erosion of superior forces by repeated attacks on outposts and supply lines. The key factor in guerila warfare is the source of supplies, since the force lacks either a home territory or supply lines to it. In revolutionary war guerila forces are only an adjunct to ◊ subversion, and supplies are extracted from the local population by covert 'administration'; when the guerillas have a home base outside the area in question, the problem is how to obtain supplies from it without having 'supply lines'. The solution may be either clandestine land lines (as in the case of North-Vietnam to Viet Cong supply lines) or air-drops (as in many instances in the Second World War). Where subversion does not work, and where the home base does not exist or cannot overcome the logistic problem, guerillas cannot survive as effective fighting forces because then they have to acquire food supplies by using force against the local population, which is politically inefficient (if practised on the general population) and tactically inefficient (if practised on enemy military or security forces).

In any case subversion is also a source of the other essential commodity needed by guerilla forces intent on more than keeping alive: local intelligence. Thus in the revolutionary war equation (guerilla warfare + subversion + national political aims), subversion is the critical factor.

'Guideline' (V 750 VK). Soviet surface-to-air missile. The Guideline, also known as SA-2 in the USA, is a Soviet two-stage anti-aircraft missile, first

seen in 1957, which has been widely exported. The entire weapon system, known as V 76 S M, consists of search, acquisition and tracking radar, a power generator, a computer and a transporter/launcher. The missile is 35 ft long and weighs 4,875 lb. Guidance is by radio command from the radar-computer unit, which tracks both target and missile, correcting its flight-path until interception; its 288-lb warhead can be detonated either by a proximity fuse or by command. Speed at burn-out is about Mach 3·5, but acceleration is quite gradual; the effective ceiling is about 60,000 ft and the maximum slant range about 25 miles. It has a solid-propellant booster and a liquid-fuelled sustainer. The Guideline is ineffective against aircraft manoeuvring at high speed, and some dozens deployed by the Egyptians have been destroyed by Israeli tactical aircraft, with the latter suffering few losses from these missiles.

'Guild'. Soviet surface-to-air missile. A S A M about 39 ft long with no booster but probably a dual-thrust solid-propellant rocket motor. It first appeared in 1960 and is probably superior to ◊ Guideline. Its operational status is unknown.

H

HA-300. Egyptian fighter project (abortive). Egyptian aircraft factories have produced a copy of a wartime German training plane (the Bücker Bu 181D) under the name of 'Goumhouria', and an exile German-Spanish jet trainer (Hispano Ha-200 Saeta) under the name of Al-Kahira. Another exile German project started in Spain was a supersonic fighter initially equipped with British turbojets. This project was abandoned by the Spanish, but the less discriminating Egyptians bought the prototype, the imported engines and the services of about 100 Spanish skilled workers and have used them to set up the project in Egypt. This appears to be directed by a number of exiled Germans who have tried to develop an engine (designated E-300); but the prototypes have flown – subsonically – with the imported turbojets. The HA-300 is a lightweight fighter of conventional appearance with a mid-wing monoplane lifting surface and swept tail surfaces. The airframe has a claimed speed tolerance of Mach 2·2 and sub-systems have already been imported from Britain and elsewhere. The HA-300 and the indigenous E-300 turbojet remained in the project stage until 1969 when the project was abandoned.

For other abortive Egyptian projects, ◇ *Egyptian Missile Programme.*

Half-track. Popular name for the tracked and lightly armoured American-made M.3 series of personnel carriers which have front steering wheels. The M.3s are ancient and offer very limited protection, but are still used by the poorer armies, including the Israeli one.

'Hard', 'Hardening'. An adjective (or verb) applied to military and other installations which are to some degree pro-tected from weapon effects, usually nuclear weapon effects. It is most commonly applied to missiles housed in underground concrete silos mounted on steel springs and fitted with armoured lids, whose associated communication and control facilities are similarly protected. Although 'hardness' is attributed to any protective structures which can resist more than 100 pounds per square inch (psi) of over-pressure, strategic-missiles silos are protected against blast effects three times as powerful. (◇ *Minutemen* missiles are generally reported to be hardened against 300 psi.) The degree of protection which a given level of hardening confers depends on the yield and accuracy of the offence. Since a strategy of ◇ *assured destruction*, or, more generally, any ◇ *second-strike* mission, requires a high degree of reliable survivability, this has to be ensured by adjusting the hardening to the evolving threat.

Satellite surveillance (to identify and locate silos), increasing accuracies and the development of multiple-warhead ICBM systems (◇ *MIRV*) have all degraded the reliability of present 300 psi harden-ing for ICBMs. Although super-hardening is now being evaluated, in-creasing the blast resistance to, say, 1,000 psi or even more is very costly and a poor response to the dynamic improvement in accuracies: the survival probability of a single missile site hardened to 300 psi against a five-megaton warhead falls from 65 to about 7 per cent as the warhead's accuracy improves from a 5,000 ft to a 2,000-ft median miss (or CEP – ◇ *Nuclear Warhead*). If the hardening is increased to 1,000 psi, the silo's survival probability against a five-megaton warhead with a 5,000-ft CEP is about 86 per cent, but an improvement in the warhead's accuracy to a 1,500-ft CEP cuts down its survival probability to 17–18 per cent. In other words, hardening ceases to be an effective protection for missile silos against a combination of current ICBM warhead yields with the accuracies expected in the 1970s. Alternatives include a variety of mobile systems such as the ◊ *Polaris* and various land-mobile configurations, but active defence by means of an ◊ *ABM* appears to be a far more promising solution. ◊ *Safeguard*.

43 Harrier. British V/STOL fighter-bomber. The Harrier, which is in squadron service with the British air force, and with the US Marine Corps, is a relatively light ground-attack aircraft which can take off and land vertically. This operationally unique capability is built into an airframe of fairly conventional layout with a cylindrical fuselage and cantilever shoulder-wing. It is powered by a single turbofan giving about 10,000 lb of thrust, and the VTOL capability is achieved by vectored thrust, i.e. the plane hovers by redirecting the jet thrust (instead of rotating the engine as in some other VTOL designs). This VTOL capability allows the Harrier to be operated from jungle clearings of little more than 50 by 30 ft, helicopter shipboard platforms and other small flat surfaces, but when it is used in this way its weapon/fuel payload is low. A much better performance is obtained by using the Harrier in a STOL configuration, for which the plane is also very suitable. No precise specifications are available, but Harrier can carry about 4,000 lb of weapons (in the STOL mode) and has a level speed of at least Mach 0·9. It has a 25-ft wing span, an overall length of 46 ft and a height of 11 ft. The gross wing area is 201 sq ft.

The Harrier is a development of the earlier P-1127 and Kestrel (F(GA)1), which was evaluated by the British, West German and US air forces, and when in service it will be the only operational VTOL fighter. It is generally thought that the STOL approach is more promising than VTOL because of the substantially smaller payload/fuel loss penalty, and because VTOL aircraft would operate advantageously only in certain specialized combat situations, where there

are only small but secure clearings, or in naval use – without carriers. A trainer version designated T 2 has also been developed.

◊ *Fighter* for general entry.

Hawk MIM-23A (ex M-3). US sur-   77 face-to-air missile. The Hawk (Homing All-the-Way Killer) is a mobile weapon system primarily intended against aircraft flying at low altitudes, The Hawk system consists of a CW ◊ *radar* which identifies moving targets even at very low altitudes, an 'illumination' radar which bounces signals off the target, and a missile-borne ◊ *semi-active radar* guidance package which picks out target reflections and homes on it. Each battery of Hawk missiles includes 6 launchers, with three missiles each, with another 18 missiles in reserve, as well as radar, power generator, communications and personnel vehicles. Hawk missiles equip the US army and several foreign countries.

Specifications (missile only): Two-stage rocket motor; CW semi-active guidance; high explosive warhead. Length: 16 ft 6 in. Body diameter: 1 ft 2 in. Wing span: 3 ft 11 in. No tail surfaces. Weight: 1,295 lb. Speed at burn-out: Mach 2·5. Slant range: 22 miles.

HE. High Explosive. ◊ *TNT*.

HEAT. High Explosive Anti-Tank. This is an armour-penetrating explosive charge also known as 'shaped' or 'hollow' charge and used in (◊ *bazooka*) rockets, recoilless weapons, anti-tank missiles, and low-velocity guns. While armour-piercing ammunition (◊ *AP*) penetrates by sheer kinetic energy and must therefore be fired from heavy, costly and delicate high-velocity guns, HEAT penetrates by focusing blast and needs no velocity at all, beyond that needed to deliver the projectile to its target.

HEAT projectiles have a copper-lined conical cavity in their explosive charge which produces a jet of liquid copper upon detonation; this jet, travelling at about 27,000 ft per second, produces very high pressures over a small area and penetrates armour quite easily. The diameter of the hole is, however, quite small and decreases with the depth of penetration, so that, though a HEAT charge penetrates a thickness of armour about 5·6 times its base diameter, ratios of 3:1 are needed to cause lethal damage and not just a very small hole. Thus an effective HEAT charge must have a calibre at least as great as conventional high-velocity rounds (◊ AP). But this is offset by the much lower weight of HEAT rounds and the fact that high muzzle velocities are not needed. Indeed, HEAT shells cannot be used efficiently with high-velocity guns which are spin-stabilized, so that if used in cannon the latter are usually medium velocity or smooth bores. ◊ AMX 30 for a spin-stabilized HEAT round which is fired from high-velocity guns.

◊ HESH for another approach to the anti-armour ammunition problem.

Hedgehog. A term used to describe both a tactical concept and an item of ◊ Anti-submarine Warfare equipment.

As tactical concept: an independent perimeter ringed by anti-tank weapons and containing artillery and anti-aircraft weaponry. Such perimeters are part of a system of ◊ defence in depth, intended to resist direct assault and to counteract ◊ Blitzkrieg tactics by (a) using artillery to interdict the passage of 'soft' transport and infantry units between each hedgehog and the next, and (b) forcing tank and other armoured units into the corridors formed between the hedgehogs. In this way, armoured units are detached from their supply and supporting infantry units and can be defeated by armoured counter-attacks when weakened by supply shortages and exposed by the lack of accompanying infantry.

As an item of anti-submarine warfare equipment: a group of depth-charge mortars arranged in a round firing pattern fitted to a traversing and elevating platform on the ship's deck; their depth-

Brief Specifications of Some Common Helicopters in Military Use

Designation	Country of origin	Year opera- tional	Primary mission	Transport capacity
SE 3160 Alouette III	France	1961	general purpose	crew plus 7 seats
SA 330	France	1969	medium transport	crew plus 12 troops and equipment or 18 passengers
Super Frelon SA 321	France	1966	heavy transport	crew plus 30 troops or 9,920 lb
Augusta Bell 204/205	—	—	—	—
Westland Scout AH-MK 1	UK	1963	light general purpose	5 seats or 1,500 lb
Whirlwind HAR Mk 10	UK	1961	military or civil	crew of 2 plus 10 troops
Wessex Mk 2	UK	1962	general purpose	crew of 3 plus 16 passengers

charge warheads are supposed to destroy the targets even when no direct hit is achieved. Now obsolete, it is still found on many older ASW ships.

Helicopter. Heavier-than-air aircraft sustained by power-driven rotary wings. This basic characteristic of helicopters means that the machine can generate lift without a ground roll, but it also means that there are energy penalties in sustaining flight which fixed wings obviate. Other drawbacks include mechanical complexity in the rotors and heavy demands on the pilot. These factors taken together mean that the capital and operating cost, pilot skill and maintenance requirements of a helicopter are far higher than those of fixed-wing aircraft on a ton/mile basis. Further, the speed and operational ceiling of helicopters are quite limited. But the vertical take-off and landing (and the hovering) capability of helicopters outweighs all these disadvantages in certain specialized missions, where hover is required or where there are no suitable airfields. The military applications of helicopters have expanded dramatically in the last twenty-five years, and a wide range of diversified uses has evolved: heavy transport, assault transport, anti-submarine warfare, air-sea rescue, mine/buoy dispensing and recovery, battlefield evacuation, tactical reconnaissance, visual command observation, and – more recently – fire support in the battlefield. ⟡ *Huey cobra* and *Cheyenne*. Their features – small frontal area, winglets for lift and weapon attachment, armour protection against small-calibre and splinter weapon effects, built-in armament in rotating turrets, higher speed and ceiling limits and advanced electronic weapon control/navigation aids – represent an attempt to overcome the vulnerabilities of 'civilian' style helicopters which were exposed during the course of the Vietnam war.

Maximum speed at sea level (mph)	*Range at sea level (miles)*	*Weight (lb)* empty	loaded	*Comments*
131	186 with 1,400 lb	2,436	4,630	new version of successful Alouette II
174	335 normal load	7,187	14,110	wide provisions for avionics
161	584 light load	14,150	n.a.*	built with US assistance
—	—	—	—	see Bell UH-IB/UH-IH
131	315 with 4 passengers	3,332 (operating weight)	5,300	there is also a naval version the WASP HAS Mk I which is heavier and slower
106	300 with standard tanks	4,952	8,000	air-to-ground missiles can be fitted
132	478 with maximum fuel	8,304	13,500	turbine-powered version of the Sikorsky S-58; nine versions in service

* not available.

Brief Specifications of Some Common Helicopters in Military Use (cont.)

Designation	Country of origin	Year operational	Primary mission	Transport capacity
Hughes OH-6A Cayuse	US	1967	light observation	crew of 2 plus 2 seats
Bell Iroquois UH-IB	US	1961	general purpose, including gunship	crew plus 7 troops or 3,000 lb
Iroquois UH-IH (UH-ID)	US	1963	assault transport	pilot plus 12–14 troops or 4,000 lb
Sikorsky (S-58) Choctaw/Seahorse CH-34A/UH-34D and others	US	1955	general purpose	crew plus up to 18 seats
Sikorsky (S-61R) CH-3/HH-3	US	1964	amphibious transport	crew of 2 plus 30 troops or 5,000 lb
Boeing-Vertol Sea Knight CH-46A/UH-46A	US	1961	twin-motor general purpose	25 fully equipped troops or 4,000 lb
Boeing-Vertol CH-47B Chinook	US	1967	medium transport	2 pilots plus 33–44 troops
Sikorsky (S-65) Sea Stallion CH-53A/HH-53B	US	1966	heavy transport	crew of 3 plus 38 fully equipped troops or 8,000 lb
Kamov Harp Ka-20	USSR	1960	anti-submarine	crew of 5–6 plus radar, sonar and weaponing or a maximum payload of 4,000 lb
MIL Mi-4 Hound	USSR	1952	basic military	crew plus up to 14 troops or 3,525 lb
MIL Hook Mi-6	USSR	1960	heavy transport	crew of 5 plus up to 65 seats or 26,450 lb of payload

Maximum speed at sea level (mph)	Range at sea level (miles)	Weight (lb) empty	loaded	Comments
143	'normal' 413, at 5,000 ft	1,156	2,700	a high performance light helicopter, large numbers being produced
138	330, light load	4,519	8,500	also produced as the AB 104B by Augusta Bell in Italy, very large numbers being produced both in US and in Italy
130 at 2,000 ft	357, light load	5,090	9,500	the UH-IH is a slight modification of the UH-ID, which has been produced in quantity. It is also produced by Augusta Bell as the AB 205
122	247 with 10% reserves	7,750	13,000	large numbers produced in eight military versions, widely exported
165	485 with maximum fuel 10%	12,423	22,050	development of earlier S-61A series, both series widely produced
166	230 with 6,070 lb of payload and 10% reserve	12,406	23,000	seven military versions widely used as assault transport in Vietnam
178	115	19,375	38,550	development of CH-47A
195	258 with 4,000 lb fuel and 10% reserve	21,780	42,000	fully loaded this is the heaviest helicopter in US service
137	250 with standard load	9,700	16,000	a specialized aircraft adapted for shipboard use. The performance data refer to the civilian version, KA-25K
130	250 with 8 passengers and 220 lb	n.a.*	17,200	this helicopter is standard equipment in the Soviet armed services and also appears in versions Mi-4P and Mi-48
186	620, with 9,920 lb of payload	60,185	93,700	heaviest helicopter in the world. It is not cost-effective except in a crane configuration. ◊ Harke

Brief Specifications of Some Common Helicopters in Military Use (cont.)

Designation	Country of origin	Year operational	Primary mission	Transport capacity
62 MIL Harke Mi-10	USSR	1960	flying crane	crew of 3 plus up to 33,070 lb of cargo over short distances

58 Hercules C-130 and variants. US military transport aircraft. The Hercules C-130 is a four turboprop military transport which has been produced in large numbers for all US military services and many foreign countries. The C-130 has been developed into several versions with progressive speed/range/payload improvements and into special purpose variants:

C-130A: Basic transport. Weight empty: 63,000 lb. Maximum payload: 36,600 lb. Range with 21,500 lb payload: 3,350 miles. Cruising speed: 356 mph.

C-130B: Developed transport. Weight empty: 69,300 lb. Maximum payload: 35,700 lb. Range with 20,500 lb payload: 3,830 miles. Cruising speed: 367 mph.

C-130E: An extended range version. Weight empty: 72,892 lb. Maximum payload: 45,000 lb. Range with 20,000 lb payload: 4,700 miles. Cruising speed: 368 mph.

Other transport versions were built in smaller numbers, including the C-130H and K with a slightly improved performance on the E. Special purpose variants include the GC-130A target drone control and launcher, the RC-130A for photographic version, the HC-130B coastguard search and rescue plane, the JC-130B satellite recovery version, C-130D for Arctic transport, the KC-130F US Marines' assault and tanker, the LC-130F UN navy Antarctic transport, the HC-130H USAF search and rescue version

(which can pick up a man after a simple kit is dropped to him), a STOL assault transport under development designated the C-130J, and the HC-130P for helicopter refuelling and mid-air parachute recovery of satellite capsules. The Vietnam war has resulted in the development of an armed version: the AC-130, with 4 20-mm multi-barrel cannon and 4 7·62-mm multi-barrel machine-guns.

All the C-130s have 4 turboprops (with ratings from 3,750 to 4,910 eshp each) and a 4-man basic crew. The troop carrier versions carry up to 92 soldiers with kit, or 64 paratroops.

HESH. High Explosive Squash-Head. A type of gun ammunition initially developed in Britain as a secondary tank gun round. (For primary rounds, ◊ AP, APDS and HEAT.)

HESH, or, as they are described in the US, HEP (High-Explosive Plastic), projectiles have a thin and soft casing which is squashed on impact so that the charge spreads out over the target and adheres to it. If the target is soft, the HESH round will have the normal blast effect (but not the fragmentation effect) of conventional thick-walled shells. On the other hand, the HESH round – unlike the conventional one – also has a good anti-armour potential. When the charge explodes in such close contact with the armour surface, it generates stress waves which – if the round is large enough in relation to armour thickness – will fracture the inside surface of the armour, throwing off lethal scabs.

Maximum speed at sea level (mph)	Range at sea level (miles)	Weight (lb) empty	loaded	Comments
124, with normal load	153, with normal load	50,525	95,790	The 'Harke', a development of the 'Hook', is fitted with very high landing gear which gives it a clearance of more than 12 ft under the fuselage. The Mi-10K has normal landing gear and is the only cost-effective version of the Mi-6 Senior

Thus the HESH round is effective for both armour and 'soft' targets. The reason why tanks still carry two kinds of ammunition, instead of the multi-purpose HESH, is because armour spacing can render it ineffectual. In fact a thin outer plate in front of the main armour is sufficient to detonate a HESH projectile, so that the stress waves are dispersed. HESH is used in British, US and German tanks as a secondary tank gun round, the primary one being ◊ APDS ammunition.

Hispano-Suiza. (Light automatic gun brand name.) Hispano-Suiza is a Swiss armament concern specializing in the manufacture of 20- and 30-mm automatic rapid-fire guns. The following types are in wide use:

HSS 673. A light 20-mm-calibre gun with a variable rate of fire (800 to 1,050 rpm) mounted on a simple two-wheel chassis/mount with limited traverse and elevation. There is a simple optical sight.

HSS 669. The same gun mounted in a more versatile and heavier platform with a more sophisticated sighting device. It is intended as a light anti-aircraft weapon.

HSS 639 is similar to the above but has better sights.

HSS 665 is a triple 20-mm with a high accuracy optical sight, a computer-operated hydraulic servo-system to point the gun and a Wankel rotative engine to provide the power. There is a 50-round magazine for each of the three guns, or alternatively 120-round magazines for the outer guns.

HSS 820. A single 20-mm gun with a special mounting to fit the US ◊ M.113 ◊ armoured personnel carrier (and other similar vehicles).

HSS 661. A 30-mm gun with the control systems of the HSS 665. This has a rate of fire of 650 rpm.

HSS 831A. Dual 30-mm mounting fitted on the French ◊ AMX tank chassis. It has a search and tracking unit consisting of ◊ tracking radar and computer as well as supplementary optical sights and manual control.

◊ Artillery.

HMG. ◊ Heavy Machine Gun.

Holy War. A war which is regarded by at least one of the parties as having been inspired, or sponsored by (or for the benefit of), a deity. A 'Crusade' is seen as sponsored by men on behalf of a religious cause; a 'Jihad' is the Islamic equivalent. The Second World War is described as the Great Patriotic War in the Soviet Union and the beneficiary was stated to be the (holy) entity of 'Mother Russia' ('Rodina') – not the Union of Soviet Socialist Republics.

Honest John MGR-1. US unguided surface-to-surface rocket. The Honest John, operational since 1960, is a large fin-stabilized rocket with a choice of nuclear, chemical or conventional warheads. The Honest John is fired from a heavy truck-launcher which has to be steered towards

91

the target, range being varied by the angle of elevation. Though there is no guidance, the 5,820-lb rocket is spin-stabilized by means of spin rockets and canted tail surfaces. It has a 12-mile range and travels at Mach 1·5. This rocket, which is due for replacement by the ⟡ *Lance XMGM-52A* missile, is currently in service at the divisional level with the American and many NATO armies under two-key arrangements.

HOT (includes MILAN). Franco-German anti-tank missile. The HOT (High-subsonic, Optically-guided, Tube-launched) is an advanced anti-tank wire-guided missile which is being developed jointly by the French Nord and the German Bölkow concerns: a smaller MILAN (Missile d'Infanterie Léger ANtichar) version is also being developed. HOT is launched from a disposable tubular container, which is also used for storage, and it is guided to target by impulses produced by a TCA unit (⟡ *Nord, Nord Harpon*) sent down the wire link. With this system the operator keeps the target in a special sight, with an IR goniometer and programmer following and correcting the missile's path. As the name suggests, HOT is faster than the first-generation missiles it will replace as well as being more accurate; further, it requires fewer operator skills.
MILAN has this very short minimum permissible range because it can be used as a ⟡ *recoilless weapon* at those ranges, i.e.

House and the Kremlin. It is supposed to act as an instant and secure channel for communication and bargaining in time of crisis. The line was so used during the June War of 1967, when (a) the US President reassured the Soviet leadership with respect to certain US air force movements, and (b) the two sides agreed to refrain from direct intervention. Similar links have been established between London and Moscow and Paris and Moscow (the 'Green Line'), but not between Paris and London, or Washington.

Hound Dog (AGM-28B). US stand-off missile. The Hound Dog, a turbojet-powered stand-off which equips the later versions of the B-52 ⟡ *Stratofortress*, is designed to deliver a thermonuclear payload at ranges of up to 600 miles, so that the B-52 bomber need not be exposed to the inner air defences. Its maximum speed is a little above Mach 2 and it can therefore be intercepted by fighters and missiles, but with a length of 43 ft and a wing span of 12 ft it is much smaller than the B-52, and its surfaces are substantially less reflective to radar. Thus its chances of penetrating (Soviet) air defences are substantially greater than those of the mother vehicle. The Hound Dog weighs 9,000 lb at launch and is believed to have a multi-megaton warhead. The ⟡ *inertial guidance* system allows alternative targeting.

Howitzer. An artillery piece which fires over a curved trajectory. ⟡ *Artillery.*

Specifications (missile and container):

	Length	Weight	Maximum speed	Range Minimum	Maximum
MILAN	2 ft 5 in	24·2 lb	400 mph	83 ft	6,560 ft
HOT	4 ft 3 in	55 lb	625 mph	250 ft	13,100 ft

without the guidance and wire-link units. The TCA unit used with both systems weighs 17·6 lb.
  ⟡ *Missile* for general entry.

'Hot Line'. A direct communication link (telephone/teletype) between the White

HQ. Conventional abbreviation for Head-quarters; used with reference to the decision-making centre of an organization, including an armed force organization. GHQ, General Headquarters, is used to indicate that the decision-making centre

concerned exercises command functions
over other HQs.

**61 Huey cobra (AH-1G). Combat heli-
copter.** A combat development of the
Iroquois UH-1 series of helicopter specifically designed for close-support missions.
The Huey cobra has a two-blade rotor, a
low silhouette, and a fuselage which is only
36-in wide. It has two small stub-wings to
assist cruising flight and which are used as
weapon attachment points. Under the
forward fuselage, the Huey cobra carries a
weapon turret containing a six-barrel
7·62-mm machine-gun (Minigun) and a
40-mm grenade-launcher. The two-man
crew sits in tandem with the pilot behind
the gunner in an elevated full-vision seat.
The gunner controls the Minigun turret,
which can be turned 115 degrees on either
side of the centreline and depressed and
elevated through 71 degrees. The machine-
gun can be fired 'slow' at 800 rounds per
minute for target location, and at 6,000
rpm. As well as firing the 40-mm grenades,
the pilot can fire the turret weapons but
only fighter-fashion, dead ahead. The pilot
also controls the firing of the rocket pods
and the release of the other external
weapon-loads. The gunner, pilot and other
sensitive parts are protected by steel
armour, another specifically combat feature
which distinguishes this helicopter from
general-purpose helicopters adapted for
combat purposes. The Huey cobra has a
simple skid undercarriage, a maximum
cruising speed of 196 mph and a hovering
ceiling, in ground effect, of 7,200 ft with
a typical combat load.

▷ *Helicopter* for general subject entry.

**Hustler B-58. US supersonic medium   50
bomber.** The Hustler, which became
operational in 1960, was until recently
the only supersonic bomber of the US
air force, but it has not been procured
in quantity. (It has now been retired.) It
is a delta-wing three-seater powered by
four turbojets; there are many original
features including the external load system.
The fuselage is relatively small and there
is no internal bomb bay. Instead there
is a large cylindrical pod carried under
the fuselage which can be fitted with any
combination of warheads, avionics, photo-
reconnaissance or fuel loads, thus giving
complete mission flexibility. The maximum speed exceeds 1,400 mph, and
the operational ceiling is about 60,000 ft.
   The Hustler's classified but limited
range (2,000 miles +), and the fact that
it is suitable neither for low-level flight
nor as a launch vehicle for ▷ *stand-off
missiles* (such as the ▷ *Hound Dog*), has
limited its procurement, the manned
bombers of the ▷ *US, Strategic Offensive
Forces* being mainly the subsonic ▷ *Stratofortress B-52s*. The B-58s will be phased
out by 1973 (or even earlier).
   Specifications: Power plant: four turbojets rated at 11,000 lb st each augmented
to 15,800 lb st with afterburning. Length:
96 ft 9 in. Height (at tail): 31 ft 6 in. Wing
span: 56 ft 10 in. Gross weight: 160,000 lb.
Maximum payload: 12,000 lb.
   ▷ *Bomber* for general subject entry.

I

CBM. Inter Continental Ballistic Missile. ⟡ *Ballistic Missile.*

ICBM, Soviet. The Soviet Union deployed its first intercontinental ballistic missiles in 1958-9, that is before the United States, but these were very large (800,000 lb of thrust), costly and vulnerable, and very few ('a handful') were actually deployed. Rather than face the costs of a realistic deployment, the Soviet leadership adopted a policy of token deployment and strategic deception (⟡ *Missile Gap*) in the context of which false claims were orchestrated to project the image of a large and capable ICBM force. In the meantime, the US reacted to the apparent threat by a real deployment programme, so that by the time of the Cuban missile crisis (November 1962) the US had about four times as many operational ICBMs as the Soviet Union. Once this attempt at turning shorter-range missiles into ICBMs by basing them near the United States had failed, the Soviet leadership initiated a real deployment programme:

Soviet ICBMs deployed:
    1960 = 30+
    Early 1965 = 200+
    Mid 1966 = 250+ US missile force stabilized at 1,000 ⟡ *Minuteman* ICBMs
    Mid 1967 = 570+
    Late 1967 = 700+
    Late 1968 = 900+
    Late 1969 = 1,200+
    Late 1970 = 1,300

Of these about 250 are large SS-9 ⟡ *Scarp* with a 25-megaton warhead (or a payload equivalent to three 5-megaton MIRVs). There is also a larger number of the less accurate ⟡ *SS-11* with a 1-2

megaton warhead, as well as some more modern SS-13 ⟡ *Savage*, the first solid-fuel ICBM deployed by the Soviet Union, which is fitted with a 1+ megaton warhead. The old SS-7 and SS-8 are still deployed but may be in the process of being phased out. The notable aspect of this deployment is the large payload of the SS-9 Scarp, which is about twenty times as large as that of the US Minuteman 1 (in terms of megaton yields). Since the payload affords plenty of scope for MIRV warheads, including a 3 × 5 megaton combination, this ICBM presents a threat to the security of the US deterrent. In view of its cost, the Scarp is necessarily seen as a ⟡ *counterforce* weapon aimed at missile silos rather than as a counter-city weapon, for which large payloads are generally inefficient.
⟡ *PVO Strany* for Soviet strategic defences, and *Safeguard* for the likely opposition. ⟡ *USSR, Strategic Offensive Forces.*

IFF. Interrogation Friend or Foe. ⟡ *Radar.*

Ikara (RN Ikara). Australian/UK anti-submarine missile. Ikara is a long-range anti-submarine weapon system developed jointly by Britain and Australia. It consists of a dual-thrust solid-propellant missile carrying a homing (Type 44) torpedo assisted by shipborne sonar devices, for target detection, and a radar/radio command guidance system. RN Ikara is basically the same system with a different computation/guidance unit developed for, and by, the British navy.

After sonar target detection has been made and the target's path predicted, the missile is launched from the ship towards the target area; continued sonar soundings and radar tracking of the missile ascertain

85

their respective positions and paths. Radio commands from the shipborne fire-control computer then correct the missile's path until the torpedo is released. When the missile reaches the target area, the homing torpedo is dropped from the missile by parachute, and proceeds to find and destroy the target.

RN Ikara and Ikara are about 11 ft long and have a 5-ft wing span.

▷ *Asroc* and *Subroc*, similar US weapons.

Indirect Approach. A strategic and tactical theme whose essence is the avoidance of frontal attacks on the enemy's forces. In a tactical sense, the indirect approach implies that the offence should disrupt the enemy's defences rather than wear them down in a slogging battle of attrition. The concept was developed and promoted by Sir Basil Liddell Hart, who advocated offensive tactics based on the acceptance cf terrain penalties. As he put it: 'Natural hazards, however formidable, are inherently less dangerous than fighting hazards.'

Col. Y. Wallach, an Israeli analyst, has described the tactics as the penetration in depth of an enemy's front line in order to cut transport and communication networks and to 'create data'. The idea is that the advancing columns of the offence would move rapidly, and this, as well as the communications failures, would confuse enemy decision-makers, and prevent the identification of their axis of advance. This would inhibit the enemy's attempt to oppose the offence effectively, even though its forces might be numerically superior. In this way, an absolutely weaker offence force can win by achieving brief but decisive local superiority at successive points as the advance proceeds, provided, that is, that the offence columns keep moving, and moving fast.

More generally, the concept of the indirect approach covers all those strategies and tactics which are based on the 'line of least expectation', i.e. anti-aircraft systems are to be attacked by commando raids rather than by aircraft, while if there is a highway from A to B which is strongly defended by the enemy B should be invested by some other route, even though it may be a longer and more exposed one.

In each case there is a price to be paid for not doing the expected thing, since risks have to be taken in doing the unconventional; but the assumption of this concept is that in war it pays to be unconventional.

Indoctrination. A process whereby ideas, methods or information are transferred to a given group of people. The word is used – with negative connotations – to describe enemy activities which if they were our own we would describe as education, information or training.

Inertial Guidance, Inertial Navigation, Inertial Platforms. Inertial guidance and inertial navigation are two applications of inertial platform systems. These are one of the most important technological innovations of the post-war period, though they are based on scientific principles known since the time of Newton. They are the standard method of guidance for ballistic missiles and advanced combat aircraft and by 1972 will also be fitted in most jet airliners. An inertial platform consists of a gyroscopically stabilized assembly on which some accelerometers are mounted. The gyroscopes have the well-known property of retaining stability *vis-à-vis* externally induced changes in orientation (relative to the fixed stars); the accelerometers have the property of detecting and measuring relative speed changes in any one direction. Thus if an object containing an inertial platform is moved, the platform will 'sense' and record the movement. It is therefore possible to navigate by using an inertial platform, since it can locate accurately its own position relative to its known starting position: given that one knows where one is at the start, maps or their equivalent (such as computer programmes) are sufficient to determine how to get to the intended objective. The enormous advantage of inertial platforms over other means of 'fixing' a position (such as astro-tracking, radio/radar beacon and visual observation) is that no reference to external objects is required. Thus inertial platforms are fully independent of the outside environment and are unaffected by jamming (▷ *ECM*), cloud cover and other forms of interference.

Inertial platforms are adapted to guidance, especially missile guidance, by having a computer coupled to the platform. The intended flight path is programmed as a series of instructions which operate the flight controls; any deviation from this flight path is detected by the inertial platform, and the computer then supplies the necessary correction data. The steady improvement in the reliability and data output/per unit weight of computers, and similar improvements in the platforms, have led to a steady decrease in the size of inertial guidance units, while accuracies have increased. The guidance package used in the ⟡ *Titan II* ICBM weighed 200 cwt; the one used in ⟡ *Minuteman 1* weighed only 150 lb. Currently available types weigh even less. The US submarine application of inertial navigation is known as SINS.

Infra-red Homing. A technique of automatic missile guidance used mainly in air-to-air ⟡ *missiles*. The mechanism consists of a detector of thermal radiation, which identifies the hottest 'spot' within its horizon (such as the efflux pipe of a jet-propulsion unit); a simple programmer, which turns the elevation/bearing data provided by the detector into flight-correction impulses; servo-systems which actuate fins and other control surfaces on the basis of these impulses. ⟡ *Sidewinder* and *Firestreak* for two first-generation 'infra-red' missiles. These are fair-weather systems, as opposed to those using ⟡ *semi-active radar homing* which is an all-weather technique. Modern infra-red homing missiles are still limited to optical visibility conditions, but have wider 'horizons' and far better target discrimination ⟡ e.g. the *Falcon AIM-4D,* and *Red Top,* which are advanced collision-course weapons, unlike the earlier types, which worked only when fired straight at the rear of the target aircraft.

Insurgency. A localized armed conflict between forces of a constituted government and other forces originating within the same national territory; the form may or may not be that of ⟡ *guerilla warfare.* If the conflict is internal but is not localized, then it is ⟡ *revolutionary war.*

⟡ *Internal war* for a classification of armed conflicts where main parties originate from the same national territory. Revolutionary war may begin as an insurgency, but unless the objective of the insurgents is to seize control over the entire national territory one need not develop into the other; the term is therefore correctly applied to cases where the insurgents are seeking limited or localized aims.

Integrated Satellite System. ⟡ *Surveillance Satellite.*

Intelligence. ⟡ *Strategic Intelligence, Tactical Intelligence.*

Internal War. Organized armed conflict between parties mainly originating from and based in what is commonly regarded as one national territory.

If the two or more parties acknowledge common nationality and wage open war with formal armies, then it is *civil war.* If the two or more parties acknowledge common nationality but the initiating party uses mainly ⟡ *guerilla warfare* and ⟡ *subversion* rather than open warfare, then it is ⟡ *revolutionary war.*

If the two or more parties are not fighting over the control of the whole national territory but for more limited aims, then it is ⟡ *insurgency.*

The terms 'sub-limited warfare', 'special warfare', 'unconventional warfare', 'internal war', 'stability operations' etc. refer to guerilla warfare or to insurgency, unless they are euphemisms for civil war or revolutionary war.

Intruder A-6. US naval attack aircraft. A carrier-based twin-jet which has been operational since 1963, the A-6 and its variants are subsonic and have a bulged frontal fuselage and swept wings. The speed and appearance of the A-6 are unexciting but this aircraft has an outstanding range and endurance and can carry a large (up to 18,000 lb) payload in all weathers, at night, and at very low altitudes. This is done by means of DIANE (Digital Integrated Attack Navigation Equipment), which with a display system presents a simulated image of the ground and air below and in front of the aircraft, and allows the pilot to select a flight and attack course which the plane then follows automatically.

The EA-6A is basically the same aircraft but has a smaller (7,200 lb maximum) payload. Instead it carries electronic counter-measures (◊ *ECM*) equipment with more than 30 antennae to 'detect, classify, record, and jam enemy radiation'. The EA-6A is intended to operate alongside strike aircraft, assisting their attack by interdicting enemy radar, as well as to collect electronic intelligence.

The EA-6B is a four-seater with more advanced ECM equipment which has lately come into service.

Specifications (A-6): Two-seater. Length: 54 ft 7 in. Height: 15 ft 7 in. Wing span: 53 ft. Gross wing area: 529 sq ft. Weight empty: 25,684 lb. Maximum internal fuel: 15,939 lb. Maximum take-off weight: 60,626 lb.

IRBM. Intermediate-Range Ballistic Missile. Ballistic missiles with ranges of between 1,500 and 4,000 miles are usually described as Intermediate-Range Ballistic Missiles or IRBMs. The ◊ *Polaris* and the ◊ *Poseidon* series fall within this category, but this range concept is usually applied to land-based systems, such as the Soviet ◊ *Skean* and the French ◊ *SSBS*.

ISS. Integrated Satellite System. ◊ *Surveillance Satellite*.

J

34 Jagdpanzer Kanone and Rakete. West German tank destroyers. These two ◊ *tank destroyers* are based on the common chassis of the ◊ *Schutzenpanzer Gruppe*.

The Jagdpanzer 'Kanone' is a turretless vehicle with a 90-mm gun fitted in a limited traverse position on the frontal plate. Thus the vehicle can engage only those targets which are within its frontal arc and it has to be turned to face targets not so placed. This limitation makes it primarily suitable for ambush or more generally defensive tactics. The turretless layout, on the other hand, has allowed the use of a large 90-mm gun almost as powerful as the one originally fitted in the M.48 ◊ *Patton* tank within a vehicle which weighs much less. The Jagdpanzer layout also results in reduced cost and height (and therefore visibility and vulnerability) and allows a large ammunition-carrying capacity (51 rounds) in a relatively small vehicle. These qualities make it very effective as a tank-killer at short and medium range. The Jagdpanzer K is 19 ft long, weighs about 25 tons and has four crew. Its road speed is about 35 mph, and its range about 300 miles.

Beyond 1,500–2,000 yds, all tank guns are considerably less effective than anti-tank ◊ *missiles*, and the Jagdpanzer Kanone is supplemented by the Jagdpanzer Rakete, whose missiles are especially effective at longer ranges. This too has heavy frontal armour, good general protection and a turretless layout, but instead of a gun it is equipped with ◊ *Nord* SS-11 missiles fitted into magazine/launchers from which they can be fired rapidly (◊ *HOT*).

42 Jaguar.   Anglo-French   fighter-bomber. The Jaguar is a relatively unambitious attack aircraft now being developed jointly in France and Britain. It is intended as a companion to the sophisticated Mach 2+ 'air-superiority' and interceptor aircraft that will be available when the Jaguar project reaches fruition in 1970–1. The development of the Jaguar is a direct result of the great sophistication and high speeds of modern fighters, which unlike previous generations of these aircraft are no longer suitable for conversion to the close-support role once their 'first-line' capabilities are obsolete.

The Jaguar is powered by two turbofans located side by side in the fuselage; the wings are of the high-lift variety, to fulfil the short take-off and landing requirement for this plane. The type of radar and other electronics have not been published, but the plane has a gyro plus computer stabilization system which makes it a good weapon platform. Five versions are being planned:

Jaguar A. French tactical support aircraft.

Jaguar S. British tactical support aircraft.

Jaguar E. French two-seat training version.

Jaguar B. British two-seat training version.

Jaguar M. French naval version.

The combat versions (A, S, M) will have a 10,000-lb weapon-load, with alternative missile (◊ *Martel,* ◊ *Nord AS.30*) and bomb options.

Specifications: Length: 50 ft 10 in. Height: 14 ft 9 in. Wing span: 27 ft 10 in. Maximum take-off weight: 28,200 lb. Maximum speed: Mach 1·7. Low-level maximum speed: Mach 1. Ceiling: 46,000 ft. Combat radius: in low-level flight: 530 miles. Maximum ferry range: 2,800 miles.

◊ *Fighter* for general subject entry.

K

Kalashnikov. Soviet rifle. ▷ *A K-47*.

'Kangaroo'. Soviet air-to-surface missiles. A long-range turbojet-powered missile with the general appearance of a swept-wing jet fighter. It is about 50 ft long, and it is believed to be guided by command from the carrying aircraft, a ▷ *Bear Tu-20*. The Kangaroo is a ▷ *stand-off missile* intended for the terminal delivery of thermonuclear warheads.
▷ the comparable U S *Hound Dog*.

Kawasaki KAM 3D. Japanese anti-tank missile. A light wire-guided anti-tank missile operational since 1968 with the Japanese army. It has a two-stage solid propellant rocket motor, and velocity control guidance. (▷ *Vigilant* for this mode of guidance.) There is a H E A T (shaped charge) warhead.
Specifications: Length: 3 ft 1 in. Weight: 34·6 lb. Cruising speed: 190 mph. Range: minimum 1,150 ft, maximum 5,900 ft.

'Kelt'. Soviet air-to-surface missile. The Kelt is a new liquid-fuel rocket missile, which is probably intended as an anti-shipping weapon for Soviet bombers. Little is known about it, except for the fact that it is about 28 ft long, and that it is credited with a range of 100 nautical miles.

'Kennel'. Soviet anti-shipping missile. A winged and turbojet-powered anti-shipping missile carried by naval ▷ *Badger Tu-16* bombers, and which has been made available to Egyptian and Indonesian airforces. The Kennel is about 26 ft long, and its range has been estimated at 50 miles. It is probably command-guided with final homing interception; also used as a surface-to-surface coastal defence weapon.

'Kipper'. Soviet air-to-surface missile. An air-to-surface turbojet-powered missile which looks like a jet fighter and is sometimes described as an anti-shipping weapon. It is carried by a ▷ *Badger Tu-16* medium bomber; it is about 31 ft long and is guided by command from the mother aircraft. ▷ comparable, but more advanced U S *Hound Dog* anti-submarine missile.

'Kitchen'. Soviet air-to-surface missile. An advanced ▷ *stand-off missile* carried by the ▷ *Blinder Tu-22* supersonic bomber. Kitchen is about 36 ft long and is reported to have ▷ *inertial guidance*.
▷ the comparable U S *Hound Dog*.

Komar. ▷ *Osa*.

Kt. Kiloton. Measure of nuclear warhead power equal to 1,000 short tons of T N T high explosive.

L

Lance XMGM-52A. US surface-to-surface missile. Lance is a 30-mile-range missile launched from a converted ◊ *M113* armoured personnel carrier chassis and intended as a divisional support weapon. As such it will replace the ◊ *Honest John* and perhaps the ◊ *Little John* rockets. It is powered by a storable liquid fuel rocket motor and guided by a simple ◊ *inertial* system; it has a small nuclear warhead, but other types of warhead are also available. An extended-range Lance is under development with a 50–60-mile range; Sea Lance XMGM-52B is a marine conversion for firing from a gyroscopically stabilized launcher.

Specifications (52A): Length: 20 ft. Body diameter: 1 ft 10 in. Launching weight: 3,200 lb. The warhead is armed after a 3-mile flight path.

Laser. Laser is an acronym for Light Amplification by Stimulated Emission of Radiation. A laser is a device which produces coherent radiation (i.e. radiation of a given frequency and with all wavefronts in phase) in the very short (optical) wave bands. It is an extension of maser (Microwave Amplification) techniques into the visible light spectrum. In 1960 the first optical maser, or laser, was constructed by using a ruby rod, which is used to amplify light which is then beamed from it. This coherent beam of light does not widen out and diffuse like normal light radiation and therefore is (a) directional and (b) highly concentrated. In theory, therefore, a great deal of heat could be projected with pinpoint accuracy over an indefinite distance – i.e. the old death-ray at last. In fact, the amount of energy required to inflict damage in this way would be out of proportion to the results, i.e. in order to destroy a missile's nuclear warhead in its ballistic

path a laser would require about 150 MW (the output of a medium-sized power station) for about a minute. But to make sure that 150 MW gets there, taking account of the likely conversion efficiency and wastage, would need several thousand MW, i.e. many large power stations. But a straight beam of light which stays thin and straight over long ranges can be used for range finding, for tracking missiles in space and for line-of-sight telecommunications, since a single beam can carry more voice channels than millions of wires – and do so more securely. (But the beam can pose a hazard to the eyes of friendly troops should any be in its path.) Laser rangefinders are now in service.

Launching Pad. Part of the hardware connected with rocket-propelled systems. Strictly speaking, the platform which, together with erector gear and fuel/communications couplings, is used to launch ballistic rockets.

Law, M.72. US anti-tank rocket-launcher. ◊ '*Bazooka*'.

Leningrad. ◊ *Moskva*.

Leopard Standardpanzer. West German tank. The first tank developed in Germany since the Second World War, the Leopard is now replacing the US M.47 and 48 ◊ *Patton* tanks in the Bundeswehr; it is the German counterpart of the main battle tanks (MBT) developed in other countries. The Leopard has a conventional layout with a driver in the hull and a three-man turret crew, like the other MBTs. It mounts the British 105-mm high-velocity gun firing APDS and fin-stabilized HEAT rounds, giving it normal MBT gunpower. It has, however, a far better overall design since it combines out-standing mobility with good armour pro-tection. The optical range-finder is also very well designed, combining both the stereoscopic and the coincidence mode, a neat trick since other MBT designs suffer the limitations of either one or the other. The simple and reliable ventilation system affords fallout protection and permits wading to turret-top height. This can be extended by using a *snorkel* tube device. The Leopard probably has the best armour-gun-mobility combination of the new MBTs, and its success is demon-strated by export sales to demanding customers.

The Bundeswehr is supplementing the Leopard with the Jagdpanzer ▷ *tank destroyers* for defensive use.

Specifications: Length: 22 ft. Width: 10 ft 8 in. Height: 7 ft 10 in. Weight: 39·3 tons. Engine: water-cooled diesel 830 bhp. Road speed: 40 mph. Road range: 380 miles. Ground pressure: 12·3 lb sq in. Ground clearance: 1 ft 6 in. Rounds carried: 56. Secondary weapons: smoke launchers, coaxial MG, turret roof MG. Crew: 4.

▷ *Main Battle Tank* for competitive type.

▷ *Armoured Fighting Vehicle* for general entry.

Light Machine-gun (LMG). Light machine-guns are magazine-fed, air-cooled, bipod-mounted automatic wea-pons firing rifle-calibre ammunition. They were developed during the First World War to bridge the gap between the machine-guns then used (▷ *Heavy Machine-gun*), which were water-cooled and weighed 100 lb or so, and the single-shot bolt-action rifles of the infantry. The LMGs, of which the first were the French Couchat, the German Bergmann, and the Anglo-American Lewis, weighed 20–30 lb and were especially suitable for small-unit ▷ *'Fire and movement'* tactics. The smallest infantry units were divided into a two-man 'fire' team, with the LMG, and an 'assault' team (the rest of the squad armed with rifles and grenades); while the 'fire' team covered their advance, the 'assault' team moved forward. Then it would take up a firing position and cover the 'fire' team as it advanced in turn.

When new lighter machine-guns were developed (▷ *Medium Machine-gun*), and when modern automatic rifles began to be issued to every soldier, the old specially designed LMG became obsolescent. Since the rifle could be turned into an LMG with the addition of a simple bipod, the LMG has been replaced by two weapon types: the medium machine-gun (minus tripod) and the standard rifle (plus a bipod).

▷ *Bar* and *Bren* for two old-style LMGs.

▷ *Medium Machine-gun*, for modes of operation, and *Rifle*.

▷ *FN* (Falo), and the Russian RPK (under *AK-47*) for the new kind of rifle-conversion LMGs.

Lightning. British Mach 2 fighter. The   **41**
Lightning is a Mach 2 interceptor with a short combat radius and a limited ground-attack capability. It has a highly swept wing and is powered by two turbojets staggered one above the other in the fuselage, fed by a frontal air-intake with a pointed centrebody which contains the radar. This and the electronic fire-control equipment give it an all-weather capability for 'blind' interception. Normal arma-ment consists of two 30-mm guns in the fuselage and two ▷ *Firestreak* or ▷ *Red Top* air-to-air missiles. Operational ver-sions include:

Mk 2: Original interceptor version.

Mk 3: Improved version with up-rated power plants giving 32,720 lb of thrust with reheat. No guns are mounted. There is provision for ferrying with ex-ternal fuel tanks and an in-flight re-fuelling probe.

T.4: Two-seat (side-by-side) training version of Mk 2/3. Some supplied to Saudi Arabia.

Mk 6: A Mk 3 with airframe modifi-cations, greater fuel capacity and im-proved weapons capability; standard UK fighter version.

T.5: Trainer version of the F.Mk 6.

F.50–57: 'Export' versions with pro-vision for carrying a wide range of under-wing and belly pack stores. Supplied to Saudi Arabia, Kuwait and other countries.

The Lightning has had limited export sales because of the high cost, the highly specialized performance and the complexity of the design from the point of view of training and maintenance.

Specifications: Length: 55 ft 3 in. Height: 19 ft 7 in. Wing span: 34 ft 10 in. Wing area: 380·1 sq ft.

◊ *Fighter* for general subject entry.

Light Tank. Light tanks are tracked, armoured and gun-turreted vehicles which have the layout and general appearance of battle tanks but weigh much less and are intended for more specialized roles. In terms of current types, these vary from the amphibian/reconnaissance of the Soviet ◊ *PT-76* to the airborne/assault of the U S ◊ *Sheridan M. 551* or the tank-destroyer role of the French A M X 13/58. This was not so in the past: light tanks were officially intended for the reconnaissance role, but were often in fact used as cheap substitutes for 'real' tanks. While the P T-76 and the Sheridan have less armour than 'real' tanks, they have compensating capabilities. On the other hand the M.41 'Walker Bulldog' simply has less of everything, including gun power, which is insufficient to fight 'real' tanks (or ◊ *Main Battle Tanks*). The M.41 is nevertheless still standard in several N A T O armies, (as is the even less adequate earlier M.24 Chaffee). Until recently the A M X 13 was also a tank-substitute of limited value (and its armour protection remains very limited), but by upgunning it from 75 mm to 105 mm it has been made into an adequate light tank/tank destroyer. Unless this class of vehicle disappears, the Sheridan M.551 will probably set future trends: a light, air-transportable tank with powerful gun/missile weaponry at least as good as that of current M B Ts.

◊ *Armoured Fighting Vehicle* for general subject entry.

Limited Strategic War. A war in which nuclear strikes are used against homeland territories but with deliberate restraint as to the number and power of warheads and/or the number and nature of targets; the recognizably limited damage so inflicted being part of a bargaining process where the nuclear exchange supplements other forms of communication. This is primarily an intellectual concept. ◊ *Threshold* and *Escalation*.

Limited War. A military conflict in which at least one party is exercising restraint in the weapons used, the targets attacked or the areas involved. The objectives of limited war are not necessarily themselves limited, since the conflict may remain on a particular ◊ *threshold* only because one party cannot, while another need not, escalate. In practice, what is limited war for a Great Power may be ◊ *total war* for a small power confronting it. In common usage a limited war is one in which a party or parties with a nuclear strike capability refrain from using it, such as the Korean war.

Little John M G R-3. U S unguided surface-to-surface rocket. The Little John is a light-weight tactical rocket which can be towed on its launcher by a jeep-sized vehicle. It is unguided but fin-stabilized; it has a 10-mile range and a firing weight of 789 lb. Accuracy is insufficient for other than area targets, therefore predicating the use of nuclear warheads, though high-explosive warheads are also available. Little John may be replaced by the ◊ *Lance X M G M-52A* missile.

L M G. ◊ *Light Machine-gun.*

Local War. This is primarily a Soviet concept: a war in which military activities are confined to a given area ('limited' and mutually recognized) within which all types of weapons, including nuclear weapons, are used. Formal Soviet doctrine anticipates the use of nuclear weapons on the battlefield in the earliest stages of a conflict with N A T O forces. (No other conflicts involving the Soviet Union are discussed in published Soviet strategic literature.) N A T O policy, however, foresees a non-nuclear 'holding' phase, and Soviet theory seems to be at variance with Soviet practice, as expressed for example in the quantity and type of weaponry deployed by the artillery (◊ *Artillery, Soviet*). This contradiction between the doctrine and the fact that Soviet artillery is prepared for sustained, non-nuclear, barrage fire is merely one aspect of the contradic-

tions in Soviet nuclear planning. These derive in turn from the contradiction between the 'offensive' posture of its highly mobile armoured forces and the 'defensive' declared policies – and, perhaps, general outlook.
  ▷ *Warsaw Pact,* and '*Forward Defence*'.

Logistics. All the activities and methods connected with the supply of armed force organizations, including storage requirements, transport and distribution. Since in modern conditions a wide range of equipment and supplies is employed in widely varied 'mixes', logistics involves a great deal of planning and calculation as well as physical activities. The aim is to provide each echelon of the armed force organization with the optimum quantity of each supply item, in order to minimize both overstocking (which restricts mobility and causes diseconomies) and shortages of essential equipment. In modern military organizations such typical operational research calculations are assisted by automatic data-processing equipment. It is convenient to divide the logistic problem into strategic and tactical: the first covering the acquisition, stockage and transport of supplies to the combat theatre, and the second their distribution within it.

M

M.16. US rifle and 0·223-cal. small arms series. The M.16, the rifle of the Vietnam war, is part of a new family of small arms designed around the revolutionary ·223-in/5·56-mm 'Stoner' cartridge. This weighs just over one half of conventional rifle ammunition and fires a small-diameter bullet at high velocities. This gives adequate range and penetration while imposing a lighter logistic load and reducing the recoil 'kick' as compared to full-power rifle cartridges. The M.16 is a short and elegant weapon which can be recognized by the characteristic combined carrying handle/rear sight and the plastic composite shoulder stock and barrel guard. It is gas-operated and has optional full and semi-automatic fire; the 20-round magazine is set in the one-piece bolt, trigger and receiver block. The same Mr Stoner has designed, apart from the M.16, a whole family of small arms known as Stoner 63 based on the same 5·56-mm cartridge and on common main components; the ◊ *light* and ◊ *medium machine-gun* have been adopted by the US army. Various configurations of the M.16 with different accessories are also in use: 'heavy assault rifle', 'carbine', 'survival rifle'. A further accessory is an under-the-barrel 40-mm grenade-launcher.

Specifications (M.16 basic): Length: 39 in. Weight (unloaded): 6·3 lb. Magazine: 20-round staggered row, detachable. Ammunition: 5·56-mm (or ·223-in) with 55 gr bullet weight and 182 gr total weight.

◊ *Rifle* for general entry and competitive types.

M.79 GL. US grenade-launcher. The M.79 grenade-launcher is a shotgun type of weapon designed to fire 40-mm high explosive, smoke or incendiary grenades.

The US Army and Marine Corps use the M.79 extensively, and at present there is an M.79 in each squad. Accurate fire to 300 yards has been achieved with regularity. The M.79 is 29·8 in long and weighs 6·45 lb.
◊ *Grenade*.

M.113. US armoured personnel carrier. The M.113 is the standard armoured personnel carrier (APC) of the US army, and it is widely used by allied and foreign armies. There are, as of now, some sixteen different major versions. The basic M.113 vehicle is mechanically simple and reliable and has excellent mobility over a wide range of terrains. It is also amphibious and can propel itself in water at 3–4 mph by using the water run-off from the tracks. At some 22,000 US dollars the M.113 is exceedingly cheap for an armoured vehicle, but then it is hardly a fighting vehicle. There is no turret, and there are no firing ports. Its armour is aluminium plate 1·25 in thick, barely enough to stop rifle bullets. Further, it has a high (7 ft 3 in) silhouette, and its vertical sides and front minimize ballistic protection.

The design of the M.113 thus reflects the Anglo-American APC concept according to which the troops are to ride to battle in the APC but not to fight from within it. European tactical concepts are quite different, and European APCs (◊ *Schutzenpanzer Gruppe* and *AMX VTT*) are provided with weaponry and thick frontal armour. Following the first experiences of combat in Vietnam a modification of the M.113 was introduced (M.113 A1 ACAV), which shows that the US army is moving towards the European APC concept. The M.113 A1 ACAV has three rather crudely improvised turrets for one 0·5-in and two 7·62-mm machine-guns.

An entirely different vehicle, the XM-701, is under development as a mechanized infantry combat vehicle. It weighs twice as much as the ten-ton M.113, and the operative word is combat: it has a turret, firing ports and thicker armour.

The basic M.113 and the improved M.113 A1 (which has a more powerful engine) can carry 12 troops in addition to the driver, and there is a hydraulically operated rear ramp. They are used as weapon carriers for the ◊ TOW missile and as carriers for battlefield radars. The M.106 is fitted out for a 107-mm mortar, while the M.125 carries an 81-mm mortar. The XM.163 mounts the M.61 20-mm AA cannon (◊ Artillery, US). Other common variants are the M.577 command carrier, the M.579 recovery vehicle, and the M.132 flame-thrower vehicle. In each case the modifications made to the basic M.113 chassis are minor.

Specifications (M.113 basic): Length: 15 ft 9 in. Width: 8 ft 9 in. Height: 7 ft 3 in. Weight: 10 tons. Engine: 209 bhp. Ground clearance: 1 ft 3 in. Ground pressure: 7·5 lb sq in. Road speed: 40 mph. Road range: 200 miles. Crew: 1 + 12.

Mace, Matador (TM-61). US tactical cruise missiles. Mace B and the older, obsolete, Matador are surface-to-surface missiles with aircraft configurations deployed by the US air force as long-range battlefield nuclear bombardment weapons. Both missiles are launched from trailer ramps by solid-fuel rocket-boosters, but sustained flight is powered by a turbojet. The speed is subsonic with a supersonic dive to the target in the last stage of flight. Matador and the first, A, version of Mace were guided by a radar-mapping system, where the guidance unit compares radar images of the ground over which the missile flies with those fed into it before launch, in order to make corrections to the flight path. The improved and longer-range Mace B has ◊ inertial guidance, so that it cannot be jammed or diverted by ◊ ECM. The Matador has a range of about 600 miles, Mace of 650 miles and

Mace B of 1,200 miles; only Mace B is retained in limited operational deployment in Western Europe and elsewhere. The ◊ Pershing tactical ballistic missile is the replacement for this series.

Machine-gun (MG). General term for weapons in which the complete cycle of loading, chambering, firing, extraction and reloading is fully automatic. Since modern rifles are also often automatic, MGs are distinguished by their capability for sustained fire and by their use of tripods or bipods. In modern conditions, 'machine-gun' means a ◊ medium machine-gun (MMG). This describes a belt-fed, man-portable weapon firing rifle-calibre ammunition and which can be used with alternative tripods or bipods (the latter is often attached, the former available).

Earlier machine-guns were heavy, usually water-cooled weapons always used on tripods, and are now described as ◊ heavy machine-guns (HMG). Apart from obsolete weapons, such as the 0·303 Vickers, HMG now describes weapons firing ammunition of greater than rifle calibre, such as the widely used 0·5 ◊ Browning and the Soviet ZPU 14·5-mm.

Because of the great weight and bulk of HMGs, light-weight full or semi-automatic rifle-calibre magazine-fed weapons were developed during the First World War. These ◊ light machine-guns (LMG) lacked the stability and belt-feed system of HMGs but were useful alongside the single-shot rifles which then equipped the infantry. Now that ◊ rifles are semi or full automatics, the LMG is no longer a specially designed weapon. Instead, ordinary rifles are converted by the addition of bigger magazines (i.e. 30 as against 20 rounds), simple bipods and selector-levers to permit single-shot or automatic operation. Though true LMGs and old rifle-power HMGs survive in old-fashioned and poorer armies, the modern infantry is equipped with three kinds of machine-guns:

(a) light-weight rifle conversion called LMGs (or automatic rifles) with simple bipods and magazine feed;

(b) portable but crew-operated (two men) belt-fed MMGs which can be used either on tripods or in a LMG configuration with built-in bipods;

(c) heavier HMGs, firing ammunition of greater calibre than the corresponding rifles (but below 20 mm/0·8 in, which is the artillery demarcation line).

Apart from these, there are hand-held automatics firing low-power (pistol) ammunition at short ranges, known as ◊ *sub-machine-guns* (SMG). Machine-carbine, machine-pistol and other terms are also used for SMGs (though not recommended).

MAD. Magnetic Anomaly Detector. This device is intended to detect submerged submarines. It consists of a magnetometer fitted in the long tail-boom typical of maritime reconnaissance aircraft, and it is supposed to detect the minute change in the earth's magnetic field induced by a submarine. Although not actually insane, MAD is effective only at very short ranges. (◊ *Anti-submarine Warfare.*)

MAG. ◊ *FN.*

Main Battle Tank (MBT). A tracked and armoured fighting vehicle intended for sustained and independent operation on the battlefield. The proliferation of vehicle types has complicated definitions (◊ *Armoured Fighting Vehicle*), but the MBT concept entails general-purpose weaponry, protection from most battlefield hazards, and cross-country mobility; it has replaced the earlier pattern of medium tanks supported by heavy tanks (e.g. M.47 ◊ *Patton* + M.103; ◊ *Centurion* + Conqueror; T.34 + T.10/JS III), as well as the even older 'Infantry' + 'cruiser' tank combination (where slow, heavily armoured 'infantry' tanks were supplemented by fast and lightly armoured 'cruiser' tanks). Though current MBT designs vary widely, reflecting different armour/weaponry/mobility combinations, they all share certain basic characteristics:

1. High-velocity guns: these range in calibre from 90 to 120 mm and take both armour-piercing (◊ *AP, APDS, HEAT*) and high-explosive rounds (for softer targets). These are supplemented by coaxial machine-guns intended mainly as anti-

personnel weapons, and turret-top machine-guns for anti-aircraft use.

2. The basic layout: with the driver in the hull, and gunner, loader and commander in the turret. The fifth crew member, a hull machine-gunner, is still present in older designs (e.g. the Sherman) but is no longer found in modern MBTs.

3. Diesel engines: sometimes with stressed piston heads (often called 'multi-fuel') giving speeds from 25 to 40 mph and road ranges from 200 to 380 miles. Mobility also depends on the ground pressure and ground clearance, with the former ranging from 10·8 lb sq in in the most mobile MBT, the French ◊ *AMX 30*, to the 14 lb sq in of the least mobile, the British ◊ *Chieftain.*

4. Protection: All MBTs have cast one-piece turrets and hull plates made of nickel/chrome/molybdenum steel of high tensile strength (e.g. 140,000 lb sq in) and medium hardness (e.g. Brinell No. 300). With a given type of armour, ballistic protection varies with the thickness and the angle of slope. The more armour, the less mobility: most MBTs compromise between the two in the 35 to 45 ton range, though, again, the Chieftain weighs over 52 tons and the AMX 30 only 33 tons. In other words all designs give protection against rifle-calibre weapons and shell splinters but differ in the degree of protection to hazards beyond this threshold.

New trends in tank design include the use of a guided-missile/howitzer combination instead of classic high-velocity guns, as on a M.60 Patton variant (the M.60 A1E1); ◊ *laser* rangefinders instead of the optical full-field coincidence or stereoscopic types (though British tanks have neither, with ranging machine-guns being used); gas turbine instead of diesels, as in the revolutionary Swedish ◊ *Strv.* 103 (S-tank), which has other even more basic design innovations and no turret. Tanks are vulnerable to many cheaper weapons developed specifically to deal with them, but their combination of armour + multipurpose weapons + cross-country mobility means that they are irreplaceable on the modern battlefield.

The following MBTs (and pre-MBT concept tanks) are separately listed: American: *Patton M.48* (includes M.47 and M.60), *Sherman M.4,* also the

Sheridan M.551 light tank; also the US–German *MBT-70*.
British: *Centurion* and *Chieftain*; also the Anglo–Indian *Vijayanta Vickers 37*.
French: *AMX 30* (and the *AMX 13* light tank).
German: *Leopard* and the US–German *MBT-70*.
Soviet: *T-54* (includes T-55 variants and T-59, the Chinese copy and T-62).
Swedish: *Strv. 103*.

⟡ *Armour* for details of ballistic protection; *Armoured Forces* for tactical concepts; *Missile* (anti-tank), '*Bazooka*', *Recoilless Weapons*, *Artillery* for armour-defeating weapons; and *HEAT, HESH, AP* projectiles.

'Marching Fire'. An infantry tactic competitive with 'fire and movement' tactics. Instead of dividing available forces into teams which alternately fire and move, all riflemen and light-machine-gunners advance in one thin skirmish line, firing at anything which could conceal or contain enemies. This tactic (pioneered in the US army by General G. S. Patton Jr) implicitly accepts casualties resulting from exposed marching, a reduced precision of fire and the impossibility of organic action. Its advantages are shock and a controllable concentration of forces. This is current US doctrine for certain phases of infantry attack.
For competitive concepts ⟡ '*Fire and Movement*'.

Marder. ⟡ *Schutzenpanzer Gruppe*.

Martel (MATRA/HSD AS.37/AJ-168). Anglo-French air-to-surface missile. The Martel (Missile Anti-Radar and TELevision) is a solid-propellant air-to-surface missile with a range of 'tens of miles' which has interchangeable TV guidance and anti-radar homing heads. The TV version is sent towards the target on a pre-set course. When the TV camera in the missile's nose-cone acquires the target, the pilot/gunner of the launching aircraft corrects its flight path by radio commands. The anti-radar version is a penetration aid for bombing missions and is intended to neutralize air defences by destroying selected radar

units. For its mode of operation, ⟡ *Shrike*.

Marut (HAL HF-24). Indian fighter. The Marut is a lightweight twin-engined fighter of conventional design which has been under development in India since 1956. Various versions have been built with a British power plant and other imported sub-systems, but no final production models have yet become operational. This attempt at an indigenous Mach 2 fighter has so far foundered on the lack of a suitable power plant, and the Egyptian turbojet development which was to be used ceased to be funded by the UAR in 1968.
The airframe is a swept-wing design with air-intakes side by side in the fuselage. The armament at present envisaged consists of four 30-mm guns plus air-to-air rockets, or up to 4,000 lb of other under-wing stores. The Indian air force has already received a number of interim models, designated Marut Mk 1 and Mk 1/a, with inadequate (4,520-lb and 4,850-lb thrust) engines, for training/combat development. These are barely supersonic in optimum conditions (Mach 1·02) at 36,000 ft, and have rudimentary avionics.
⟡ *Fighter* for general subject entry.

Massive Retaliation. An interpretation of past US policy as announced by Secretary Dulles in January 1954 (' . . . the way to deter aggression is for free communities to be willing and able to respond vigorously at places and with means of our own choosing . . . '). The essence of this interpretation is that any aggression, even if 'minor' or carried out by proxy, would be met by a nuclear strike on the Soviet Union. This policy had the attraction of economy, since all US interests anywhere in the world were to be protected *vis-à-vis* the Soviet Union by strategic nuclear weapons only – thus obviating the need to maintain large non-nuclear forces.
This (interpretation of) US policy has been criticized on the grounds that it would be ineffectual, since the threat would not be credible. For example, the Soviet leadership would not be deterred from organizing infiltration in, say, Laos, since they would refuse to believe that US response to this would take the form

of a nuclear first strike against the USSR, thus initiating a general nuclear war. In other words, US dependence on this policy would lead to a series of unacceptable choices between capitulation or mass destruction over each local crisis (◊ *Credibility*). Massive retaliation was never central US policy, and was in any case associated with the diplomatic technique of ◊ *brinkmanship*. It is essentially an intellectual concept, an attempt to equate the balance of power with the ◊ *balance of terror* in order to save money on the non-nuclear elements of US armed force organization. ◊ *Controlled Response* and *Assured Destruction* for current formulations of US strategic policy.

67 **Matra R.530. French air-to-air missile.** The Matra R.530 air-to-air missile is used on the ◊ *Mirage III* and on the ◊ *Crusaders* of the French navy. It is a two-stage solid-fuel rocket with cruciform tail control surfaces and delta wings, with either of two interchangeable guidance units: ◊ *infra-red homing* and ◊ *semi-active radar*. The 60-lb high-explosive warhead is fitted with a proximity fuse.

Specifications: Length: 10 ft 9 in. Body diameter: 10 in. Wing span: 3 ft 7 in. Launching weight: 430 lb. Maximum speed: Mach 2·7. Range: 11 miles.

MBT-70. US–West German battle tank. A main battle tank (MBT) jointly developed by American and West German military authorities, and intended to go into production by 1970. The MBT-70 has a conventional external appearance, except for the very large turret, but the armament and mechanical arrangements are quite original. The crew consists of only three men, rather than the usual four, and the driver sits in the turret in a counter-rotating capsule. There is a new hydro-pneumatic suspension which allows the whole chassis to be raised (for fast cross-country movements) and lowered (for concealment). There are two alternative weapon options: a conventional 120-mm high-velocity gun, or a 152-mm howitzer/missile-launcher which can fire ordinary shells or the ◊ *Shillelagh* guided missile. There are an automatic loader (which explains the three-man crew) and a stabilization device to allow accurate moving fire

as well as a ◊ *laser* rangefinder and night-vision aids. Armour protection is good all round, and there are radiological, biological and chemical agent filtration devices. There is also, for the first time, an integrated computer system for both fire control and navigation. Inevitably, the question arises of how such an expensive system can be justified, given the threat of anti-tank devices which are cheap, reliable and effective and can be operated by a single man.

Specifications: Width: 11 ft 6 in. Height: variable, between 6 ft 6 in and 7 ft 6 in. Weight: 50 tons approximately. Engine: either 1,475 hp air-cooled multi-fuel, or 1,500 hp water-cooled diesel. Road range: 300 miles approximately. Estimated speed: 40 mph. Laser rangefinder. Secondary weapons: 20-mm cannon, 7·62-mm gun, smoke and/or grenade launchers.

◊ *Main Battle Tank* for current competitors.

◊ *Armoured Forces* for tactical concepts.

Medium Machine-gun (MMG). Modern MMGs are air-cooled, belt-fed, and man-portable and have alternative tripods or bipods. Weights vary from the 36·4 lb of the US M.60 to the 50 lb of the German MG 42/59 (including tripod but no ammunition). The cyclic rate of fire varies from the 500 rpm of the ◊ *Browning* M.1919 A4 to the 1,200 rpm of the MG 42/59. Different modes of operation include:

(a) Gas, where expansion gases from the bullet just fired travel into a cylinder and push back a bolt-piston assembly, this having various attachments which extract the spent cartridge, feed in the next one and, pushed forward by a spring arrangement, fire it.

(b) Recoil, where the direct 'kick' energy is used to achieve the movement of the bolt relative to the barrel.

(c) Delayed blow-back, where expansion gases are allowed to push back the bolt after a catch is released following the exit of the bullet.

The first system is used in the M.60, the ◊ *FN GPMG* and the Soviet SGM RPD and RP 46. All the Browning MMGs are recoil-operated, as is the MG 42/59 and the heavy ·303 Vickers.

Delayed blow-back is used in only one MMG, the French M.1952.

Since water-cooling has been abandoned, all modern MMGs have quick-change barrels for sustained fire. The feed relies on fabric belts (as in the *Browning* M.1919 series), metal belts (as in the MG 42/59 and the Soviet SGM RPD and the RP 46 series), or disintegrating links as in the more modern FN GPMG and the M.60 machine-guns.

The higher cost and heavier weight of MMGs are justified by their capacity for sustained automatic fire, and by their accuracy, due to the stable tripod mounts. These qualities are used in the offensive to 'prepare' and 'cover' the advance. While automatic rifles (in the light machine-gun role) deliver short bursts at close range, the MMG can remain back and deliver heavier volumes of fire at long range. In the defensive, MMG fire from prepared shelter ('Nests') will delimit the perimeter of the area defended. In modern MMGs, the heavy tripod mount can be left behind in 'fire and movement' tactics and the MMG taken forward and the bipod used instead. The MMG then operates in the light machine-gun mode, or even as a personal weapon. If a soldier can use an M.1919 A 6 or an M.60 as a personal weapon, as some do, he is pretty good.

▷ *Machine-gun* for general subject entry.

MG. ▷ *Machine-gun.*

'Microwave Radar'. A ▷ *radar* operating in the higher part of the electro-magnetic spectrum. The 'microwave region' covers the UHF, SHF and EHF frequencies:

because atmospheric absorption is too high (since it is near the optical range).

Midas. ▷ *Surveillance Satellite.*

Mig-21. ▷ *Fishbed.*

Mig-23. ▷ *Foxbat.*

MILAN. ▷ *HOT.*

Mine. An explosive device which is detonated by contact, magnetic proximity or electric command. Mines are very important weapons of passive defence, and they come in all shapes, sizes and functions. Very large (e.g. 25 kilos of high explosive) land mines detonated only by heavy pressure or impact are known as anti-tank mines; these are designed so that the weight of lighter objects, including men, do not detonate them. Anti-personnel mines have fragmentation cases. Some are designed so that the explosive splintering element is projected upwards, thus extending its killing range. Land mines can be detected by magnetometers, but this is a slow process. Plastic mines, being non-magnetic, are immune and may be mixed in with (larger) iron mines to impede mine clearing. All land mines have a pressure sensor which is above or just below the ground. And manual mine-clearing means finding these – without stepping on one. More safely, mines can be detonated by special-purpose tanks. Underwater mines can be anchored to protect harbours or to interdict sea lanes, but most modern mines are simply laid on the seabed. Their acoustic (sonar) or

Frequency	Wavelength	Term
300 Mc. to 3 Gc.	1 metre–10 cm.	UHF
3 Gc. to 30 Gc.	10 cm.–1 cm.	SHF
30 Gc. to 300 Gc.	1 cm.–1 mm.	EHF

The higher the frequency, the better the definition of the radar 'image' which can be obtained, but the upper end of the EHF region is not used in radar sets

magnetic sensors are supposed to detonate them when a ship passes overhead. Such mines can be dropped by aircraft. The countermeasure is magnetic detection or pre-emptive explosion. Minesweepers have wooden hulls, and even steel ships can be de-magnetized by 'degaussing'.

94 Minuteman. LGM-30 series. US intercontinental ballistic missile. The Minuteman series are three-stage missiles with solid-propellant motors designed as simplified and low-cost weapons for launching from 'hard' silo sites. They need very brief pre-launch preparation (32 seconds). The first production model was assembled in April 1962, and since then 1,000 Minuteman missiles (in three versions) have become operational. A new version (Minuteman 3) with multiple warheads (◊ *M I R V*) has also been developed. LGM-30A (Minuteman 1) is 53 ft 9 in long, weighs about 60,000 lb and delivers a megaton-warhead over 6,000 miles at a speed of about Mach 22. Like all modern I C B M s, it has an ◊ *inertial guidance* system which gives good accuracies. Countdown can be stopped at any time up to six seconds to blast-off; the warhead is armed only after the missile is airborne. Only one wing (159 missiles) was equipped with ĻG M-30A. LGM-30B, an early improvement of LGM-30A, is a little longer (55 ft 11 in), heavier, and, above all, more precise. 650 LGM-30Bs were produced. Partly as a result of reliability problems these were phased out in favour of the LGM-30F (Minuteman 2). LGM-30F is 59 ft 10 in long and 70,000 lb in weight and has a 7,900-mile range. Instead of the swivelling quadruple nozzles of the two earlier types, it has a liquid-injection single nozzle system. LGM-30F is more accurate, and a larger number of alternative targets can be preset in the inertial guidance system. The Minuteman 2 has a two-megaton warhead. It became operational in 1966 and is scheduled to replace all previous Minuteman missiles, building up a force of 1,000 missiles in six wings. A Minuteman 3 with MIRV multiple warheads is currently being produced. The silo-launcher for each Minuteman is about 80 ft deep and 12 ft in diameter; the surface area of each site is about 2 or 3 acres. Each flight of 10 silos has a launch-control centre about 50 ft below the surface, protected by shock-absorbers and blast

walls. This is manned by two Strategic Air Command officers. To start a countdown, two separate levers at two separate centres must be activated.
   ◊ *US, Strategic Offensive Forces* for other missiles systems. ◊ *Ballistic Missile.*

Mirage. French fighter series. The   **4** Mirage III is a single-seat delta-wing Mach 2 fighter initially developed as a high-altitude interceptor. It has an elongated fuselage and a particularly 'clean' appearance with only the side-by-side air-intakes to interrupt the smooth fuselage line. The sixteen versions marketed so far have been evolved from three basic variants:
   Mirage III-C: The basic model with a 13,225-lb thrust turbojet, an optional rocket velocity booster, and electronic fire control. Additionally, it has a limited air-to-ground capability. Weaponry in the air-to-air role consists of a single ◊ *Matra R.540* missile and (optionally) two 30-mm guns and a pair of ◊ *Sidewinder* missiles. Ground-attack armament can include four 1,000-lb bombs or a ◊ *Nord A S.30* missile and two bombs. Mirage III-Cs have been delivered to Israel, under the III-CJ designation, with modified – reduced – electronics. A South African version is designated CZ. The Mirage III-C is 48 ft 5·5 in long and weighs 13,040 lb empty.
   Mirage III-E: A slightly longer plane with a more powerful, 14,110-lb, thrust turbojet, an improved nose radar, a downward-pointing doppler radar for low-level 'all-weather' flight, and a navigation computer. This version is intended primarily for long-range all-weather ground attack. The Australian version, III-O, has gyro stabilization and different navigation equipment. The Swiss version, III-S, has an U S fire-control system for the ◊ *Falcon* HM-55 missile. A further version, the Mirage 5, has been designed with Israeli help for export as a ground-attack plane with a larger payload/longer range and much simpler avionics. These have been sold to Peru, Belgium (who make it under licence) and Israel, the last-mentioned order being paid for but not delivered. The Mirage III-E is 49 ft 3·5 in long and weighs 14,375 lb empty.
   Mirage III-R: A reconnaissance version of the III-E with no radar, a self-

contained navigation system, and camera windows in the nose.

Mirage III-V: A much enlarged experimental vertical take-off and landing version.

Mirage III-B: A two-seat training version of the III-C with air-to-ground armament.

Mirage F.1: A new aeroplane which retains certain features of the III series. It has a single turbojet rated at 15,400 lb of thrust with afterburning. The fuselage is similar to that of the III series, but there is a swept wing and tailplane, instead of the delta wing. The F.1 weighs 16,425 lb empty and can carry a greater fuel/weapon/avionics load than the III series. Top speed in ideal conditions is Mach 2·2, and maximum range exceeds 2,000 miles. The F.1 is expected to replace the III series in French service.

Mirage F.2: A substantially larger plane, with the swept-wing layout of the F.1, under development since 1964.

Mirage G: A variable-geometry version of the F.2, which reached the flight stage in 1967 and is to be developed into a range of operational types.

Specifications (Mirage III-C): Overall length: 48 ft 5 in. Height: 13 ft 11 in. Wing span: 27 ft. Gross wing area: 375 sq ft. Maximum take-off weight: 26,015 lb. Maximum level speed with normal mission load: 925 mph at low level, 1,430 mph (Mach 2·15) at 36,000 ft. Climb to 36,000 ft with two ◊ Sidewinders: 6 minutes and 30 seconds. Maximum combat radius with ground attacks: 560 miles; radius in optimum conditions: 745 miles.

◊ Fighter for general subject entry.

Mirage IV. French light bomber. The Mirage IV is a twin-jet, two-seat supersonic bomber designed as a nuclear delivery weapon for the French ('Force de Frappe') 'independent' deterrent. ◊ France, Strategic Offensive Forces for the operational status and deployment of this force.

In the context of current Soviet air defences (◊ PVO Strany) the Mirage IV is a weapon of limited capabilities; it can deliver an 80-kiloton fission ('atomic') bomb on targets in European Russia but cannot do so in the one way which could

ensure a high penetration probability: supersonic low-level flight most of the way. Airframe modifications for low-level flight have been made but, even with in-flight refuelling, the plane lacks the range for this role. In any case, the small size of the Mirage IV force (45 in all) and their limited electronic countermeasures equipment would ensure almost 100 per cent interception by Soviet defences, unless these were also meeting a major US attack – in which case the impact of a small number of 80-kt warheads would hardly be noticed. ◊ Trigger for the strategic doctrine used to justify this force.

Specifications: Power plant: two 15,400-lb turbojets side by side in fuselage. Crew: 2. Air-to-ground radars including doppler. Navigation computer. No inertial platform. Length: 77 ft. Wing span: 38 ft 10 in. Weight empty: 31,965 lb. Combat take-off weight: 69,665 lb. Maximum speed: Mach 2·2. Service ceiling: 65,000 ft.

◊ Bomber for general subject entry.

MIRV. Multiple Independently-targetted Re-entry Vehicle. The destructive capacity of a nuclear warhead per unit yield (kilotons or megatons) falls off sharply as the total yield increases. For example, three 5-megaton warheads are far more effective against almost all targets than a single 25-megaton warhead; 10 small 50-kiloton warheads are similarly superior to a single 10-megaton warhead, though the latter's yield is twenty times as large as that of the ten small warheads. This sort of ratio is especially important with small and dispersed targets such as missile silos, and these considerations contributed to the development of MRV (multiple re-entry vehicles, which fall in a fixed cluster) and MIRV payloads for strategic missiles. In this way a single booster can dispense several warheads, each of which is separately guided to its target. The US ◊ Minuteman 3 is reported to carry three 0·2 megaton MIRVs, while the ◊ Poseidon is to deliver about ten MIRV warheads of some tens of kilotons each (or fourteen, according to another estimate).

MIRV warheads have been developed with the limited intent of improving

ICBM cost-effectiveness, but their strategic implications are such that their inventors would perhaps like to put the genie back into the bottle. This is due to the fact that with MIRV payloads there is a built-in advantage to the side which shoots first, so that – other things being equal – their deployment is likely to be exceedingly destabilizing. The strategic implications of MIRV payloads are illustrated in the examples below. In both it is assumed that the single-shot kill probability of an ICBM warhead against an enemy ICBM in its hardened silo is 80 per cent (though any number less than 100 per cent will do just as well).

Case 1. Both sides have 1,000 single-warhead ICBMs.

A strikes first at *B*'s ICBMs attempting to disarm *B*. 800 of *B*'s ICBMs are destroyed, but 200 survive. These are more than adequate to allow *B* to retaliate against *A*'s cities, or exact a price for refraining from doing so. It therefore does not pay either side to strike first, and the strategic balance is stable.

Case 2. Both sides have 1,000 ICBMs with three MIRV warheads on each ICBM.

A strikes first and allocates about 1,500 warheads in two waves to destroy all of *B*'s ICBMs. *A* now has 500 ICBMs left to dissuade *B* from using any remaining ICBMs and to dispose of bomber and submarine bases. *B* will therefore be disarmed by *A* and dissuaded from using its remaining forces by *A*'s threat of utter destruction. *A* has therefore eliminated *B* as a strategic rival by striking first, and *B*'s only defence is to strike before *A*. Thus it pays (both sides) to strike first, and the strategic equilibrium is unstable.

In other words, MIRV payloads are going to end the relative strategic balance of the mid and late sixties unless restabilizing countermeasures are adopted by both sides (such as an ◊ *ABM*).

Missile. An unmanned and disposable vehicle which is guided to, rather than aimed at, its target. The first operational missiles, the German V-series, were 'strategic', i.e. intended for homeland-to-homeland use, but since then a wide range of diversified missile weapons has been developed for a wide variety of missions. These vary in range from a few dozen feet to many thousands of miles, and their warheads are intended for targets ranging from single vehicles to large urban conurbations. (The conventional functional classification is listed at the end of this entry.) Missiles consist of three main elements: the power plant, the guidance unit, and the payload.

Early missiles, from the V-2 to the 'first generation' of nuclear delivery vehicles (such as the US Atlas), were generally powered by liquid-fuelled rocket motors. The liquids used were highly unstable and could not be stored for long inside the missile. Thus rather lengthy fuelling was required before launch. Storable-liquid (SL) fuels were introduced in the next generation of missiles, such as the US ◊ *Titan II* and the Soviet SS-9 and many shorter-range missiles. Such 'SL' rocket motors are still used on many missiles including most Soviet ICBMs, but most missiles are propelled by solid-fuel motors, which are generally more reliable and economical than liquid-fuelled ones (though their specific energy per unit weight tends to be lower). Since most missiles require a powerful initial thrust for lift-off and a substantially weaker sustaining thrust, different 'stages' are generally used: a powerful 'booster' and a longer acting but less powerful 'sustainer'. More recently, however, dual-thrust rocket motors have been evolved; these are single-unit motors which can modify the thrust to fit different phases of flight. But rocket motors did not power one of the earliest operational missiles, the German V-1. This had a pulse-jet engine, an air-breathing motor of the ramjet family. Ramjet- and turbojet-propelled missiles are less common than rocket-propelled ones, and are limited to short and medium ranges; the British ◊ *Bloodhound* and ◊ *Sea Dart* and the US ◊ *Bomarc B* surface-to-air missiles, the US ◊ *Mace B* tactical missile and the Soviet ◊ *Kennel* anti-shipping missile are all 'air-breathing'.

Guidance is what makes a missile what it is, and several different techniques are used. Long-range missiles are generally ◊ *ballistic missiles* (with a high curved trajectory to their target) and are separately discussed under that heading. The rest of this section covers mostly non-ballistic or 'cruise' missiles, i.e. those which fly a relatively straight, aircraft-like course to their target.

◊ *Infra-red homing* is used in many air-to-air and some anti-aircraft types, such as ◊ *Sidewinder,* ◊ *Red Top* and ◊ *Redeye.* Since it relies on optical radiation this form of guidance is (a) short range and (b) dependent on reasonably clear skies.

◊ *Semi-active radar homing* units consist of a narrow-beam ◊ *tracking radar* receiver which 'locks on' the target and a computer which generates flight-correction signals for the servo-controls. In semi-active systems, the target is 'illuminated' by a radar transmitter on board the launching ship or aircraft, or, in the case of surface-to-air missiles such as the ◊ *Hawk,* on the ground.

Active radar homing, where the missile has a complete transmitter-receiver set, is a method of guidance suitable only for rather large missiles. This too is a short-range method, but unlike infra-red homing it is usable in all weathers (and at night). But this type of guidance can be used only against targets which have a clear radar 'image' when opposed to their back-grounds, such as aircraft or ships.

Radio-command is generally used in air-to-ground systems, where the target is invisible to radar because it blends into the ground image. In the simplest form, the operator observes visually both target and missile and directs the latter by joy-stick controls transmitted to the missile by radio or along a wire link. The French A S.20/A S.30 (◊ *Nord*) and the US ◊ *Bullpup* use this method of guidance. Many ground-to-ground anti-tank mis-siles use a variation of this theme, where the signals are transmitted from the 'pilot' to the missile by a thin wire un-coiled from the missile in flight. More sophisticated radio-command methods employ remote observation by TV – as used in the US ◊ *Walleye* – or infra-red-assisted tracking of the missile, as in the US ◊ *Tow.* Radio-command systems which do not rely on visual observation are employed in longer-range anti-aircraft and anti-shipping missiles (and in ballistic missile defence (BMD)

systems). In these, the missile is flown by remote control on the basis of data pro-vided by radar tracking both target and missile. Radar is also employed in two more guidance methods: beam-riding, where the missile 'rides' or tracks a pencil point beam directed at the target, as in the British ◊ *Seaslug,* and compar-ative radar-mapping. This is a far more complicated guidance system, and its use has been correspondingly limited (◊ *Mace, Matador*). A high-definition air-to-ground radar maps the ground over which the missile is flying and compares the ground 'image' with a recorded image stored in the memory of its computer. The com-puter compares the two images and, if the ground image picked up deviates from that stored in its memory, it actuates servo-controls which make suitable adjustments. Comparative radar-mapping therefore requires the collection of the initial data by radar-reconnaissance, and this may not always be possible. (◊ *Inertial Guidance* for the technique used in ballistic missiles.)

The payload is the pay-off in a missile as in other vehicle weapon systems. Strategic missiles and ◊ *stand-off missiles* (i.e. those launched from a bomber at long ranges to avoid having to penetrate inner air defences) are equipped with ◊ *nuclear warheads* (◊ *Ballistic Missile*). Very short-range missiles are generally equipped with high-explosive warheads, usually fitted with a proximity fuse (◊ *Radar*). In between, many 'tactical' missiles are generally equipped with inter-changeable nuclear, high-explosive and chemical warheads. Some anti-aircraft and anti-shipping missiles, though quite small, are also equipped with nuclear warheads, in order to reduce the accuracy require-ment for a 'kill', or to deal with particular-ly 'hard' targets, such as bridges (◊ *Nu-clear Warhead* and *MIRV*). Electronic countermeasures equipment is an in-creasingly important part of aircraft and missile payloads. It is mainly intended to deceive or confuse enemy radar surveil-lance and tracking. Missiles which rely on telecommunications or radar guidance/homing systems are themselves subject to ECM and sometimes equipped with counter-ECM (or ECCM) devices. ECCM are usually built into the guidance system in the shape of filters or selectors. One missile whose payload consists entirely of ECM equipment is the US ◊ *Quail*, a ◊ *'penetration aid'* for US strategic bombers. A direct form of ECM is the use of special-purpose anti-radar missiles, such as the US ◊ *Shrike*, which 'rides' enemy beams to destroy the set at the other end. Ballistic missiles equipped with inertial guidance are immune to ECM since they are fully independent of ground links, but if anti-ballistic missile (◊ *ABM*) systems are deployed, ICBMs will probably be equipped with ECM in order to deceive enemy radars. These would take the form of decoys, chaff and possibly echo distortion or simulation equipment. Confusion measures, such as various forms of 'jamming', are much less likely to be employed, since the power required to jam ABM radars is likely to be very great indeed.

The speed of missile weapons varies from the few hundred miles per hour of the wire-guided anti-tank missiles, such as the French SS.*11*, to the 16,000 mph plus of long-range ballistic missiles. But speed is generally not a critical variable since guidance systems determine the limits and usefulness of speed performance. But in ABM missiles, such as the ◊ *Spartan*, ◊ *Sprint* and ◊ *Galosh*, acceleration is, however, a critical variable; Sprint has by a wide margin the most rapid acceleration of any missile now performing (though it is not yet in operational service).

Accuracy is usually a far more important variable than speed. This is usually stated as the CEP, **Circular** Error Probable, defined as the **radius** around the aiming point within which half the warheads will impact. CEPs for long-range ballistic missiles are classified, are generally thought to be as little as 0·3 mile or less for the latest ICBMs, such as the ◊ *Minuteman 3* and ◊ *Poseidon*. 'Battlefield' systems with non-nuclear warheads have (and need to have) CEPs of the order of a few yards, though the use of nuclear warheads on some air-to-air and surface-to-air missiles allows 'kills' with far higher CEPs. ◊ *Nuclear Weapon* for some yield/CEP calculations.

The conventional classification for ballistic missiles is given under that entry;

non-ballistic missiles are usually classified as follows:

Air-to-Air (AAM): Used as aircraft-to-aircraft weapons to supplement or replace guns.

Air-to-Surface (ASM): A larger category, which includes ◊ *stand-off missiles*, anti-radiation missiles such as *Shrike* as well as ordinary ASMs.

Surface-to-Air (SAM): Anti-aircraft missiles from the man-portable *Redeye* to the long-range *Bomarc*; also includes projected anti-missile missiles such as *Sprint*, *Spartan* and *Galosh*.

Surface-to-Surface: Includes non-ballistic nuclear bombardment missiles (◊ *Mace B*), as well as the separately listed ballistic family. Also includes quite small non-nuclear anti-tank missiles for use on the battlefield, and naval missiles such as the Soviet *Styx*.

Anti-submarine missiles are a new but growing family: ◊ *Asroc*, *Subroc* and *Ikara*.

◊ also *Honest John* and the Soviet *Frog* series for unguided nuclear-warhead rockets; ◊ *Walleye* for a guided but unpropelled weapon.

Missile Gap. The presumed inferiority of the United States in strategic missile deployments with respect to the Soviet Union. This was a slogan of the 1960 Presidential election, and it played a major role in US domestic politics (and administration) over the period 1957–61. During 1959–60 US intelligence projections of the number of Soviet ICBMs deployed was revised downwards by more than 97 per cent, but by then the US missile deployment programme was under way. There is now little doubt that this error was due to a deliberate and highly successful Soviet campaign of strategic deception. During the period 1957–63 every Soviet statement on the subject sought to create the impression that the Soviet Union deployed large numbers of highly effective ICBMs. In fact, at the beginning of 1960 the Soviet Union had only 'a handful' of operational ICBMs deployed, but the deception allowed Soviet leaders to exert a good deal of leverage with their few ICBMs. It was only large-scale photographic reconnaissance by means of surveillance

satellites that finally exposed the deception and allowed the United States to determine the true level of Soviet strategic capabilities.
▷ *Soviet Strategic Offensive Forces* for their current deployments.
▷ *ICBM, Soviet* for the evolution of their ICBM force.

MLF. Multilateral nuclear force. An American plan of the early sixties which envisaged the deployment of a mixed-manned NATO force of twenty-five surface vessels equipped with ▷ *Polaris* intermediate-range nuclear-warhead missiles. The MLF plan provided for a US veto power on missile firings from the ships and for the participation of as many NATO powers as wanted to join. France rejected the plan since it was not an independent deterrent. Britain rejected it partly because its cost would have provided less leverage than the maintenance of its already deployed independent deterrent. The Soviet Union attacked the plan because of German participation and much propaganda hay was made with appropriate slogans about German fingers on nuclear triggers. In June 1964 the United States and Germany decided to proceed without France, but the decision was later reversed. The British plan, ▷ *Atlantic Nuclear Force*, was a package deal which provided for German investment, American veto power and a costless British participation. Both plans are defunct, though the problems to which they were addressed are very much alive.
▷ *European Security.*

MMG. ▷ *Medium Machine-gun.*

Mortar. ▷ *Artillery.*

I **Moskva and Leningrad. Soviet missile-equipped helicopter carriers and command vessels.** In the last few years the Soviet navy has been extending the range of its operations into new areas and especially the eastern Mediterranean. Its fleet of modern vessels (▷ *Destroyers, Soviet*), the reactivated Marine infantry, and the speeches of Soviet admirals all indicate the intention of establishing a naval presence around the Soviet empire, in the best nineteenth-century tradition –

showing the flag and all that. But these ships have no air cover, neither aircraft carriers nor (as yet) land-based fighters if operated outside the Black Sea. In other words, except for the limited protection of surface-to-air missiles and the symbolic protection of AA guns, the vessels are entirely incapable of surviving a local conflict with a coastal power which has an operational air force. The Israeli air force, for example, could sink the entire Soviet fleet in the Mediterranean at a relatively low attrition cost and in a very short time. The Soviets know this, but aircraft carriers are very expensive and the Soviet leadership may be sufficiently prudent to refrain from establishing bases in quasi-countries such as Syria.

The *Moskva* and *Leningrad*, beautiful, modern ships with clean lines and effective armament, are being deployed within this strategic context. They cannot provide air cover but they do offer a platform for some elegant helicopter displays to back up any symbolic 'vertical envelopment' manoeuvre by Soviet Marines. Such ships will certainly impress the Iraqis and Yemenis (if the Suez Canal should be reopened), and they may even impress the Egyptians.

Each ship can carry up to 30 medium-sized helicopters (though the complement is likely to be a more practical 20) and is equipped with three twin launchers for ▷ *Goa* SAMs as well as two twin 57-mm AA guns (indispensable for salutes, etc.), five trainable tubes for AS torpedoes (21-in), and ASW mortars.

Other specifications: Displacement: 15,000 tons standard, 18,000 full-load. Length: 645 ft. Steam turbine propulsion. Maximum speed: 30 knots.

MRBM. Medium-Range Ballistic Missile. ▷ *Ballistic Missile* and *Missile.*

MSBS. French submarine nuclear delivery system. The MSBS (Mer-Sol Balistique Stratégique) is a two-stage solid-fuel medium-range ballistic missile intended for underwater launch from the ▷ *Redoutable* class of French ballistic missile submarines. Three of these submarines are supposed to complement the surface-to-surface weaponry of the French 'independent' deterrent. The following

specifications are available. Body diameter: 4 ft 11 in. Nuclear warhead: 0·5 megatons; a 22,050–lb thrust first stage and a 8,820-lb thrust second stage. Steering by gimballed nozzles in the first stage and fluid injection vector control in the second stage.
▷ *France, Strategic Offensive Forces* for other French systems.
▷ *Trigger* for the implicit strategy the MSBS is intended to implement.

MSR. Missile Site Radar. A large and sophisticated radar, computer and communications unit which is one of the components of the US ballistic missile defence system. The MSR units operate in conjunction with the long-range ▷ *PAR* units. The latter alert the MSRs and supply them with the predicted path of the incoming attack, but the MSR, located within the defended area, would discriminate between warheads, decoys and spent missile parts and guide the ▷ *Spartan* and ▷ *Sprint* missiles until interception was achieved. The MSR is therefore intended to fulfill three basic missions:
1. identification and tracking of the warheads;
2. launch and tracking of the Spartan and Sprint missiles;
3. computation of flight-correction data and its transmission in the form of command impulses to the missiles until intercept.
The computers of the MSR would therefore have to 'choose' between alternative targets to identify the real warheads and also 'choose' whether to use the long-range Spartan or the short-range Sprint. The many tracking beams needed by the MSR could not easily be supplied by conventional single-beam mechanically slewed radars, and instead the ▷ *phased array* technique is used, where the antenna is fixed, and directional-beam emission and control are achieved by electronic means. Each MSR is to be housed in a moderately blast-resistant (or ▷ *hardened*) building with four large circular 'windows' for the 13·5-ft-diameter phased array. The MSR housing will be 120 by 120 ft, and forty feet of it will emerge above ground.
▷ *ABM* and *Radar*, general subject entries.

Mt. Megaton. Equivalent to one million short tons of TNT high explosive; a measure of the power of a ▷ *nuclear warhead*.

Mustard Gas. Mustard gas is the common name for a liquid toxic agent which becomes a gas only at 271° C (it is also known as Yperite). First used by the Germans during the First World War at Ypres in 1917, its variants are still deployed for military use. Mustard and its modern derivatives are almost colourless but have a faint garlic or mustard smell if impure. Mustard penetrates normal clothing, causing severe burns, and it inflames the eyes and irritates the lungs. Short exposure to an atmospheric concentration of one part in 100,000 causes acute vomiting and fever. The effect is delayed so that an effective dose can be absorbed before the agent is detected. Modern derivatives, T, Q, and HN3, are lethal in smaller doses. No adequate treatment exists, and mustard is the most severe blistering agent in existence. (Q causes incapacitating blisters after 3 milligrammes are absorbed.)
▷ *Chemical Warfare.*

Mya-4. Soviet bomber. ▷ *Bison.*

N

Napalm. An acronym for *NA*phthenic acid and *PALM*etate. The former is a petroleum product, while the latter is extracted from palm oils. Both are cheap and readily available. Napalm is an incendiary used as a filler for canisters for air-to-ground bombing and as a fuel for flamethrowers. The jelly composition, the extremely high temperature produced on ignition and its low cost make napalm an ideal weapon against 'soft' targets such as humans and most kinds of primitive dwellings. It is also extremely effective against armoured vehicles, because their armour operates as a heat conductor which broils the crew inside. Since it 'flows' as a liquid it can also be used against concrete bunkers and similar structures (though most modern fortifications have overhanging lips above the firing slits intended to prevent the fluid from flowing inside). Napalm is not suitable for use against 'hard' structures such as bridges. The problem with napalm is that its density is low so that tanks filled with it are bulky in relation to their weight and this reduces the speed of carrying aircraft. The intense heat wave produced by napalm can 'kill' armoured vehicles even if only a near-miss is scored, so that it is far superior to an explosive bomb for this mission. The fact that napalm is an efficient weapon gives it no moral qualities; its users may perhaps be castigated for waging war, but the use of napalm is in itself ethically meaningless.

Narrow-beam Radar. ◊ *Tracking Radar.*

NATO. North Atlantic Treaty Organization. Formed to implement the provisions of the treaty signed in 1949, its structure has not changed significantly since then. The members are Belgium, Britain, Canada, Denmark, West Germany, Greece, Iceland, Italy, Luxembourg, the Netherlands, Norway, Portugal, Turkey and the USA. France signed the Treaty but is no longer an effective member of the Organization; Iceland, Luxembourg and Norway impose various limits to their participation.

There are a North Atlantic Council consisting of permanent ambassadors of the 14 nations with its HQ in Brussels, a Defence Planning Committee, and a Secretary-General with a permanent multi-national staff. Each member – except Iceland – sends a permanent representative to the Military Committee, which supervises the NATO command structure. This consists of two major territorial commands, ◊ *SHAPE* for Europe and ◊ *ACLANT* for the Atlantic, with the smaller ◊ *ACCHAN* for the Channel area. SACEUR, the commander of SHAPE, and SACLANT, the commander of ACLANT, participate in the US Joint Strategic Planning System and have direct planning control over some nuclear strategic delivery systems (British V-Bombers and ◊ *Polaris* and US Polaris/ ◊ *Poseidon* submarines). A meeting of Defence Ministers known as ◊ *NDAC* deals with general policy with respect to the use of nuclear weapons, but detailed contingency planning for the use of tactical nuclear weapons is carried out by a more permanent body, known as ◊ *NPG*. Total peacetime NATO forces include about 1,100,000 combat and support troops, just over 7,000 ◊ *main battle tanks*, and about 3,000 combat aircraft. ◊ *Warsaw Pact* for the supposed opposition.

◊ *European Security* for the underlying political problems of NATO.

NCO. Non-Commissioned Officer.
Any rank within the separate and mostly
subordinate hierarchy of soldiers who do
not pass through or into the officer system.
This division derives from aristocratic
traditions and is unrelated to personnel
requirements in modern armed force
organizations.

**NDAC. Nuclear Defence Affairs
Committee.** A policy-making body for
nuclear planning established in 1966 within ◊ *NATO*. Membership of NDAC is
open to all Treaty signatories, but France,
Iceland and Luxembourg do not take part.
NDAC operates through meetings of
Defence Ministers, and the NATO
Secretary-General is ex-officio its Chair-
man. Detailed contingency planning is
actually carried out by NPG.

Nerve Gases. The most toxic of all
chemical agents with military potential,
known as anticholinesterases. They are
liquids, and they act by causing contrac-
tion in body muscles, including involun-
tary and opposed muscles. When muscle
vibration is induced, drooling, jerking,
twitching, staggering, convulsions and
asphyxia result. Death occurs after very
short exposure.

The first nerve 'gas', Tabun, was dis-
covered in Germany in 1937, and it was
followed by the more toxic Sarin and
Soman. All these are odourless and colourless and can penetrate through the skin as
well as by inhalation. 4 grammes of Sarin
on clothing (the military clothing of a
moving soldier) is a lethal concentration;
the inhalation of 25 milligrammes per
cubic metre per minute causes a 50 per
cent lethality. 13,000 tons of Tabun were
available in Germany by 1945, and a Sarin
plant was on a stand-by condition. Both
the USA and the Soviet Union captured
stocks and plants and have developed these
agents further. UK scientists have de
veloped a new series of nerve 'gases' known
as V-agents. 'VE' and 'VX' are both
liquid and apparently lethal in very small
concentrations ('one small drop on the
skin'). Like other liquid agents, nerve
gases would be used in aerosol form
for maximum efficiency, and both the
USA and USSR are thought to have

developed suitable shells and missile
warheads.
◊ *Chemical Warfare.*

Neutrality. Claimed or declared non-
involvement in an armed conflict. Neutrality is invariably respected by belligerent
powers if (a) the neutral state refrains from
breaches of its status and (b) no belligerent
regards the benefits of an attack on the
neutral power as greater than the cost of
such an attack.

Neutron Kill. A method of destroying
objects, including enemy nuclear warheads in the atmosphere, by means of a
nuclear detonation. The ◊ *Sprint* interceptor missile of the ◊ *Safeguard* ◊ *ABM*
programme is intended to destroy incoming warheads by means of the neutron
effect of its 1–2-kiloton warhead as well as
by blast. The neutron emission of the
interceptor's warhead is intended to trigger
fission within the enemy warhead, thus
generating very high temperatures which
damage the warhead and prevent a normal
detonation. Among other ◊ *penetration
aids*, the shielding of nuclear warheads
against the neutron effect of interceptors
has been prospected. This is unlikely to be
deployed as a protection device since
effective shields would absorb far too
much of the payload to justify the effort.
◊ *Nuclear Warhead.*

**Nike-Hercules. US surface-to-air
missile.** The Nike-Hercules is an anti-
aircraft missile intended to intercept
strategic bombers at medium to high
altitudes. The missile is powered by a
solid-fuel rocket motor and is launched
vertically from a fixed ramp. Its maximum
slant range is about 75 miles and its speed
and ceiling (150,000 ft) exceed those of any
likely target, but this missile is obsolescent
since it cannot intercept low-level intruders or highly manoeuvrable aircraft.

The Nike-Hercules has a nuclear warhead and is guided to its targets by radio-
command from the ◊ *SAGE* air defence
system. It is 41 ft 6 in long and weighs
over 10,000 lb.

**NORAD. North American Air De-
fense Command.** A joint USA–Canada
organization with its combat operations

centre under Cheyenne Mountain near Colorado Springs. It controls:

1. A chain of radar and communication systems for detection and tracking, which includes the Ballistic Missile Early Warning System (◊ *BMEWS*), the Distant Early Warning Line (◊ *DEW*) and AEWCF, a fleet of Super Constellations with radar and communication gear, which are to be replaced by ◊ *AWACS*.

2. 15 interceptor squadrons of the US Aerospace Defense Command: 11 squadrons of F-106 ◊ *Delta Darts*, three F-101B ◊ *Voodoo* squadrons and a residual squadron of F-102A Delta Daggers. These aircraft are equipped with ◊ *Sparrow*, ◊ *Sidewinder* and ◊ *Falcon* air-to-air missiles, and the ◊ *Genie* nuclear rocket. These forces are supplemented by 17 Air National Guard squadrons, mainly equipped with Delta Daggers, and by three Canadian Voodoo squadrons. There are also ◊ *Nike-Hercules* and ◊ *Hawk* surface-to-air missiles for point defence, operated by the US Army Air Defense Command, and about 170 ◊ *Bomarc B* long-range nuclear interceptor missiles operated by the air force ADC.

Both the interceptors and the radar chains are coordinated by the ◊ *SAGE* system which is supplemented by the ◊ *BUIC* system.

The US Defense Department has been reducing the establishment of NORAD, since Soviet bomber forces present a diminishing threat, and NORAD forces are now much smaller than those of its Soviet counterpart, PVO Strany, with about 700 aircraft versus the latter's 3,500 or so.

◊ *PVO Strany, CAFDA* and *UK, Strategic Defensive Forces* and the same for China for NORAD's counterparts. ◊ *USSR, Strategic Offensive Forces* and the same for UK, France and China for the potential opposition to NORAD.

◊ *Air Defence System* for general subject entry.

Nord. French guided-missile brand name. Nord-Aviation makes a wide range of small battlefield missiles of which the following are the most common:

AS.11   See SS.11 below.

AS.12   See SS.12 below.

AS.20   A dual-thrust solid-propellant air-to-surface missile with a 66-lb war-head and visual guidance where the pilot controls the missile by radio commands. The AS.20 is 8 ft 6 in long, weighs 315 lb and has a Mach 0·7 minimum and Mach 1·7 maximum speed. The range is 4·3 miles.

AS.30   A scaled-up AS.20 with a 510-lb warhead. There is an optional ◊ *infrared* aiming/guidance system. It is 12 ft 9 in long, weighs 1,146 lb and has a Mach 0·45 minimum and a supersonic maximum speed. Range is about 7 miles. There is a lightweight (838-lb) modification known as AS.30L. Another modification is the AS.33 which has ◊ *inertial navigation*. This permits the pilot to leave the target area once the missile is set on its course.

ENTAC   An anti-tank wire-guided missile. This is a simple and lightweight first-generation anti-tank missile guided by joystick, with the operator keeping the missile's flare in his sight and on target. The maximum operator-to-missile distance at launch is 360 ft. The HEAT (shaped charge) warhead weighs 9 lb. Entac is 2 ft 8 in long, weighs 26·9 lb and travels at 190 mph. Minimum range is about 1,300 ft and maximum range is 6,000 ft.

SS.11 (5210), SS.B1.   A wire-guided battlefield missile guided by joystick on visual/manual control. The operator keeps the missile's flare in his sight and on target by making corrections sent down the wire to the missile in flight. The original version has been replaced by an improved B1 type with transistor equipment. There are various optional warheads with HEAT anti-tank, APHE penetration-explosion and fragmentation anti-personnel charges. Both the original SS.11 and the SS.11B1 are made in air-to-surface versions, where a stabilized sight has been developed for helicopters and slow aircraft. The length is 3 ft 11 in, and it weighs 66 lb. The missile travels at 300 mph and has a minimum range of 1,650 ft and a maximum range of 9,840 ft.

HARPON   This is a SS.11B1 missile with automatic guidance. The system consists of a goniometer which follows the missile's tracer flare, a computer which calculates correcting impulses for the missile, and an optical sight which is used by the operator to obtain the initial programming 'fix' on the target, guidance being automatic after this point. This guidance unit (designated TAC) has allowed the minimum range to be reduced to about 1,200 ft.

SS.12 (and AS.12) An up-scaled version of the SS.11B1, also suitable for naval use, with a larger, 66-lb, warhead. This missile is also being adapted for the Harpon guidance unit. The SS.12 is 6 ft long, has a 7 in body diameter, and weighs 167 lb. The speed at impact is 210 mph, and it has a maximum range of 19,650 ft in the surface (SS.12) and 26,250 ft in the AS.12 versions.

◊ Missile and ◊ Radio-command Guidance.

For the Nord-Bölkow HOT and MILAN missiles, ◊ HOT.

NORTHAG. A NATO command organization. ◊ SHAPE.

NPG. Nuclear Planning Group. A policy-planning body within NATO which is responsible for the formulation of detailed rules covering the use of tactical nuclear weapons. Since the ◊ Warsaw Pact forces have a net superiority in the Central European theatre (assuming a prior Russian mobilization) a full-scale attack on NATO would have to be met by using tactical nuclear weapons. Further, localized nuclear strikes against NATO territory would also call for a tactical nuclear response. The NPG, which was set up in 1966, is to determine when and how such a nuclear response is to be delivered. Apart from the nominal member, France, all NATO members except for Iceland and Luxembourg participate in nuclear planning, but only seven NATO members sit on the NPG at any one time, with the smaller powers taking their place in rotation. The NPG is supposed to be supervised by the ministerial meetings of the Nuclear Defence Affairs Committee (◊ NDAC), but evidently a more direct supervision has been exercised since the NDAC appears to be inactive.

NPT. Non-Proliferation Treaty. Signed on 1 July 1968, this USA–USSR inspired multilateral treaty is intended to prevent the proliferation of nuclear weapons by forbidding the manufacture or transfer of fissile materials, production plant and certain ancillary equipment as well as the weapons themselves. The text agreed to by the USA and the Soviet Union was tabled at the Eighteen-Nation Disarmament Committee (ENDC) at the beginning of 1968, which endorsed it with very minor changes. It was then presented to the General Assembly of the UN as a motion, which was passed with 95 votes for, 4 against and 21 abstentions. The treaty was then signed in Washington, London and Moscow on 1 July 1968. Since then another 80 countries have signed the Treaty, but it has to be ratified by 40 states before it can come into force. The US Senate ratified the treaty in March 1969, but the Soviet Union has not done so as yet. In the case of the United States, ratification is a constitutional requirement; in the case of several other countries (including the Soviet Union and Pakistan) it is intended to provide an 'out' in case a particular country refuses to sign the treaty (including West Germany and India). France and China (CPR) have not participated in the preparatory work and are not expected to sign the treaty. Certain non-nuclear powers are also expected to delay or avoid signature, including the most likely nuclear countries: India, Israel and Japan. In the case of the first two, a *quid pro quo* in the shape of a nuclear umbrella or territorial guarantee is being asked for; in the case of the latter two it is more a matter of status, fears of negative effects on their nuclear engineering industries and resentment against the signatory nuclear powers and their monopoly (or rather oligopoly) status. These feelings were externalized at the Conference on Non-Nuclear Weapon States held in Geneva between 29 August and 27 September 1968.

◊ Nuclear Proliferation.

Nuclear Free Zone. A stated area within which no nuclear warheads may be stored or deployed. A specialized form of ◊ arms control which, like other forms of ◊ arms limitation, can be very useful to a party which has a local superiority in some other groups of weapons – i.e. a party with a local superiority in non-nuclear forces can improve the local balance of military strength by securing the agreement of other (nuclear) parties to institute a nuclear free zone in the area in question (for example, the 'Rapacki plan' for a European Nuclear Free Zone).

Nuclear Proliferation – or the nth country problem (or rather the $n + 1$ country problem). So far five countries have developed nuclear weapons: the USA, Britain, the Soviet Union, France and China. All of these have developed both the plutonium-239 and U-235 technology and have produced both fission and fusion weapons, though only the two super-powers have developed a full range of weapons. ($\lozenge$ *Nuclear Weapon* for the full nuclear 'family'.) Since the possession of nuclear weapons is in any case difficult to achieve and since it is generally thought of as conferring significant political advantages, the possessors of nuclear weapons have adopted policies of non-dissemination. At the same time, they have exported 'peaceful' reactors, most of which can be used to produce plutonium-239. The policy of non-dissemination followed by each power in turn as it joined the nuclear 'club' consists basically of three elements:

(a) no sales of nuclear weapons, fissionable material or technical data for direct weapon manufacture;

(b) no publication of technical data, specially to do with weapon manufacture, and security arrangements at their own nuclear weapon installations;

(c) contractual arrangements backed up by inspection to ensure the return of plutonium produced in reactors supplied or fuelled by 'club' countries.

India is, however, on the verge of producing nuclear weapons and more than twelve 'non-nuclear' countries produce plutonium in sufficient quantities for weapon production. Many more have research reactors useful for training personnel, including Formosa, Columbia, Congo (Kinshasa), Yugoslavia, Romania, Poland, Indonesia, Iran, South Korea, the Philippines, Portugal, Turkey, Venezuela, Thailand, and South Vietnam. Though very few of these facilities have been developed independently, and though 'safeguards' are applied, this represents pro-

liferation on a massive scale. All members of the 'club', except for China, have been competing to spread these resources and information to all comers. In this way competitive pressures (and major policy errors, such as the US 'Atoms for Peace' programme) have subverted the officially restrictive policies. In practice, the intermediate stages of weapon production, including plutonium-producing reactors, have been treated as if their technology was fully insulated from that required for a weapons programme. Inspection, mainly intended to secure the return of weapon-grade plutonium to the supplier of uranium fuel for reactors, is of course a chimera, since a country determined to produce nuclear weapons will do so even if contractual agreements are thereby violated. (Will Canada declare war on India when the latter infringes the return-of-plutonium provisions?)

Nuclear reactors are rated in megawatts thermal, MW(th), in the case of small, 'research' facilities, and megawatts electrical, MW(e), in the case of large energy-supply reactors. For each MW(th) of yield a reactor fuelled on natural uranium will produce about 0·25 kilo of plutonium-239 per year; each MW(e) is similarly equivalent to about one kilo of plutonium-239 per year. Since ratings of 200, 200 and 660 MW(e) are common for reactors such as those exported to Italy, Japan and West Germany, large numbers of bombs (at, say, 7 kilos of plutonium each) could be produced by a number of countries; even quite small countries have reactors rated at 25 MW(th), which can produce at least one bomb-equivalent per year. Thus non-dissemination has entirely failed for plutonium weapons.

The technology required to produce U-235 (indispensable for fusion weapons) is on the other hand more demanding – but the development of gas centrifuges ($\lozenge$ *Nuclear Weapon*) may change this state of affairs. Of course the $n+1$ country problem cannot be isolated from the security and political problems that lead countries to make the very heavy investment required for a nuclear weapons programme. Canada for example has all the resources required for the production of a very large number of fission bombs as well as those required to sustain an accelerated fusion bomb programme. And yet Canada is not likely to become a nuclear power. This is of course due to the fact that Canada (a) faces no particular threats from any power, nuclear or otherwise, and (b) is· protected by formal and informal security arrangements against any conceivable threat to its security. Thus Canada has no military-security reason to develop and deploy nuclear weapons; it may still do so in the future because of political/prestige reasons, but these pressures are less likely to become operative, and if they so operate a Canadian prestige programme is not likely to cause any major instabilities. Consider on the other hand the case of Israel. Its leaders are indifferent to political/prestige considerations such as those which moved the French, for example, but they do face an immediate and grave threat from a number of Arab countries, partly acting as Soviet client-states. Israel is one of the very few countries which has not been admitted to any military (or political) alliance system. It is also the only country involved in a conflict where its opponents are aiming for politicide, rather than territorial adjustments or political changes. Thus Israel, whose nuclear resources are a tiny fraction of those available to Canada, is likely to develop nuclear weapons, and this is likely to cause major instabilities. But then 'instabilities' are a small matter when the inner core of security, national survival, is exposed to a continuing threat. Nuclear proliferation, and its control, is therefore

inseparable from the problem of small-country security in the nuclear age.

Nuclear Test Ban Treaty. ◊ *Test Ban Treaty.*

Nuclear Warhead. Although the power or yield of nuclear devices is measured in units of 1,000 tons of ◊ *TNT* (◊ *Nuclear Weapon*), the effect of a nuclear explosion on actual targets is not proportional to the energy yield, and there are several effects which do not occur with non-nuclear explosives to a significant extent:

(i) Blast and shock, most usefully measured in pounds per square inch of overpressure, or psi (and in the velocity of the winds produced, and the size of the crater produced by groundbursts).

(ii) Heat, conveniently measured by the radius up to which specific materials or environments are ignited in given conditions.

(iii) Radiation, or immediate radiation, due mainly to gamma rays and neutron emission, which for convenience are measured in Röntgen units, in terms of the distance at which certain effects are experienced by unsheltered humans (or electronic equipment).

(iv) Residual radiation or 'fallout', which consists of the radioactive debris, including weapon parts, which is thrown up to various altitudes and eventually descends to the ground. Local fallout takes the form of a cloud whose main axis corresponds to the direction of the wind.

But from the point of view of the user,

Nuclear Weapon Damage Effects

Type of target	Relevant nuclear weapon damage effect, and type of burst for optimum results
(i) Missile silo and other small and hardened facilities.	Blast, as measured in psi of overpressure. Groundburst.
(ii) Urban population, industry centre, and all 'soft' facilities.	Heat, as measured in temperatures produced at various radii from point of impact. Airburst.
(iii) Missile warhead, and other electronic systems.	Immediate radiation (neutron and X-rays); as used in the ◊ *Sprint* and ◊ *Spartan* missiles of the ◊ *ABM* system. Not applicable.

the effects that matter are determined by the nature of the target; in each case only one or two effects are useful, while others are often undesirable (◊ *Nuclear Weapon*). The primary and most relevant 'effects' are tabulated opposite for three types of targets:

1. *Nuclear Warheads versus Missile Silos* Missile silos and ◊ *hardened* facilities are the most difficult targets for nuclear warheads since they are both small and specifically protected against the primary weapon effect, overpressure. The two critical variables for the effectiveness of a nuclear warhead are the CEP and the energy yield. The CEP (Circular Error Probable) measures the accuracy of the delivery system (the radial distance from the aiming point within which half the warheads impact). The interaction between CEP and energy yield as applied to a group of targets hardened to 300 psi (typical stated figure for the US Minuteman sites) is measured in terms of the 'kill probability' in the table below:

livered with a CEP of 1,000 ft, that warhead has a 96 per-cent probability of destroying a hardened silo. The US has or will soon have warheads of this yield and accuracy. More serious from the US point of view is the development of large-payload Soviet missiles that could, for example, deliver three 5-megaton warheads with CEPs of 2,000 ft or so (◊ *ICBM, Soviet*). And this raises the second issue that will multiply the destabilizing effect of really accurate delivery systems: ◊ *MIRV* multiple independently targetted warheads. While the low CEPs of forthcoming delivery systems nullify hardening, *MIRV* warheads will reverse the other axiom of strategic stability: the built-in advantage of defence over offence. Each missile equipped with *MIRV* warheads will be able to destroy two or more missiles in their silos. Thus the one who shoots first can pre-empt the retaliatory force of the other. In other words, each side's ◊ *assured destruction* capability could be nullified, and with it the very basis

Single-Shot kill probability of stated warheads against targets hardened to 300 psi (nearest percentage point)

Accuracy, CEP in feet		5 mt	1 mt	500 kt	50 kt
(a)	10,000	9%	3%	2%	0%
(b)	5,000	34%	12%	8%	2%
(c)	2,000	93%	55%	40%	11%
(d)	1,000	99%	96%	87%	35%
(e)	500	99%+	99%+	99%+	82%+

Until recently lines (c) and below were purely theoretical, since accuracies of this sort were unattainable by ◊ *ballistic missiles*, the main strategic weapon of the 1960s. There is increasing evidence that such accuracies are increasingly obtainable, or have been already attained by the ballistic missiles of both the USA and the Soviet Union. The strategic balance has rested so far on each side's inability to destroy the other's hardened missile silos; each side could therefore follow a relaxed strategic policy since each side had a survivable and therefore valid deterrent. But when a one-megaton warhead can be de-

of deterrence. For further discussion and examples, ◊ *MIRV*. Also ◊ *ABM*.

2. *Nuclear Warheads versus Cities*
While hardened targets such as missile sites are best destroyed by groundbursts, against large soft targets such as cities or military bases airbursts are more efficient, with the optimum height of the detonation depending on the yield of the warhead. The key weapon effect against large soft targets is the heat effect, since the perimeter of the blast and immediate radiation effects are encompassed by the lethal heat effect perimeter. The primary heat effect is compounded by the combination of

winds (produced by the shock wave) and secondary fires which produce the so-called 'fireball'. The likelihood and extent of the ignition of the environment depend on a number of factors, including the materials involved, the spacing between them, and the weather. The table below summarizes immediate effects relevant to city-type targets:

Nuclear weapon effects on civilian targets

	Groundburst		Airburst at optimum height	
Energy yield:	1 megaton	10 megatons	1 megaton	10 megatons
Crater, depth	230 ft	500 ft	does not occur	
diameter	950 ft	2,600 ft		
Complete destruction of brick structures at	2·7 miles	6 miles	3·5 miles	8 miles
Light damage at	7·2 miles	15·5 miles	13 miles	26·5 miles
Lethal winds at	4 miles	9 miles	6·5 miles	14 miles
Blistering burns at	9·4 miles	23·5 miles	11 miles	26·5 miles
Ignition of fabrics at	5·6 miles	14·5 miles	6 miles	17·5 miles

But in reality single large warheads would not be used against cities since *MIRV* warheads would be much more effective. The wasteful overkill which occurs at the centre of the detonation would be avoided by using several smaller warheads whose combined yield would be lower but whose effect would be much greater.

3. *Nuclear Warheads versus ballistic missile warheads; ballistic missile warheads as anti-ballistic-missile defence weapons, i.e. ABM.* The two missiles developed for the US ⟡ *Safeguard ABM* system under construction, the Sprint and the Spartan, are intended to destroy incoming warheads mainly by means of immediate radiation. But the uses of immediate radiation against electronic systems, whether those of incoming warheads or those of ABM radars, cannot be easily tabulated. Two types of radiation effect seem to be useful for destroying incoming warheads: hard ⟡ *X-ray kill* as used, outside the atmosphere, by the Spartan missile, and ⟡ *neutron kill*, as used in the atmospheric Sprint missile. The use of ⟡ *blackout*, produced by heat and beta-ray radiation, has been prospected as an anti-ABM countermeasure.

4. *Nuclear Warheads as Global Weapons: Fallout*
Fallout is an effect which is not generally wanted in weapon systems. Nuclear warheads intended for use against missile silos as part of an ABM system or for battlefield purposes are generally made as 'clean' as possible, since their fallout effect is a positive drawback. But there is a require-ment for very 'dirty' weapons as part of the assured destruction element of an offensive force ⟡ *Nuclear Weapon*. The radioactive effect of nuclear warheads is conventionally divided into three elements:

Immediate radiation, the effect of the direct radiation produced during the first minute following an explosion.

Local fallout, the effect of the cloud of radioactive elements thrown up by an explosion whose fireball touches the ground.

Global fallout, or the long-term radioactive effect attributed to a detonation whose debris goes up to the upper atmosphere eventually to scatter and descend.

Radiation affects the complex biochemical processes of the body. Since it is cumulative under certain conditions, and since nuclear explosions produce unstable isotopes whose 'half-life' (radioactive life-span) is variable, no precise measure of radioactive effects on humans is possible. The table opposite indicates certain ranges for immediate radiation. The units used, Röntgen (r), are an approximate combined measure of different radiation elements, though gamma-rays are the most important. It is known that two different doses of radiation within a 24-hour period are cumulative without 'discounting', i.e. 1or + 2or = 3or. But the same doses spaced out over longer periods must be discounted, i.e. 1or + 2or (one week later) = less than 3or.

ing the structure of the atoms themselves; energy differences are therefore far greater, so that far more energy is liberated per unit weight. Apart from the power (or 'yield', measured in 1,000-ton units of ◊ TNT) of such explosions they cause

Radiation (Röntgen units)	Effect on unsheltered humans of 'average' health
Up to 150 r	Only long-term cumulative.
150–250 r	Nausea and vomiting within 24 hours. Sickness.
250–350 r	Nausea and vomiting within 4 hours. Some deaths in 7 to 30 days.
600 and over r	Nausea and vomiting immediately. Death within one week.

Seven hours after a nuclear explosion 10 per cent of radiation remains, and even after a week 1 per cent persists.

Taking quite small weapons, in the kiloton range, these immediate readings are experienced at the stated distances from 'ground zero':

certain effects which chemical explosions do not produce at all, such as lethal radiation.

There are two ways of altering the structure of atoms: splitting very heavy ones (fission), or fusing together certain very light ones (fusion). The former pro-

Weapon yield, kilotons	5	20	20	40
200 r registered at (metres)	1,200	1,300	1,400	1,600
1,000 r registered at (metres)	900	1,000	1,100	1,250

Nuclear Weapon. A nuclear explosive device uses one or both of the two energy-liberating nuclear phenomena, 'fission' (as in the 'A' or atomic bomb) and 'fusion' (as in the 'H' or thermonuclear bomb). The effects of nuclear weapons, including their tactical and strategic implications, are discussed under ◊ Nuclear Warhead; this entry covers the actual physical processes and the manufacture of nuclear explosive devices.

Chemical, i.e. non-nuclear, explosions are produced by altering the atomic composition of certain compounds: different compounds have different energy charges, so that in going from one to another energy can be liberated. If the process is rapid, and if the energy differential is great, the process is explosive. Nuclear explosions, on the other hand, are produced by alter-

duces far less energy per unit weight, but even the so-called 'nominal' bombs produce yields equivalent to twenty thousand tons of TNT, one of the most powerful of chemical explosives.

Two materials have been used in fission devices: U-235, a rare form of the rather common uranium atom, and plutonium-239, which is produced artificially in nuclear reactors. A certain amount of sufficiently pure U-235 or plutonium-239 will spontaneously initiate a chain reaction leading to fission. The amount sufficient to do this, the 'critical mass', has been unofficially estimated at 16 to 20 kilos of U-235 and at 7 kilos of Plu-239, assuming that the materials are in a compact 'heap'. A weapon consists therefore of two or more shaped blocks of U-235 or Plu-239, each of which is sub-critical; a chemical explosive device compresses them into a single compact block which exceeds the critical mass of the material. Of course, safe military use requires extensive shielding and packaging as well as elaborate control equipment. (US nuclear weapons have sustained fragmentation, chemical explosions, fires, mechanical impact and electrical charges without detonating.)

U-235 is found in pure natural uranium, but there are only seven parts of it out of a thousand, the rest being U-238, which cannot be used to initiate fission. Since U-235 and U-238 are chemically identical, they cannot be separated by chemical means; since they are in other respects exceedingly similar, only very sophisticated physical techniques can be used to extract U-235 from the rest of natural uranium. Three processes have been used: (i) gaseous diffusion, which is the only practicable method so far and which requires vast quantities of energy as well as a very expensive and technologically demanding plant; (ii) electromagnetic separation, which was tried during the 'Manhattan' programme but which could not be used to produce useful quantities of U-235, and (iii) gas centrifuging. This process was also tried during the 'Manhattan' project, but the speeds required in the centrifuge were then unattainable. Recent technological advances, however, have apparently changed the situation, and U-235 separation may now be possible by this method. Since this process is technologically less demanding (and far cheaper) workable gas centrifuge separation would considerably lower the technical barriers to nuclear proliferation (gas centrifuge separation is also far easier to conceal than gaseous diffusion).

Fusion makes use of the lightest element, hydrogen. Certain types of hydrogen (deuterium and tritium) can be fused at very high temperatures, thus producing much more energy than was required to initiate the process. In theory, the heat required could be produced by other means, but so far fusion devices have been triggered by a U-235 (fission) detonation. Thus 'H' or 'thermonuclear' bombs are really fission–fusion bombs.

· Fission alone can be brought about by using Plu-239. This is simpler and cheaper to produce than U-235. Further, while the technology of gaseous diffusion is mostly useless for purposes other than weapon manufacture, most of the plutonium-239 production stages are part of the important civilian reactor technology. A controlled chain reaction in a reactor fuelled with natural uranium will produce plutonium by transforming some of the U-238 into the former material; if thorium is used it will be turned into U-233. This too is apparently fissile, but it has not been used in weapons so far. The production of plutonium is a normal side-effect of operating a natural uranium reactor; but a chemical separation plant is required in order to extract the pure (weapon grade) plutonium-239. These separation plants are useful for civilian purposes too since the material extracted can be used to fuel further reactors, but in general their acquisition by non-nuclear powers (in the weapon sense) indicates an interest in weapons. The chemical separation plant is complex and expensive, but its cost is probably no greater than 1–2 per cent of the cost of a gaseous diffusion plant. Plutonium is therefore much cheaper than U-235, but only the latter has so far proved suitable as a trigger for fusion. Plutonium on the other hand is suited for low-yield light-weight weapons.

Apart from fission and fusion, a third kind of weapon has been developed, which uses both techniques and represents an economical approach towards large weapon construction. U-238 is not fissile under normal conditions but is fissionable at the very high temperatures of fusion. U-238 is cheap, and can be used to make the weapon casing, since a fusion bomb requires in any case a thick casing and U-238 is suitable for this purpose. A fission–fusion–fission bomb is therefore very economical; it is also very 'dirty', i.e. it produces a lot of fallout, since while fusion in itself is relatively clean (with the fallout coming from the fission trigger) the fission of the thick U-238 casing produces a great deal of fallout.

A typical nuclear weapon 'family' is therefore composed of the following elements:

(a) Fission devices based on plutonium-239 suitable for lightweight low-yield weapons, 1 to 20 kilotons.

(b) Fission devices based on U-235

suitable for larger, medium-yield weapons, perhaps 20 to 200 kilotons.

(c) Fusion devices with fission U-235 triggers; suitable for weapons with yields of 0·5 to 20 megatons. (These weapons will produce fallout in proportion to the size of the fission trigger.) ('Clean' weapons are especially desirable for a ▷ *counterforce* attack.)

(d) Fusion devices with fission U-235 triggers and with outer cases made of U-238 for secondary fission, i.e. fission–fusion–fission weapons. The largest nuclear weapon ever detonated, a 57-megaton Soviet device, was apparently of this type. The U-238 casing creates a great deal of fallout, and this was a very 'dirty' weapon indeed. ('Dirty' weapons are especially suitable for a ▷ *counter-value* attack aimed at cities, as part of an ▷ *assured destruction* posture.)

NUDETS. A US nuclear detonation warning system. A matrix of seismic sensors spread over the entire US territory which is intended to estimate the point of impact and power of nuclear-detonated nuclear warheads.

O

Orbital Bomb. A satellite containing a ◊ *nuclear warhead* which circles the earth in a low orbit and which can be commanded to descend on a particular target. No such weapons are now known to be operational, and their deployment in orbit would be prohibited under the terms of the Outer Space Treaty of 1966 (though anything less than a full orbit is permitted, hence ◊ *FOBS*).

Osa and Komar. Soviet missile patrol boat classes. The *Osa* and *Komar* classes are small and very fast boats which are used as platforms for the ◊ *Styx* non-nuclear surface-to-surface cruise missiles. Though primarily intended as an anti-shipping weapon, the 15–18-mile range Styx is also suitable for the offshore bombardment of coastal targets. The boats are equipped with search radars and programming units to set the missiles on their path; since final interception is probably by ◊ *semi-active radar homing*, the radar unit includes an 'illumination' capability. The Israeli destroyer *Eilat* was sunk by Styx missiles fired from *Osa* or *Komar* vessels.

Because their missiles are lethal and the boats themselves are so elusive, the *Osa* and *Komar* classes present a new and effective threat to larger vessels. But the range and stability of these boats limit their use to calm seas within easy range of friendly harbours, and so their capabilities can be regarded as extending those of shore-based anti-shipping missiles. Nevertheless, the threat presented by the *Osa* and *Komar* vessels has led to the development of countermeasures, including detection equipment for the missiles, which being relatively slow can be interdicted by anti-aircraft missiles and even artillery – if controlled by fast response radar systems.

Each *Osa* carries four Styx missiles in large covered launchers, disposed in overlapping pairs and facing forward; each *Komar* vessel carries two launchers. Neither boat carries spare missiles.

Defensive armament includes four 25-mm AA cannon in the *Osa* and two in the *Komar*, and both types of boat are diesel-powered; since the power plant in both is the same, the *Osa* class is slower because larger.

Outer Space Treaty. Formally known as the Treaty on the Exploration and Use of Outer Space, this treaty was signed by Britain, the Soviet Union and the

Specifications:

| | Tonnage | | Dimensions in feet | | | | Maximum speed |
	standard	full-load	length	width	draught	Main engines	knots
Osa	160	200	131·5	23	6·5	3, each 4,800 bph	35
Komar	75	100	88	21	6	3, each 4,800 bph	40

United States in January 1967. The
adherence of almost all other countries,
except for France and China, followed
shortly thereafter. The terms of the
treaty prohibit the deployment of weapons
in outer space, including both natural and
artificial celestial objects. Any satellite
containing a weapon would therefore
violate the treaty, though only if it makes
a full orbit before returning to earth
($\triangleright$ *FOBS*). The treaty does not cover
the use of satellites for military communication and intelligence purposes.
$\triangleright$ *Surveillance Satellite.*

P

Pacifism. ◊ *War*.

Pacific Security Treaty. ◊ *ANZUS*.

PAR. Perimeter Acquisition Radar.
A radar, computer and communications
unit intended to detect and acquire incoming ballistic warheads as part of the
US ◊ *Safeguard* programme of ballistic-
missile defence (◊ *ABM*). The mission
of the PAR begins with the search for
incoming warheads at the (radar) horizon,
which for minimum-energy ICBMs is
over 1,000 miles away; once a target is
acquired the PAR is to follow its trajectory and to detect the dispersal of missile
parts, decoys and warheads while its
data-processing unit predicts the targets'
trajectories and intended target. The
PAR units are not intended to control
the defence interceptors but the target
data is communicated to the higher
frequency shorter range ◊ *MSR* unit,
which performs the weapon-control
function. The PAR is based on the ◊
phased-array technique, where a large
number of beams are generated electronically and 'steered' across the sky on the
basis of a computer-controlled time-
phasing system. The large antenna faces
(one or two in each PAR, depending on
its location) are therefore fixed, since no
mechanical movement is needed to scan
the skies (as opposed to classic radar,
where only one beam is generated and it
has to be slewed mechanically by physic-
ally pointing the antenna).

The PAR units can therefore track a
large number of targets simultaneously
and do so at long range thanks to the
relatively low UHF frequency on which
they operate (though this increases a
unit's vulnerability to nuclear ◊ *blackout*).
Each PAR is to be housed in a moderately
blast-resistant building 211 ft by 209 ft
and 130 ft high with large circular antenna
'windows' whose diameter is to be 112 ft.

Partial Test Ban Treaty. ◊ *Test Ban
Treaty*.

**Patton. US battle tanks M.47, M.48
and M.60.** The US tanks brought into
service since the wartime ◊ *Sherman M.4*
include the M.46, M.47 and M.48
'medium' tanks and the M.103 'heavy'.
Though the US armed forces have made
a very limited use of 'heavy' tanks, it was
only with the M.60 that the main battle
tank (MBT) concept was adopted. The
M.46 and M.60 series are the product of
continuous development, originally based
on the M.26 heavy tank chassis. This had
the conventional tank layout: driving
compartment at the front, a fighting
section with rotating turret in the middle,
and an engine compartment at the rear.
Unlike current MBTs, the M.26 also had
a machine-gunner in the bow, which,
with the driver and three-man turret
crew, resulted in the five-man layout of
most wartime tanks. The M.26 gun was a
90-mm giving fire power considered
adequate until the introduction of the
latest MBT generation of tanks. In 1948,
the M.46 was produced as an interim
model. It was an M.26 with better engines
and transmission, and other minor improvements. In 1950 another 'interim'
model was derived by adding a new and
characteristic elongated turret to the
M.46, a new engine and minor modifications. This, the first 'Patton', is the
M.47, which is still standard issue in
many NATO and Asian armies. Weighing 44 tons, the M.47 has adequate
armour and gun power, but the five-
man layout, the very limited range and

15

the great complexity of its fire-control system render it obsolete as compared to modern MBTs.

In 1953 the Patton M.48 was brought into production. It has a family resemblance to the M.47, but the hull is cast in one piece and the elimination of the machine-gunner reduced the crew to the now usual four. The turret, a great improvement on the M.47, is turtle-shaped and mounts an improved gun of the same 90-mm calibre.

The M.48 A1 of 1954 added a combination anti-aircraft machine-gun and commander's turret, on top of the main gun turret. The M.48 A2 of 1955 introduced a petrol injection engine which somewhat improved the inadequate range. The M.48 A2 and the M.48 A3, which feature

there is a potentially superior system in the alternative M.60 A1E1. This is a 152-mm combined howitzer-launcher which fires either (large) conventional rounds or the ◊ *Shillelagh* missile. The M.60's weight, at 47 tons, is almost the same as that of the M.48 A2 and armour protection is of the same order, so that the new weaponry and other improvements have been compensated by internal weight saving. The M.60 (gun version) has one major defect: the height, which renders it more visible – and therefore more vulnerable – than any other MBT. Also the full-field coincidence optical range finder is inefficient in weak light (but ◊ *Laser* range-finders will soon be available), and the machine-gun turret is perhaps unnecessarily large.

Specifications:

Gun		Rounds: Type and number	Weight	Length	Width	Height
M.48 A2	90-mm	APCR and APC (60)	47 tons	21 ft 6 in	11 ft 11 in	10 ft 2 in
M.60 A1	105-mm	APDS and HEAT (63)	47 tons	21 ft 9 in	11 ft 11 in	10 ft 6 in

	Engine, bhp	Speed, mph	Range, miles	Ground pressure	Ground clearance
M.48 A2	Petrol 865	30	160	12·2 lb sq in	1 ft 6 in
M.60 A1	Diesel 750	30	300	11·3 lb sq in	1 ft 6 in

minor modifications, are among the most common tanks in NATO armies. It is a reasonably well armoured and adequately armed tank, but it is of great mechanical complexity and has a limited range.

The M.60, which went into production in 1959, is the current MBT of the US army and also equips certain NATO armies. It resembles the M.48 A2, but the diesel engine and the 105-mm gun mounted in a smaller turret have produced a modern MBT whose performance far exceeds that of the M.48. The British 105-mm gun with 63 APDS and HEAT rounds is a very good weapon indeed, but

◊ *Main Battle Tank* for competitive types.
◊ *Armoured Fighting Vehicle* for general subject entry.
◊ *Sheridan M.551* for US ◊ *Light Tank*.

Penetration Aids. Generic name for devices and tactics intended to assist the penetration of aircraft and missiles against defence systems (◊ *Air Defence System* and *ABM*). Six main classes of 'pen aid' have been prospected:

(a) Use of chaff, jamming or nuclear ◊ *blackout* to hide the delivery vehicles or warheads from the defence radars, or as a second-best, to postpone target identification.

(b) The saturation of defence radars by the use of multiple warheads (including MIRV payloads) and target simulations such as decoys, balloons or false radar echoes (◊ ECM).

(c) The shielding of missile warheads against weapon effects (◊ Neutron Kill and X-ray Kill) and the use of manoeuvrable re-entry vehicles (or ◊ stand-off missiles in the case of strategic bombers).

(d) Active defence against interceptors by means of anti-ABM missiles or anti-SAM missiles in strategic bombers.

(f) Salvage fusing. This is intended to detonate the offence weapons and especially ICBM warheads when interception appears probable, or, more practically, upon sensing the first effects of interceptor weapon effects. (This reduces the effectiveness of the offence since its weapon is not yet on target, but it may achieve a partial kill.)

A further evasion measure specifically intended against ballistic-missile defences is the low trajectory ◊ FOBS flight pattern for strategic missiles. This reduces the height and therefore the radar visibility of incoming warheads, thus complicating the defence's mission.

All these penetration aids exact a price in terms of the delivery vehicle's payload, which could otherwise be used for weapons. Further, their effect against an ABM system is likely to be limited, since lightweight decoys, chaff, balloons and fragmented missile parts are filtered out by the atmosphere as the warhead and its pen aids re-enter. Nuclear blackout can be very effective against an ABM radar operating in relatively low frequencies, but its effect against a system of radars would probably be unreliable. There appear to be no pen aids which are both cheap (in terms of payloads and development costs) and effective.

◊ Safeguard, MIRV and Phased-array Radar.

90   **Pershing MGM-31A. US tactical missile.** The Pershing is a surface-to-surface missile deployed by the US army and the West German air force. It has solid-propellant rocket motors and two stages apart from the final, warhead, stage. Three powered tail fins, jet deflection and second-stage air vanes control elevation and direction; the range is controlled by varying the length of the coasting period between first-stage burnout and second-stage ignition. The complete system consists of an erector-launcher vehicle which also carries the complete missile, a programmer-test station and power generator, both carried in one vehicle, and a battery operations centre and a radio communications vehicle for long-range transmissions. All four vehicles are, in the improved Pershing 1A version, wheeled and articulated and have good mobility (the mobility and air-transportability being essential requirements in this system). See also the less efficient ◊ Sergeant.

Specifications (missile only): Length: 34 ft 6 in. Diameter: 3 ft 4 in. Launching weight: 10,000 lb. Range: 115 (minimum), 460 (maximum) miles. Nuclear warhead.

PETN. ◊ *TNT.*

Phantom II. F-4 and RF-4 series.   36
US Mach 2 fighter. The Phantom is generally considered to be the most effective multi-mission fighter now operational. It is a large two-seat twin-jet machine with complex swept wings and tail surfaces, which was initially developed as a carrier-based interceptor. Since the initial F-4B (operational) version came into service in 1962 nine variants have reached operational status. These share the very high speed, long range and large maximum payload of the F-4B detailed in the specifications below, but have different engines and avionics/fuel/weapon-load options. The F-4D is the current USAF fighter-bomber version. This has dual control, nose 'fire-control' radar, weapon-release computer and an ◊ *inertial navigation* system. This, like its predecessor the F-4C, is a 'close-support' tactical bomber. The RF-4C is a reconnaissance version for the USAF. It has a lengthened nose and carries side-looking radar for filming high-definition radar maps, an infra-red detector to locate enemy forces at night, and several cameras with in-flight processing. The RF-4B is a Marine Corps

reconnaissance version. The F-4E is an 'air-superiority' version for the USAF with a 20-mm multi-barrel gun and an improved fire-control system and more powerful engines. The F-4E can carry up to 8 tons of bombs as well as four ◊ *Sparrow* missiles. The F-AJ is the current naval version. It has a doppler ◊ *radar* and modified wing structure. It is intended primarily as an interceptor but has a good ground-attack capability. It has replaced in production an interim model, the F-4G. The F-4K is an export version for the British navy. It has been modified structurally to fit the smaller British carriers and has more powerful British engines and avionics. It is used with the ◊ *Sparrow* and ◊ *Martel* missiles. The F-4M is the British air force version similar to the K and retaining the folding wings and arrester gear of the naval version. The F-4 has also been sold to West Germany, Israel and Iran.

Specifications (F-4B): Length overall: 58 ft 3 in. Height: 16 ft 3 in. Wing span: 38 ft 5 in. Gross wing area: 530 sq ft. Weight empty: 46,000 lb. Maximum take-off weight: 54,600 lb. Maximum level speed, with external stores: Mach 2+. Combat ceiling: 71,000 ft. Take-off run: 5,000 ft. Combat radius: as interceptor, 900 miles; as ground-attack, 1,000 miles. Ferry range: 2,300 miles. ◊ *Fighter.*

Phased-array Radar. A radar technique used in the US ◊ *Safeguard* ◊ *ABM* system and some sophisticated air defence systems. In conventional radar, the electromagnetic radiation used to detect, locate and track objects in space is fed into a single-beam antenna which is moved mechanically in order to project the beam in a given direction. In phased-array radar millions of beams are produced each second, each in a direction determined by time phasing on the basis of computer control. The large antenna is fixed and contains a very large number of radiating elements; their emission is phased so that a wave of radiation is sent in a particular direction. As in conventional radar a relatively low frequency (i.e. low UHF) is preferable for the search function over long ranges, while a higher frequency is preferable for the tracking function at

somewhat shorter ranges. These differences are embodied in the two radars of the US Safeguard system ◊ *PAR* and ◊ *MSR*. The US *Long Beach* guided-missile cruiser (◊ *Cruisers, US*) is equipped with a phased-array radar, as are the larger carriers of the US navy. ◊ *Radar* for the general principles and other applications.

Phoenix XAIM-54A. US air-to-air missile. A 1,000-lb air-to-air missile being developed for the US navy, Phoenix represents the next generation of air-to-air missiles with built-in self-testing of components, module construction and improved all-round performance. The missile is fired by command from an onboard radar-computer tracking and missile-control unit; final interception is by active ◊ *radar* homing. The Phoenix was intended for the F-111B; when this was abandoned it was switched to the Grumman F-14A, now under development. Its range has been estimated at 60 nautical miles.

Phosgene (US code - CG). A colourless gas, known as CG, which caused 80 per cent of First World War gas deaths. The gas acts only through inhalation, the effects being: coughing, retching, frothing at the mouth, asphyxia, and death. It has 'new-mown hay' smell and gives a metallic taste to tobacco. It was extensively stockpiled by all parties in the Second World War. ◊ *Chemical Warfare.*

Platoon. An army formation subordinate to the battalion and compromising a number of squads or sections. Normally the smallest unit with an organizational identity, it varies in size from the 12 men of a Soviet army tank platoon to the 40 plus men of a US army infantry platoon.

Polaris UGM-27 series. US nuclear delivery system. The US navy's contribution to the US strategic offensive forces consists of 41 nuclear-powered submarines each carrying 16 ballistic missiles of the Polaris series. The initial Polaris A1 (UGM-27A), which became operational in 1960, has been phased out; the A2 (UGM-27B) equips about 10 submarines, while the remainder have the

95

Polaris Missiles – Specifications:

	Length	Diameter	Weight	Range (miles)	Speed	Warhead (megatons)
UGM-27B (A2)	31 ft	4 ft 6 in	30,000 lb	1,700	Mach 10	0·7
UGN-27C (A3)	31 ft	4 ft 6 in	30,000 lb	2,875	Mach 10+	0·7

improved A3 (UGM-27C). The new ◊ *Poseidon* missile with MIRV multiple warheads will eventually equip 31 of the 41 ships.

The system consists of a nuclear-powered submarine (◊ *Polaris Submarines*) with an ◊ *inertial navigation* (SINS) unit which determines the exact position of the vessel at any moment in time; a control and communication system which is supposed to provide contact between decision-makers and the vessel; and the missile itself, which is also inertially guided. Navigation errors plus guidance errors are thought to limit the Polaris (though not necessarily Poseidon) to a counter-city role, but continuous improvements in both may have invalidated this assumption.

Polaris submarines are based in Holy Loch (Scotland), Rota (Spain), as well as in South Carolina and Guam. The successive improvements in missile range have reduced basing constraints. Reliability is of a high order, with 14 of the 16 missiles being available for firing 100 per cent of the time. But command and control problems have limited the strategic flexibility of this force.

The Polaris is launched under water by a gas-steam injection device; as soon as the missile leaves the water, the first stage ignites and the inertial system stabilizes it on the programmed trajectory.

Three years after the first firing of the A2 in 1960, the A3 was ready for mass production. As well as the substantial increase in accuracy, its range was also improved without resulting in an increased weight or a payload loss penalty.

◊ *US, Strategic Offensive Forces* for other delivery systems.

the ◊ *US, Strategic offensive forces* belong to three classes, the *George Washington* (5), the *Ethan Allen* (5), and the *Lafayette* (31). These share the same basic nuclear propulsion system – pressurized-water S5W reactors coupled to geared single-shaft turbines rated at 15,000 shp. The design arrangement for the ◊ *Polaris* (and now ◊ *Poseidon*) missiles consists of 16 tubes set vertically into the hull aft of the fin-shaped 'sail' which has wing-like diving planes. The whale-like appearance of these submarines is enhanced by the prominent vertical tail fin. Another common feature is the SINS (Ships ◊ *Inertial Navigation System*).

The *George Washington* class were converted to (before completion), rather than designed for, the missile role. They have 'whale' (or 'Albacore') hulls and originally carried the A1 Polaris missile, now replaced by the A3. They also have six 21-in torpedo tubes for the ◊ *Subroc* anti-submarine missile. The first underwater firing of the ballistic missile was in July 1960, from the name-ship *George Washington*. Initially the underwater launch of the missile was by compressed air, but a more advanced steam-injection method has been developed This class also pioneered the SINS inertial navigation unit. The accuracy of the Polaris missile is a function of its own navigation error plus the submarine's navigation error. Accurate position finding is therefore of crucial importance: various methods are used, but the SINS unit is the main device. These submarines also require adequate communication links with their command and control shore bases and low frequency and VLF radio.

The *Ethan Allen* class, built in succession to the *George Washington* class, have the new 'tear-drop' hull design and are longer and generally improved. Like the

6 **'Polaris' Submarines.** The 41 ballistic-missile submarines which are part of

former series, they are being refitted with improved Polaris A3 missiles.

The latest, *Lafayette*, class vessels, of which 31 have been built, are the largest undersea craft ever built. Their A2 and A3 Polaris missiles are to be replaced by the new Poseidon series, with MIRV multiple warheads and longer range. A new steam missile ejection system was introduced with the later *Lafayette* class. A small rocket motor is ignited, and its very hot gases produce steam instantaneously, which ejects the ballistic missiles from their cylinders.

All these vessels are deployed on three-month cruises with alternative crews for each cruising period. ◊ *Submarine, US*. For British 'Polaris' submarines, ◊ *Resolution*.

improve *A*'s bargaining position at *A-B* negotiations if *B*'s demands are convincingly presented as extreme, thus stimulating internal opposition to them.

Poseidon.◊ US submarine-launched 96 ballistic missile. The Poseidon is intended to replace the ◊ *Polaris* missiles in thirty-one of the forty-one ◊ *Polaris Submarines* now deployed by the US navy. The Poseidon, which is 34 feet long and has a diameter of 6 feet, has been credited with twice the payload and a longer range than the Polaris A3. (The latter has a 0·7-megaton warhead and a range of 2,850 miles.) The Poseidon has already been tested with a MIRV payload; this has been reported to consist of

Polaris Submarines – Specifications:

| | Main class features | | |
	George Washington	Ethan Allen	Lafayette
Displacement (short tons)			
surface	6,010	6,900	7,250
submerged	6,700	7,900	8,250
Length (feet)	382	410	425
Width (feet)	33	33	33
Draught (feet)	29	30·7	31·5
Missile type (in final arrangement)	Polaris A3	Polaris A3	Poseidon
Torpedo tubes	6	4	4
Speed (knots)			
surface	20	20	20
submerged	28	28	28
Crew	112	112	140
First of class completed	1959	1961	1963

Common features: PWC reactor, designated S5W, driving geared turbines with a single shaft to the propeller. 16 vertical tubes for ballistic missiles. Torpedo tubes are intended for the Subroc anti-submarine missile.

Political Warfare. The manipulation of political forces within the enemy camp. It includes the use of ◊ *subversion* and other covert operations but is mainly based on ◊ *psychological warfare*. It may be associated with a parallel armed conflict or a ◊ *confrontation*, but some mild forms of political warfare (minus the covert operations) are a normal adjunct to international relations, i.e. a speaking tour by *A*'s ambassador in *B*-land may

more than ten separate warheads each with a yield of some tens of kilotons. ◊ *US, Strategic Offensive Forces*.

P.O.W. Prisoner of War. The phrase describes individual status with reference to pseudo-legal concepts (◊ *Geneva Convention*) or to presumed 'usage, custom or convention'. P.O.W. status is usually claimed for armed force personnel in captivity; it is also claimed by parties waging ◊ *Revolutionary War*. In the latter case it is usually not accorded until the guerilla forces have established safe areas where captured live soldiers can be held for bargaining purposes.

PPI. Plan Position Indicator. A form of radar display which gives realistic horizontal-plane radar images of surrounding radar-reflective objects and ground features. Frequently used in navigation units. ◊ *Radar*.

Pre-emptive Attack. An attack launched in the belief that an enemy attack has already entered the executive phase, i.e. that the decision has already been made. Unless the attack actually reduces or eliminates the effect of the imminent attack, it cannot be called pre-emptive. An example of a successful pre-emptive attack is the Israeli air strike of 5 June 1967.

Program 437. A limited capability anti-satellite system, based on obsolete ballistic missiles fitted out as non-ballistic S A Ms.

Propaganda. A term applied to enemy statements. War propaganda is a phrase used to describe such statements in the context of armed conflict. If the conflict is presumed to be impending, 'war of nerves' is the appropriate term for enemy statements.

'Atrocity propaganda' is a suitable term to describe enemy statements about one's own security measures and their much regretted effects. For the use of manipulative information techniques in the context of conflict, ◊ *Psychological Warfare*.

Proximity Fuse. Also known as 'Radio Proximity Fuse' or 'V T' fuse. An electronic (radar) or infra-red device small enough to fit into guided weapons or shells, which detonates the warhead within a pre-set distance of a reflecting surface, which is presumed to be the target. The pre-set distance is within the 'kill-radius' of the warhead. ◊ *Radar* for the technique employed.

P S I. Pounds per Square Inch. A measure of overpressure, used to measure blast effects from detonations, including nuclear ones. ◊ *Nuclear Warhead*.

Psychological Warfare. All actions designed to influence enemy personnel (including the political leadership and non-combatants) in order to serve the manipulators' purposes. The tools of psychological warfare are: the presentation and distortion of images (◊ *propaganda*); the coordination of military and/or diplomatic action in order to create particular images; the modulation of existing pressures within the enemy camp in order to affect morale, discipline or the efficiency of decision-making. 'Brain-washing' refers to the use of ordinary manipulative techniques applied to the minds of captive subjects, i.e. prisoners or a television audience. Psychological warfare is always important in conflict situations; since ordinary military and/or diplomatic activities have a psychological warfare 'angle' these can be modulated to maximize benefits in that area. Such modulation is often more effective than propaganda, whether 'hard' or 'soft', direct (the source is acknowledged) or 'black' (the real source is disguised).

P T-76. Soviet light tank. In service since 1955, the P T-76 is the standard Soviet heavy reconnaissance vehicle, and it has been supplied to a large number of satellite and foreign countries. The basic P T-76 chassis is also used in the ◊ *B T R-50 (P)* armoured personnel carrier and in the ◊ *A S U-85* airborne assault vehicle. The P T-76 is a true amphibian with water-jet propulsion, light armour and a low turret mounting a 76-mm gun. At 14 tons it is too heavy for efficient airborne use, and in any case the gun is inadequate for the assault role. But the low ground pressure, powerful engine and boat-shaped hull make it very mobile in varied terrains and conditions and in water; it can swim fast and without needing preparation – unlike many Western 'amphibians'. The P T-76 is 22 ft long, 10 ft 1 in wide and 7 ft 2 in high. There is also a P T-85 version fitted with an 85-mm gun.

◊ *Light Tank* for competitive types.
◊ *Armoured Fighting Vehicle* for general subject entry.

Pulse Radar. ◊ *Radar.*

PVO Strany. The Soviet air defence command. The PVO (*protivo vozdush-naya oborana strany* = anti-air defence of the country), a separate service of the Soviet armed forces founded in 1954, is organized into a large anti-aircraft command (PSO) and a small ◊ *ABM* 'anti-rocket command' (PRO). Apart from a hundred or so ◊ *Galosh* sites around Moscow, the PRO's systems appear to be in the development stage (as opposed to the US ◊ *Safeguard* now being deployed).

There are 14 command centres for the PSO in the Soviet Union as well as six command centres in Eastern Europe whose operational status is unknown, except for the fully operational East German PSO. Apart from the ◊ *Tallinn Line* (or 'Blue Line') system, which is thought to have a limited ABM support capability, there are a large number of radars in PVO (including airborne early-warning systems) coordinated in a target-detection and weapon-control system. The weapons deployed include a large number of anti-aircraft guns: radar-controlled twin 23-mm, single 57-mm, 85-mm, 100-mm and 130-mm guns as well as some self-propelled 4-barrelled 23-mm and twin-57-mm guns. Many of these are obsolete by any standards, and Western air defence systems (◊ *NORAD*) deploy no anti-aircraft guns at all. Interceptor fighters are mainly ◊ *Fishbeds,* ◊ *Firebars* and ◊ *Fishpots* as well as the more advanced ◊ *Fiddlers,* ◊ *Flagons* and ◊ *Foxbats;* only the latter have large avionics/fuel payloads and speeds in excess of Mach 2. There are some 3,300 aircraft in all, but the total number includes obsolete (in this role) Mig-19 and the even older Mig-17 aircraft.

The number of surface-to-air missiles is not known, but these include some 8,000 of the older ◊ *Guideline* as well as the more modern ◊ *Griffon,* ◊ *Ganef* and Gainful. The ◊ *Goa* is in limited deployment and the ABM Galosh appears to be a semi-experimental deployment.

◊ *Air Defence System* and *US, Strategic Offensive Forces.*

Q

Quail. US air-launched diversionary missile. Quail is part of the electronic countermeasures (▷ *ECM*) equipment of the B-52 ▷ *Stratofortress* strategic bomber. It is a miniature turbojet-powered aircraft which flies at the same speed as the B-52 and is equipped with ECM gear which simulates the B-52's radar 'signature'. Quail is 12 ft 10 in long and has a 5 ft 5 in wing span (once it has been dropped from the bomber's weapon bay and its wings have unfolded). It weighs 1,200 lb. Unlike the active ECM devices carried by the B-52, Quail is intended to degrade hostile radar by dilution, rather than by deception. Its range – which determines how long the dilution can be maintained – is not known.

R

Radar and its application. Radar (*RA*dio *D*etection *A*nd *R*anging) devices operate by transmitting a particular form of electromagnetic energy and processing that portion of it (the 'echo') which is radiated back. Radar techniques are used to detect and locate targets, to control weapons, for navigation and – in a 'secondary' form – for aircraft identification. In these and other applications the radar is often the most important single element of sophisticated weapons systems and takes up a substantial part of military equipment budgets. Radar techniques are based on the exploitation of three main properties of electromagnetic 'waves':

1. Electromagnectic waves travel at the velocity of light so that the time interval between the transmission of a signal and its returning echo divided by two and multiplied by the speed of light gives the range of the reflecting object.

2. Electromagnetic waves can be transmitted in beams of varied shape through appropriate antennas, including pencil-point beams which can accurately locate and follow moving objects.

3. Electromagnetic waves undergo a shift in frequency when they are reflected by an object moving relative to the radar equipment. The shift in frequency (the doppler effect) is proportional to the relative speed of movement of the reflecting object, so that the speed can be calculated.

Radar devices in their most elementary form consist of an oscillator which produces electromagnetic radiation, a waveguide that carries it to the antenna (which beams the 'waves' and collects returning echoes), a receiver which selects and amplifies the echo, and a display or signal unit which makes the data available in the required form. (◊ *Phased-array Radar* for

another basic technique.) Radars have been operated at frequencies ranging from 25 to 70,000 megacycles, but most ordinary sets operate in the microwave region, in the UHF, SHF and EHF bands, that is, from 300 to 35,000 megacycles.

The choice of frequency is determined by the intended use of the radar system, since the size of the antenna required for a given narrowness of beam increases as the frequency falls; small antennas (such as those mounted in aircraft) can achieve directional precision only with high frequencies (i.e. 10,000 mc). But the higher the frequency, the greater is the degree of atmospheric absorption, which reduces the effective range. Thus long-range search radars, such as those used in early-warning systems (◊ *BMEWS*, ◊ *DEW* and the ◊ *PAR*), have large antennas and lower frequencies (i.e. 600 mc and less), and these achieve both range and directionality.

Radars in Air Defence
Search radars have large antennas and operate in low frequencies to scan the airspace perimeter and detect targets. Since the intensity of the returning echo is greatest when the receiving antenna is pointing directly at the reflecting object, the angle of elevation of the antenna and its bearing when the echo is strongest give the position of the detected object. But in order to determine the range of the target the transmitted signal must be 'coded' or marked in some way so that the time interval for the round-trip can be determined precisely. There are two ways of doing this:

(a) by transmitting the signal in short bursts or 'pulses' and leaving a silent gap within which the returning echo is isolated. This is known as pulsed radar.

(b) by changing the frequency of the transmission at a known rate so that the echo is paired to the right signal by comparing the frequencies. This is known as frequency modulation and is used on the second type of radar, Continuous Wave or CW radar.

Once the target's angle of elevation, bearing and range are known, the next step is to follow it, predict its path, and direct weapons against it. To do this, tracking, or 'narrow beam', radars are needed, which often operate at higher frequencies since the higher the frequency the 'straighter' the beam. These tracking radars produce a 'pencil' beam which can rapidly scan the area of the detected target and, having 'acquired' it, 'lock' on to it. Apart from the elevation, bearing and range, the tracking radar can also produce target speed data by means of the doppler effect. Information about the target is then passed on to a third item of electronic equipment, which unlike the search and tracking units is not a radar but a computer. This records and predicts the path of the target and produces instructions which are fed into the shooting part of the weapon system. If anti-aircraft guns are used, the tracking radar-programmer system actuates servo-controls which point and fire the guns; if surface-to-air missiles are used, both target and missile are tracked and the computer generates flight instructions which are sent to the missile in order to keep it on an interception course. In so-called 'all-weather' fighter systems, the links are more complex. While the target is being tracked, the fighter is flown towards it on the basis of the data produced by the programmer, the aircraft's nose-cone AI (airborne-interception) radar then 'locks' on to the target and keeps the fighter in line with it. When the target comes within the range of the aircraft's weapons, a signal is given to the pilot or else routed directly to the missiles or guns,

Many air-to-air and surface-to-air missiles use semi-active homing guidance, yet another form of radar weapon control. The missile is equipped with a radar receiver set which can collect, though not transmit, electromagnetic radiation. The target is 'illuminated' by a ground or airborne radar, and the missile homes on to the echoes, which bounce off the target and are picked up by the missile's receiver. The echoes are then put through a programmer which generates signals to the autopilot in order to keep the missile on an interception course. In active homing types, as the name suggests, there is a transmitter as well as a receiver.

Radars in Navigation

Returning echoes are usually displayed on a cathode-ray tube. This can be designed to show range, bearing and elevation as spots or 'blips'. In order to get a picture of what is happening all round a PPI (Plan Position Indicator) is much better and is generally used for naval 'blind' navigation and air traffic control. In the PPI display, the sweep is a radial which starts at the centre of the screen, though it can be offset if necessary. The distance between the centre of the screen and its rim represents the range of the set, and the sweep rotates in step with the rotation of the antenna. PPI screens are usually coated with a material which has an 'afterglow' effect so that the echoes remain visible until the sweep has gone through a complete circle and shows them once again. Thus for a ship navigating in crowded waters or seeking a target out of visual range or in poor visibility, the display PPI gives an accurate synthetic picture of the coastline. But the picture is only two-dimensional, so that an aircraft flying at 30,000 feet and a ship on the surface will both appear as blips with no indication of their very different elevation. A further display showing elevation is therefore required, or, if the radar is intended purely for sea navigation, the antenna must be shaped so as to project a beam which sweeps only at very low altitudes.

Radars in Airborne Navigation

The FM CW Radio Altimeter, a device not usually thought of as a radar, produces altitude data by bouncing a signal off the ground below and computing the 'range' of the ground by multiplying the time interval for the one-way trip of the echo by the speed of light. The doppler effect is used in a similar manner to give the ground speed of an aircraft; CW signals are bounced off the ground and the shift in frequency is processed by a programmer to give the relative speed, or 'ground speed'. SLAR, or side-looking airborne radar, used for ground-mapping and other purposes, consists of a high-frequency set with a small antenna. SLARs can produce good images of the ground below in any weather, and these can be filmed to provide a linear picture for reconnaissance purposes. The SLAR technique is also used in low-altitude, high-speed navigation in order to identify ground features, since the forward image becomes a 'blur' at high speeds. The TFR, terrain-following radar, is a high-definition radar pointed forward, which feeds data on the ground ahead to a programmer which generates instructions to the aircraft servo-controls so as to avoid obstacles by appropriate adjustments in the flight-path. Since air defence radars are ineffectual at very low altitudes, there is a premium on flying as low as possible, and TFR-type units are being installed on new aircraft so that these can contour-fly to target even over terrain which is quite irregular.

Radar-bombing techniques have been in use since the Second World War. In their simplest form, the bomber is tracked by friendly radars, and when it is over the target (a fixed point of known location) the free-fall bombs are released, thus achieving accuracy even at night and in bad weather; but current techniques are a good deal more complex and often rely on ◊ *inertial guidance* rather than radar for a 'fix'.

There are many things that radar cannot do; it cannot give very high definition images, it does not discriminate between colours, and, more important, it cannot pick out a target on the ground as it can in the air or over water (so that homing techniques cannot be used against ground targets). Above all, most radars are limited in range to the horizon, since the beams are straight. Recently, the need to detect missile firings at long range has stimulated the development of radars that can 'look over the horizon'. The US 'OTH' operates in the HF range of frequency (i.e. 3 to 30 mc, or much lower frequencies than ordinary radar) to take advantage of the ionospheric reflection, or 'scatter', of transmitted signals. These OTH radars use the ionosphere as a mirror to reflect back to Earth (at an

angle) the transmitted signal, with a reverse effect for the echo. As the process is repeated, the signal is reflected from the ionosphere to Earth, and vice versa until both it and the echo travel over long ranges.

The poor reflectivity of most ground features and their interference ('clutter') have limited the application of radar in the detection of ground targets, but at least two types of battlefield radars have been developed: mortar-locating sets, which track a mortar shell so that its ballistic path can be used by a programmer to compute the location of the mortar (or howitzer) which fired it, and battlefield surveillance sets. For the reasons mentioned, these cannot pick out fixed targets with any precision but (a) they can identify movements at long ranges or in 'blind' visibility (or rather invisibility) conditions by using the doppler effect, and (b) they can give the range of a target which has been identified, and towards which a narrow beam is pointed. (◊ *Laser* for an alternative way of doing this.) All these applications rely on primary radar, i.e. on the use of echoes bounced off a reflecting object; but IFF (Identification Friend or Foe) uses 'secondary' radar. In this, an 'interrogator transmitter' on the ground sends a signal which is picked up in the aircraft, and this automatically switches on a transmitter (transponder) which sends an answering signal in a code which is different for every aircraft, so that the ground controller can identify the aircraft (or rather the blip on the screen). This IFF system is used, among other things, for air defence systems, when the ground controllers have to be able to discriminate between friendly interceptors, 'Angels', and enemy bombers, 'Bandits'. Proximity fuses (VT), which produced spectacular improvements in the effectiveness of anti-aircraft (and other) artillery in the last war, are also radar devices. The VT fuse has a simple amplifier connected with the equally simple oscillating detector and antenna; when the output of the amplifier is of sufficient magnitude (i.e. when the target comes within the 'kill-radius' of the shell) the firing circuit (usually a thyratron) triggers off the detonation.

Radar techniques are of crucial importance in determining the effectiveness of many weapon systems. Since they mostly rely on transmissions which can be interfered with or confused, electronic countermeasures and their natural concomitant, electronic counter-countermeasures, have become important (▷ ECM). ▷ also ABM (Anti-Ballistic Missile) defence for certain radar applications.

Radio-command Guidance. A form of missile guidance in which steering instructions from an operator/launcher are relayed to the missile's servo-controls in the form of radio or wire-carried impulses. ▷ Missile for general entry on guidance and homing systems.

Radome. The protective cover on a ▷ radar or antenna. In aircraft this is sometimes seen as a black nose-cone.

Rapacki Plan. A Soviet plan for the denuclearization of central Europe marketed by Poland and named after the then (1957) Polish foreign minister, Adam Rapacki. This proposal sought to prohibit the storage, manufacture or deployment of all nuclear weapons on the territories of Poland, Czechoslovakia and both Germanies. Since at that time NATO defence planning was based on the early use of nuclear weapons in the event of a Soviet attack on Western Europe, the plan was rejected. Given the size of the two sides' conventional forces in the theatre, acceptance of the plan would have altered radically the balance of power in central Europe in favour of the Soviet Union.

74 Rapier. A limited British battlefield anti-aircraft missile system intended to provide low-altitude defence in both ground and seaborne roles. The system consists of a detection radar which spots incoming aircraft, an IFF device which discriminates between friendly and hostile aircraft, an optical tracker, and a mounting for four missiles which are automatically aligned on the target. After an alert, the operator sees the target and follows it by manual tracking; when the target comes within range the computer alerts the operator and he can then fire the missile. Deviations between the missile path – which carries rear flares – and the target

is measured by the computer, which then relays commands to the missile in order to keep it on a collision course. The system, including the generator, is fitted into three small units which can be towed by a jeep-sized vehicle, the whole being air-droppable; is ineffective against supersonic aircraft, it can be operated by a single man, and is rugged and easy to maintain. The missile itself needs no servicing, and the other components are easily replaceable. Rapier can be mounted in armoured personnel carriers.

▷ Missile (surface-to-air) for general subject entry and competitive types.

RCT. Regimental Combat Team. Used with reference to a composite force of more than battalion size. It may or may not include armour but is intended to be self-sufficient for some defined tasks.

RDX. ▷ TNT.

Recce. ▷ Reconnaissance.

Recoilless Weapons. Weapons where the recoil effect is reduced and counteracted by vents in the breach which deflect up to four fifths of the expansion gases to the rear. Recoilless weapons can therefore fire artillery-sized ammunition without generating artillery-sized recoil, or any recoil at all. Thus the heavy mountings and recoil-absorbing devices needed with artillery weapons are not required; with internal stress and muzzle energies low, the barrel can be much thinner and lighter than that of a full-recoil gun firing the same weight of shell. Naturally, the effective range is also much smaller than with full-recoil weapons: for calibres of about 100-mm range a recoilless rifle (RR) will have a maximum range of 1,000–2,000 yds, as opposed to a full-recoil gun range of about 25,000 yds. But the ratio of weapon weights is also of the order of 1 to 20 even counting the wheeled mounts used with some RRs, and cost ratios are also in this range. Recoilless weapons are used whenever the range is secondary to the size of shell delivered and where high velocities are not required, or quite simply for economy. RRs are in fact primarily intended for anti-tank use, where their 'hollow charge' (◊ *HEAT*) shells penetrate armour without needing the kinetic energy required by conventional solid shot fired from full-recoil artillery.

The US 106-mm RR and two earlier US models which can be fired from the shoulder, the 75-mm and 57-mm, are widely used in the West; the Soviet B-11 107-mm RR and the older B-10 82-mm RR are equally popular in the 'East'. Another widely used type is the one-man 84-mm 'Carl Gustav', manufactured in Sweden and used by, among others, the British army. The largest recoilless weapon, the British Wombat 120-mm, is not widely used.

These are all properly called recoilless rifles since their barrels are rifled as in the case of normal artillery. But some light recoilless weapons have smooth barrels and project a HEAT charge attached to a reduced calibre 'stick' which fits into the barrel. The West German 'Panzerfaust' and the Soviet RPG both consist of a small calibre tube (about 40-mm) within which the charge and the 'stick' are set; when the weapon is fired, the 'stick' and the much larger HEAT charge affixed to its end are projected out. These are short-range one-man weapons fired from the shoulder.

Reconnaissance. The collection of visual, photographic and other sensory data (infra-red, electronic) about enemy forces in or on a given terrain, or about the terrain itself. Reconnaissance activities are performed overtly by military personnel or automatic equipment, though the same are also sometimes used in clandestine intelligence gathering. Within each branch of the armed forces of most countries there are reconnaissance units especially trained for the task and provided with specialized equipment. Sometimes standard combat units are used in reconnaissance missions, the only difference being that their primary role is to detect and appraise the enemy and the terrain, rather than to engage the former or occupy the latter. Whether the units are specialized or not, reconnaissance operations are usually conducted by selected personnel with light and highly mobile equipment; in many armies, reconnaissance units are also elite units.

Reconnaissance operations provide one of the main sources of raw data for ◊ *tactical intelligence,* and are especially important in the conduct of mobile operations, where conditions are fluid and data are a prime resource.

For special-purpose or adapted equipment primarily intended for reconnaissance missions, ◊ *Fighter* and *Armoured Car.* ◊ *Surveillance Satellite.*

Redeye (FIM-43A). An American anti-aircraft weapon of the 'bazooka' type which is fired from the shoulder and is suitable for one-man operation. It consists of a simple missile with infra-red guidance and rocket motor which is housed in a tube for both storage and launch. This is the first weapon which gives some sort of protection against all but the fastest low-flying aircraft and which can be carried about as a personal weapon by a single infantryman. The tube is about 4 ft long and about 3 in diameter, and it weighs just over 28 lb. When an aircraft is sighted, the operator follows it in an optical sight and prepares the missile for launching by pressing a button. When the missile is ready, a buzzer alerts the operator, who can then fire the missile, which is ejected by a small explosive charge, and at a safe distance away the rocket ignites. The infra-red homing device then guides the missile to the target. Redeye is very cheap at $5,000 US and has already been exported. Since it is intended as a battlefield weapon against fighter-bombers and helicopters, its limited ceiling is no disadvantage; but supersonic aircraft taking evasive action will not be vulnerable to it while infra-red systems of this type are low on reliability. ◊ *Missile* (surface-to-air) for general entry and competitive types. 75

Redoutable. French nuclear-powered ballistic-missile submarine. The *Redoutable,* launched in 1967 and due for completion in 1970, is the first of a class of French 'third-generation' nuclear-delivery vehicles planned for the seventies. They are patterned on the US ◊ *Polaris submarine* and are intended to carry 16 French underwater-launched missiles. These, designated ◊ *MSBS,* are somewhat larger than Polaris missiles but will have rather inferior capabilities even if the present development problems can be overcome. The French, unlike the British, received very little American assistance for the development of this system, and it shows it. The ship is larger, slower, and more detectable than those of the US *Lafayette* class (all US classes are discussed under ◊ *Polaris*), but has no known countervailing advantages.

Specifications and performance data (estimated): Displacement: 7,900 tons surface, 9,000 tons submerged. Length: 420 ft. Width: 34 ft 8 in. Draught: 32 ft 8 in. Weapons: 16 MSBS ballistic missiles with a 1,900-mile range; 4 torpedo tubes in the bow of 21 ft 7 in diameter (not NATO standard). Power plant: 1 pressurized water-cooled reactor; 2 turbo-alternators; 1 electric motor 15,000 hp, single shaft; auxiliary diesel. Speed: 20 knots surface, 25 knots submerged. Crew: 135.

◊ the equivalent British *Resolution* class.

For other French submarines, ◊ *Submarines, French*.

66   Red Top. British air-to-air missile. The Red Top is a 'second-generation' infra-red homing air-to-air missile developed from the ◊ *Firestreak* and used by the RAF and RN since 1963. The improved infra-red homing device allows for a wider range of interception configurations, notably a collision course one, while first-generation missiles of the Firestreak type were usable only on pursuit-course attacks. Red Top is 11 ft 5 in long, and has an 8 in diameter. The maximum speed is about Mach 3. The warhead weighs over 60 lb and is fitted with a proximity fuse (◊ *Radar*).

Re-entry Vehicle. ◊ *REV*.

Regiment. An army formation subordinate to the ◊ *division* usually parallel to the ◊ *brigade* and comprising a number of battalions. It is the standard sub-divisional unit in the Soviet, Warsaw Pact, and Chinese armies where, in the infantry, it consists of about 2,000–2,500 men. In the British army the regiment is the traditional focus of loyalties and social activities, though the operational link goes from brigade to battalions.

REM. Röntgen Equivalent Man. A measure of the effects of radiation, other than gamma or X-rays, derived by multiplying the energy yield of the radiation by a ratio which expresses that radiation's effect on man. ◊ *Nuclear Warhead*.

Resolution. Class of British nuclear-powered ballistic missile submarine. Four *Resolution* class submarines, each equipped with 16 ◊ *Polaris* A3 missiles with multiple (but not MIRV) warheads, form the second generation of ◊ *UK strategic offensive forces*. The missiles and some other parts are based on American components or designs, but the ship and the missile war-heads are locally produced. The main external difference between these and the ◊ *Polaris* type submarines is in the diving planes, set on the bow instead of on the sail, and the 'whale'-type forward hull, similar to that of the first *George Washington* class of US ◊ *FBMS* submarines.

The *Resolution* class vessels have a crew complement of 141 with two crews alternating at three-month intervals. The power plant, a presssurized water-cooled nuclear reactor, is rated at about 15,000 shp, and it feeds geared turbines which operate the single-shaft drive. Assuming a normal maintenance cycle, no more than two vessels will be operational at any one time when the whole force is completed; this very small figure raises questions about the kind of strategy to which such a force can be harnessed. ◊ *minimum deterrence* under *Deterrence, Trigger* and *Catalytic War*.

For specifications see opposite.

For other British submarines, including nuclear-powered hunter-killers, ◊ *Submarines, British*.

REV. Re-Entry Vehicle. The last stage of an extra-atmospheric vehicle whose outer surface is resistant to heat in order to survive descent through a planetary atmosphere. As used in strategic literature, REV includes the warhead (invariably a nuclear weapon), ◊ *penetration aids* and other objects, such as terminal guidance devices, not known to be deployed.

Revolutionary War. ◊ *Guerilla warfare* + ◊ *Subversion* = Revolutionary War.

Specifications for the whole *Resolution* class submarines:

	Laid down	*Completed*
Renown	June 1964	1968
Repulse	March 1965	1969
Resolution	February 1964	1968
Revenge	May 1965	1969

Displacement, submerged: over 7,500 long tons. Length: 425 ft. Width: 33 ft. Draught: 30 ft. 16 Polaris A3 missiles set in vertical tubes in the hull aft of the conning tower. 6 torpedo tubes, 21-in diameter. Crew: 141.

An armed conflict between organized forces originating within the same country and fighting over its total control, where the initiating party relies mainly on guerilla action and subversion rather than formal warfare. The initiating party operates by setting up a rival state structure which embodies a political ideology and which is intended to replace the older structure. The administration of the territory is taken over on an area-by-area basis, and this is not the outcome of local victories but rather the actual instrument of war. In revolutionary war, the winning party out-administers rather than out-fights the loser. The mechanism operates by underground 'administration' which collects taxes, conscripts and information – all of which can be extracted from the population even if the opponent is in apparent military control over the area in question. These resources are then fed to the guerilla arm and used by the latter to erode the military, or 'daylight', control. This in turn facilitates the spread of subversion and the cover administration. The key factor in the cycle is subversion and not the guerilla action, and concentration on fighting the latter is usually inefficient. In any case, conventional warfare against guerilla forces is uneconomic and requires a very high degree of manpower superiority in order to succeed. For antidotes to revolutionary war ◊ *Counter-insurgency Warfare*. For a classification of various forms of internal conflict ◊ *Internal War*.

Rifle. The main personal weapon of the infantry, which fires a small-diameter elongated bullet through a rifled barrel, propelled by smokeless powder held in a metal cartridge. Most rifles used in the two World Wars were bolt-action weapons with a 5- to 10-round magazine which fired a bullet of 0·3 in or more (7·6 mm +) weighing more than 150 grains, and produced muzzle energies of 2,500 foot pounds or more. As well as a powerful kick, these cartridges also meant long range (2,500 yards plus) and a large logistic load, especially when used in machine-guns. (The light and medium machine-guns use rifle ammunition.)

Modern rifles, of which the first were the US M1 Garand and the Russian 7·62-mm Tokarev (both introduced before 1939), are mostly semi-automatic and gas-operated (◊ *Medium Machine-gun* for modes of operation). The exception is the delayed blow-back CETME G3. Semi-automatic rifles as delivered in the infantry of modern armies can easily be modified for full-automatic operation, have 20-round (or larger) magazines, and fire bullets of about 7·62-mm calibre (such as the NATO standard) of 150 grains weight or less, with higher muzzle velocities (2,700-2,800 fps as against 2,600 +) and lower muzzle energies (2,150-2,250 foot pounds) than the Second World War types. ◊ *FN 7·62* and ◊ *CETME G3* are two important modern types.

From such weapons ◊ *Light Machine-gun* substitutes have been developed by the addition of simple bipods, an automatic or single-shot selector lever, and sometimes a heavier barrel and larger magazines or belt-feed conversion, in such weapons as the FN FALO and the Soviet RPK (for which ◊ *AK*). Though these rifles can deliver burst fire, they are still inefficient for close fighting since they use unnecessarily heavy and powerful cartridges. The ◊ *Sub-machine-gun* cartridge is, on the other hand, not sufficiently powerful for fire at any but the closest ranges. The Germans, whose wartime rifle (the Mauser M 98/37) fired the particularly heavy and powerful 7·92-mm ammunition, developed the first intermediate rifle ammunition, the 7·92 Kurtz (i.e. short-7·92), which weighed 120 grains – versus the 198 grains of the 7·92 Standard – and its muzzle energy was correspondingly smaller. With the 7·92 Kurtz they developed the first 'assault rifle' (MD 44), that is a rifle firing smaller cartridges with optional semi- or full-automatic fire.

After the war, both the British and the Russians took up the idea of an automatic rifle firing a reduced-power bullet, on the assumption that 500 yards was the maximum range that the infantry could actually use and that anything more powerful than 'intermediate' cartridge was wasted in a personal hand-held weapon. The British developed EM-2, a general multipurpose rifle + sub-machine-gun + light machine-gun with a short ·28-in 120-grain cartridge to go with it. This had to be abandoned when the NATO standard 7·62-mm cartridge was imposed for rifles and machine-guns. The Russians, however, developed their cartridge. This, a 7·62-mm (short) weighing 122 grains, is the standard Russian cartridge used on the SKS carbine, the AK rifle and the ◊ RPD and RPK light machine-guns. The US armed forces, whose experts had rejected the EM-2 intermediate for the NATO standard 7·62 mm, built up the M.14, M.14M and M.60 family of weapons to use it. But some years later they adopted a tiny 0·223-in cartridge of only 55 grains which produces high muzzle velocity (3,300 fps versus 2,800 for the NATO cartridge) and a low total muzzle energy (1,328·foot pounds versus 2,150). A variety of weapons were developed by the designers of the 0·223-in cartridge, but eventually the elegant ◊ M.16 rifle was adopted as well as the LMGs and MMGs to go with it. The M.16 permits accurate fire (hand-held) at up to 500 yards, while in a tripod-mounted MMG the cartridge is effective up to 800 yards. This is less than full rifle power but just as good for most uses, and the Vietnam war has led to the gradual shift of the Army and Marines to the new, small, light and ultra-fast bullet. With both the USSR and the USA converted to the intermediate cartridge, full rifle power will not long survive as a standard cartridge.

Rio Treaty. Inter-American Treaty of Reciprocal Assistance. A treaty signed in 1947 by almost all Latin American countries and the United States. Ecuador and Nicaragua did not sign the Treaty but do participate in the executive Organization of American States (OAS). Under the terms of the Treaty, all signatories are pledged to intervene on behalf of any member attacked by an outside power. The precise form of this intervention is not defined, and it would be subject to a decision of the conference of ministers of foreign affairs which can be convened within the framework of the treaty. This regional collective security treaty is supplemented by detailed provisions for the peaceful settlement of internal disputes embodied in the Charter of the OAS. Cuba was expelled from the latter (and from Treaty membership) in 1962.

Rocket. A vehicle or projectile driven by the reaction effect of a high-velocity stream of gas. ◊ *Missile* for a discussion of the various kinds of rocket motors. Most military rocket systems are guided (and therefore discussed under missiles of various sorts), but small unguided rockets are standard air-to-ground armament, and multi-barrel rocket-launchers are used as a form of ground ◊ *artillery*. (◊ *Artillery, Soviet*.) Unguided rockets are also used as naval weapons, to provide barrage fire for landing operations.

RPD. Soviet light machine-gun. The RPD (*Ruchnoi Pulemet Degtyarev*) is a ◊ *Light Machine-gun*, bipod-mounted, belt-fed and gas-operated, which is used in the Soviet army as a standard squad weapon (i.e. 1 RPD to 8–10 rifles) and has been exported to a large number of countries. There are also Chinese and North Korean copies, which have also been widely exported. Like other post-war Soviet infantry small arms, the RPD uses the 7·62-mm 'intermediate' cartridge (◊ *AK-47*) of less than full rifle power, and the RPD weighs only 15·6 lb, which makes it the lightest belt-fed gun in service anywhere in the world. This weight excludes the 100-round metallic link-belt which is carried in a drum suspended from the main assembly. The RPD is a weapon of unusual simplicity and great reliability, but in spite of this it is being replaced in Soviet services by the RPK, a bipod modification of the AK rifle. The RPD is 40·8 in long, and has simple post-sights in front and tangent sights in the rear and a cyclic rate of fire of 650–750 rpm.

RPK. Soviet light machine-gun. ▷ *AK-47.*

S

SAC. Strategic Air Command. A US air force command which comprises the land-based ◊ *Minuteman* and ◊ *Titan* missile forces and the B-52 ◊ *Stratofortress* bombers of the ◊ *US strategic offensive forces*. SAC became important within the US armed forces during the fifties, when deterrence was based on manned bombers only. Unless the ◊ *A.M.S.A.* project materializes, the residual manned-bomber force will disappear in the mid-seventies.

Safeguard. US anti-ballistic-missile programme. The Safeguard programme of ballistic-missile defence is intended to protect the ◊ *Minuteman* missiles, bomber bases and command centres of the American retaliatory forces. It is also intended to provide a 'thin' defence of US population centres against a small and/or unsophisticated nuclear strike. Although the programme is flexible, it is based on a system whose components are now frozen for production and deployment: the ◊ *PAR* search and acquisition radar, the ◊ *MSR* missile control and designation radar (both of which include data-processing systems), the ◊ *Spartan* long-range interceptors, and the ◊ *Sprint* terminal interceptors. A picket line of PARs is intended to detect ICBMs (coming over the Poles) and SLBMs (from the oceans) and to perform a limited degree of target designation; the MSR tracking unit would then take over and refine the discrimination between enemy warheads and other objects, including decoys; it would then launch and control the Spartan for extra-atmospheric interception and, if necessary, the MSR would follow up this long-range intercept with a terminal, atmospheric, intercept by means of the Sprint missile.

The PAR + MSR + Spartan + Sprint module can provide an effective defence against a sophisticated ICBM strike against the ◊ *hardened* retaliatory forces because a low-altitude terminal intercept is possible since fallout and other near-miss weapon effects can be ignored. In protecting cities, these effects cannot, of course, be ignored, and enemy warheads must be intercepted at long range, by means of the Spartan. Such intercepts are likely to be ineffectual against a large and sophisticated force because the defence radars would not benefit from atmospheric filtering of lightweight decoys, and, more generally, because the defence radars would have less time to track and discriminate enemy warheads. Depending on the number of Sprints and Spartans deployed, the system becomes more area or more 'hard point' oriented, but the same system can provide, as Safeguard does, both a thin area defence and a thick hard-point defence. In the event of a Russian attack on US deterrent forces, the Spartans would be used to disperse decoys, and other lightweight ◊ *penetration aids*; the Spartan's powerful detonations would also keep the offence 'honest', i.e. it would discourage an attempt to overload the system by a massive one-wave attack by presenting the counter-threat of multiple kills against the incoming warheads. This is possible because the Spartan's kill-radius is large (several miles) and because minimum-energy ICBM trajectories would have to come over a fairly narrow and predictable front in order to strike at US missile silos. The Sprints would then be launched against warheads by now isolated from their surrounding penetration aids; in the event of a large-scale attack against US Minuteman silos all the Sprints would be concentrated to defend selectively the MSRs, their silos and a part of the Minuteman force. This 'preferential defence' would increase very considerably the number of Russian warheads required to destroy most of the Minuteman force, so that a ◊ *counterforce* strike force would be expensive to deploy (and detectable well before it became operational), and its effects would not be reliable from the Russian point of view.

The only reliable counter to Safeguard would be an exhaustion attack, i.e one where the offence sends in a number of warheads which exceeds the number of interceptors deployed and then follows up with a second wave aimed at its real target: the US deterrent forces. This involves (i) very high deployment costs; (ii) the possibility of being overtaken by the deployment of more defence missiles; (iii) the risk of a US strike when a large number of offence warheads detonate on US territory but before its own Minuteman force is entirely destroyed.
◊ *ABM* and *ICBM, Soviet.*

SAGE. Semi-Automatic Ground Environment System. The air intruder tracking and defence weapon-control units of the ◊ *NORAD* ◊ *air defence system.* SAGE controls the interceptor aircraft and the ◊ *Bomarc B* missiles of this antibomber defence. The thirteen SAGE centres, giving overlapping coverage of the entire US territory, are supplemented by the emergency ◊ *BUIC* centres.

'Sagger'. Soviet anti-tank missile. A compact and more recent version of the ◊ *Snapper* wire-guided anti-tank missile with the same ◊ *HEAT* (shaped charge) 11·5 lb warhead. It is 2 ft 6 in long.

Saladin FV 601. British armoured car. In service since 1956, the Saladin is based on a 6 by 6 chassis also used in the ◊ *Saracen*, Stalwart and several other vehicles. This is a heavy ◊ *armoured car* with three large equidistant wheels on each side; apart from the driver it has a two-man crew in the square turret, which mounts a 76-mm gun. Like other gunarmoured cars, Saladin is intended for the 'cavalry' role, which, in many armies, is fulfilled by ◊ *light tanks.* The six driven

wheels, and the four-front-wheel steering, make it a very mobile vehicle. Although the gun fires both high-explosive and armour-piercing shells, it is inadequate against ◊ *main battle tanks.* The Saladin is to be replaced by a light tank known as the ◊ *Scorpion.* This mounts the same 76-mm gun.

Specifications: Drive: 6 × 6. Steering: front four wheels. Length: 16 ft 2 in. Width: 8 ft 5 in. Height: 7 ft 10 in. Weight: 11·4 tons. Engine: 170 bhp. Road speed: 45 mph. Road range: 250 miles. Armament: 76-mm QF gun, coaxial 0·3-in MG and turret-top MG. Crew: 3.

SAM. Surface-to-air missile. ◊ *Missile.*

'Samlet'. Soviet shipborne surface-to-surface missile. A naval version of ◊ *Kennel.*

SAMOS. Satellite and Missile Observation System. ◊ *Surveillance Satellite.*

'Sandal'. Soviet medium-range ballistic missile. A liquid-propellant ballistic missile with a range of about 1,200 miles. It has small tail fins and control vanes in the rocket efflux, like the wartime German V.2. Sandal is a more developed version of the earlier Shyster, which was probably a direct V.2 development. Guidance is by initial radiocommand, probably assisted by ◊ *inertial guidance* in the final flight path correction stage. Sandal has been operational since 1959; some were deployed in Cuba in 1962. It is 68 ft long, and has a body diameter of 5 ft 3 in, and its weight is estimated at 60,000 lb. The warhead is estimated at one megaton.

Saracen FV 603. British armoured personnel carrier. This 6 by 6 wheeled and lightly armoured vehicle has been in service since 1952. It shares a common chassis with the Saladin heavy armoured car and the Stalwart supply carrier. Though it is officially called an armoured personnel carrier, it has in fact largely been used in internal security roles. This is just as well, since it lacks sufficient protection for battlefield roles – other than light reconnaissance – for which it is much too heavy (10 tons laden) and elaborate.

It carries a total of twelve men, including the crew, and it mounts a small hand-operated machine-gun turret. The engine is in the front, and there are rear doors and firing ports in the superstructure. ◊ *Saladin* for the almost identical automotive details. The Saracen is 15 ft 11 in long and weighs about 10 tons.

◊ *Armoured Personnel Carrier* for competitive types, including the British tracked carrier ◊ *Trojan FV 432*.

◊ *Armoured Fighting Vehicle* for general entry.

Sarin. ◊ *Nerve Gases.*

'Sark'. Soviet submarine-launched missile. The Sark, which first appeared in 1959, is a storable-liquid-fuelled ballistic missile which is about 48 ft long and has a base diameter of 5 ft 9 in. The Sark carries a one-megaton warhead, and the range has been estimated at about 400 miles. The Sark cannot be launched under water and is generally inferior to its US counterpart, the ◊ *Polaris* series.

◊ also *Serb.*

'Sasin'. Soviet intercontinental ballistic missile. A two-stage ICBM operational since 1963. The missile appears to use a storable liquid fuel propellant. It is about 80 ft long, and has a maximum body diameter of 9 ft. Its range has been variously estimated between 5,000 and 6,500 miles, and it is thought to have a five-megaton warhead.

◊ *ICBM, Soviet.*

'Savage'. Soviet intercontinental ballistic missile. A three-stage ICBM operational since 1968. The missile is thought to use solid-fuel propellant. It is about 66 ft long and has a base diameter of 6 ft 6 in. Like ◊ *Scrag* it has the stages separated by open-work truss structures. Range has been estimated at 6,000 miles, but lower figures (4,000–5,000) have also been suggested. It is thought to have a one-megaton warhead.

◊ *ICBM, Soviet.*

'Sawfly'. SS-N-6 Soviet submarine-launched missile. The Sawfly is a more advanced version of ◊ *Serb*. It has a

claimed range of 1,500 nautical miles and is reported to have solid-fuel rocket motors.

'Scamp'. Soviet intermediate-range ballistic missile. Scamp is only a little larger than the US *Polaris*, and it is carried on a tracked erector/launcher. It has a claimed range of 2,500 miles, so that if it were in fact operational it would be a true mobile strategic delivery system. It is almost certainly solid-fuelled. It is 40 ft long and has a 5 ft 7 in body diameter. Not operational.

◊ *ICBM, Soviet.*

'Scarp'. Soviet intercontinental ballistic missile. This very important item of Soviet strategic weaponry is a two-stage ICBM propelled by storable liquid fuel. The Scarp is very large, some 100 ft long and 9 ft in diameter, and has a correspondingly large payload. Alternative warheads proposed for it are three 5-megaton ◊ *MIRVs*, or a single 25-megaton. The Scarp has already been tested in the Pacific with MRV (i.e. multiple but not independently-targeted warheads). At that time it was noticed that the fixed aiming pattern of the MRV happened to correspond to the silo pattern of the US ◊ *Minuteman* missiles. The Scarp is therefore seen as a ◊ *counterforce* weapon, instead, that is, for use against the US deterrent rather than as a retaliatory weapon. What makes the Scarp important in strategic terms is the fact that 500 of these could knock out the entire US ICBM force. Since other Soviet forces (including an ABM) could take care of the bombers and *Polaris* submarines which make up the rest of US strategic offensive weaponry, a large-scale Scarp deployment would clearly threaten the stability of the ◊ *balance of terror*. By the spring of 1968 some 200 of the Scarps were deployed, and after a nine-month pause deployments were resumed at the rate of 4–6 missiles per month. It was this threat which led directly to the American response, the ◊ *Safeguard* ABM system. ◊ *ICBM, Soviet.*

Schutzenpanzer Gruppe. West German armoured vehicles. A family of ◊ *armoured fighting vehicles* based on a common chassis which includes two types of ◊ *tank destroyers* (◊ *Jagdpanzer*), a 120-mm mortar carrier and the Spz.Neu (Marder) ◊ *armoured personnel carrier*. The common chassis is fully tracked, well armoured, very mobile and fitted with night-vision, air-filtration and wading devices. The Spz.Neu is the first armoured personnel carrier designed in Germany since the war, and it is a true combat vehicle intended to implement German doctrine on the armoured infantry which is trained to fight on the move. The frontal armour is thick and well sloped, and the crew can use four rifle-calibre weapons as well as two machine-guns and one 20-mm cannon from behind the armour protection. In contrast, the current British and American APCs (◊ *Trojan* and ◊ *M.113*) are designed as 'battle taxis', and the crew must expose itself to be able to use its weapons – or even to observe what is going on – while their thin vertical armour plates offer very limited ballistic protection. The mortar carrier is also intended as a mobile combat weapon, and the 120-mm weapon can be fired from within the vehicle. These vehicles are intended to replace the HS 30 and M.113 APCs which currently equip the West German army.

Specifications (Spz.Neu): Length: 21 ft 9 in. Width: 10 ft 2 in. Height: 6 ft. Full load weight: 20 to 25 tons. Maximum road speed: 43 mph. Road range: 370 miles. Ground pressure: 10·6 lb sq in. Ground clearance: 1 ft 4 in. Armament: 20-mm automatic cannon in cradle mounting with coaxial 7·62-mm machine-gun; 'anti-aircraft' 7·62-mm MG in separate turret in the rear. Crew: 10; commander and driver alongside the engine in the hull, 2 gunners in the 2 turrets, 6 crew in centre facing outwards.

Scorpion. British light tank. The British army has not deployed a ◊ *light tank* for over twenty years, using instead gun-armed wheeled armoured vehicles, such as the Saladin. Perhaps it is the shift from a colonial to a European orientation which has caused the changeover.

The Scorpion is a small tracked vehicle, built out of aluminium alloys and fitted with a largish turret. Its armament consists of the same short 76-mm gun as is used on the Saladin, which is intended to fire ◊ *HESH* ammunition. Such ammunition would be effective against fairly thick armour, but is hardly a tank-killer. The Scorpion has a three-man crew, and it is powered by a militarized motor-car engine. Its armour protection is necessarily limited, but the poor design of the turret appears to weaken still further the vehicle's ballistic protection.

As well as the Scorpion, which is intended as a general reconnaissance vehicle, several other fighting vehicles are to be produced on the basis of this eight-ton chassis. The variants include the Scimitar, which is to be fitted with the 30-mm Rarden cannon used on the ◊ *Fox*, the Striker, missile carrier for the ◊ *Swingfire* anti-tank missile, and the Spartan. This last is a cut-price armoured personnel carrier which is to carry only four (equipped) soldiers as well as its three-man crew.

Specifications: Length: 14 ft 4 in. Width: 7 ft 3 in. Height: 6 ft 10 in. Weight: 17,500 lb. Engine: 195 bhp. Road speed: 45 mph. Range: 350 miles. Armament: 76-mm QF main gun with 40 rounds of ammunition, coaxial 7·62-mm machine-gun with 3,000 rounds. Two multi-barrel smoke dischargers.

'Scrag'. Soviet intercontinental ballistic missile. A three-stage ICBM first displayed in 1965. The missile uses liquid propellant and may carry a 30-megaton warhead. It is about 120 ft long and has a base diameter of 9 ft 5 in. The range has been variously estimated at 5,000, 5,500 and 6,000 miles. It is not operational.

◊ *ICBM, Soviet*.

'Scrooge'. Soviet intermediate-range ballistic missile. Scrooge is described by the Soviets as a mobile ballistic missile. It has an estimated range of 3,500 miles, but its operational status is not known. Like ◊ *Scamp* it is carried on a tracked erector/launcher. Its dimensions are not known but to judge from those of the container, it must certainly be much larger than Scamp. The container is 62 ft in length and 6 ft 6 in in diameter.

◊ *ICBM, Soviet*.

93 **'Scud'. Soviet tactical missile.** Unlike the smaller ◊ *Frog* series, Scud is guided, and it is a battlefield missile comparable to the US ◊ *Sergeant* (MGM-29A). It has a cylindrical structure with cruciform tail fins; the sustainer motor is a liquid-propellant rocket, and there is no booster. It has been operational since 1957 and is thought to carry a kiloton-range warhead. It is 35 ft long and has a weight of 10,000 lb at firing. The speed is estimated at Mach 5, and the maximum range is about 500 miles. It is carried on and launched from a converted tank chassis. Two new versions of Scud have been displayed. The Scud series is believed to have a simple form of ◊ *inertial guidance.*

84 **Seacat/Tigercat. British surface-to-air missile.** A British subsonic anti-aircraft missile, known as 'Tigercat' in the air-portable land version of this originally marine system. The Seacat missile is propelled by a solid-fuel rocket and is fitted with a proximity fuse. Guidance is by command, with control impulses conveyed by radio. There are several different variants of the control unit, from simple manual joystick control to full-scale radar tracking. The Mk 20 system used in the British, Australian and Brazilian navies has manual control, where the operator keeps the target in the cross-hairs and corrects missile deviations by joystick moves which are converted into radio impulses by a shaper-transmitter. The Dutch, German and Swedish navies have adopted the Seacat missile and have combined it with a radar tracking guidance unit which operates by following both target and missile, computing and correcting any deviations. Other versions include a variety of optical and electronic systems. One of the advantages of the Sea-cat is its compatibility with varied control gear, and it can be deployed on even the smallest ships. The missile weighs about 120 lb, and its range is about 4,000 yds.
◊ *Missile* (surface-to-air) for general entry and competitive types.

83 **Sea Dart CF-299. British surface-to-air shipborne missile.** The Sea Dart is a medium-range shipborne anti-aircraft missile powered by a ramjet and guided by semi-active radar homing. It is smaller than the earlier ◊ *Seaslug*, but its all-round performance is reported to be superior; notably it can intercept incoming high-speed targets at very low altitudes, as well as having the Seaslug's high-altitude capability. The body is cylindrical, with a large nose air-intake for the ramjet; the warhead, guidance package and control elements are disposed concentrically around the ramjet duct. There are 4 interferometer aerials instead of the dish-shaped antenna usually used with radar homing units, and this permits the use of a full-diameter ramjet. The Sea Dart is propelled from its twin launcher by a solid-fuel booster: a tracking radar follows the target and a computer/guidance unit corrects the missile's path by command; final interception is achieved by the missile's homing unit, which locks on impulses bounced off the target by the shipboard 'illumination' radar. Sea Dart is 14 ft 3 in long and has a 1 ft 4 in body diameter.

Seaslug. British surface-to-air shipborne missile. A shipborne anti-aircraft missile for medium-range interception. It has a solid-fuel rocket sustainer assisted by four wrap-around solid-fuel boosters; it has fixed cruciform wings and pivoting tail control surfaces. The Seaslug is guided by beam-riding, where the shipboard radar acquires the target, plots the range, bearing and height, and then provides a beam which the missile 'rides' to target. The warhead is high-explosive with a proximity fuse. The Mk 2 version has longer range and a better low-level performance, and can also be used against surface targets. The Seaslug Mk 1 is 19 ft 8 in long and has a body diameter of 1 ft 4 in. Its effective range exceeds 20 miles. Unlike its US counterpart ◊ *Terrier* and ◊ *Tartar* this missile has a good record of availability and consistent performance, but its low speed renders it obsolescent.

SEATO. South East Asia Treaty Organization. Formed as a result of the Manila Treaty of 1954, the members being the USA, UK, France, Australia, New Zealand, the Philippines, Thailand and Pakistan. Members are 'to consult' in order to prepare for joint action against aggression directed at them or at the non-member 'protocol' states (Cambodia, Laos and South Vietnam). France and Pakistan are generally considered as 'detached' from the alliance. No central command structure exists, but there is an *ad hoc* council. The Vietnam war, and the resultant intervention of Australia, New Zealand, the Philippines and Thailand, with expeditionary contingents, has not been handled within the framework of SEATO. No troops are permanently committed to SEATO, but US and UK forces in the area are earmarked for it. For other multilateral Western alliances, ◊ *NATO, CENTO* and *ANZUS*.

Second-Strike (Capability, Strategy). A strategic nuclear strike mounted in reply to an enemy nuclear attack after the latter has already occurred. The less ambiguous term 'strike-back' is generally recommended but little used.

If a party can mount a nuclear strike after having suffered the effects of one, it has a 'second-strike capability'; this is usually thought of as consisting of ◊ *hardened* or otherwise protected missiles and associated control facilities. But vulnerability depends as much on the enemy's capabilities as on the number and nature of delivery systems. If it has an effective second-strike capability, a party may decide to use it only in reprisal for a prior nuclear attack made upon it; such a choice is described as a 'second-strike strategy'. But though such a strategy implies the possession of a 'second-strike capability', having the latter does not imply a 'strike-back only' strategy. ◊ The *Assured Destruction* policy of the USA, which being a contextual second strike is a more useful concept.

Self-propelled Artillery. Artillery weapons mounted on wheeled or tracked chassis which are lightly or partly armoured (except for anti-tank weapons mounted on such vehicles, which are listed under ◊ *Tank Destroyer*). The general class of weapons described as 'self-propelled' artillery ranges from fully tracked tank-like vehicles, such as the ◊ *Abbot FV 433*, to stripped-down chassis with guns (or howitzers) mounted

upon them, such as the US M.107 175-mm gun and the M.110 howitzer (◊ *Artillery, US*).

In Western armies the trend is towards converting all but the lightest artillery weapons to self-propulsion, with fully enclosed light armour envelopes for medium-calibre weapons and unarmoured tracked chassis for the heavier calibres. But the Soviet armed forces, within which the artillery branch still retains its traditional importance, have largely opted to retain the classic two-wheel towed chassis configuration. This, of course, is far cheaper but the choice appears to reflect other considerations. It seems that Soviet planners still believe in the value of traditional large-scale barrage fire with non-nuclear warheads. This contradicts their own strategic doctrines (◊ *Local War*), which anticipate the early use of nuclear weapons on the battlefield. Large-scale barrages require large quantities of ammunition, and the Russians' system of supply vehicles and unprotected artillery caters for this. Western equipment affords greater protection to the gun and the crew, but it is not practical to carry vast quantities of ammunition within the armour envelope; this squares with Western doctrine, which foresees small local barrages and 'sniping' but not the large volumes of (non-nuclear) fire to which the Soviet armed forces are orientated.

The only self-propelled items in the Soviet inventory are the 'airborne' tank destroyers (◊ *ASU*) and the anti-aircraft armoured fighting vehicles, ◊ *SU-57-2* and *SU-23-4*. These have independent mobility because they are deployed with tank units and are based on converted tank chassis.

◊ *Artillery, British, French, West German, Soviet, US.*

◊ *Artillery* for general subject entry.

Semi-active Radar Homing. A radar guidance system used on many air-to-air and surface-to-air missiles. This is an 'all-weather' technique, unlike the other main form of 'homing' guidance: ◊ *Infra-red Homing.* The system consists of a tracking radar unit in the missile, which receives signals bounced off the target by a ground-based 'illumination' radar which is part of the launching unit. (Since no impulse transmission is required from the missile itself, this is a fairly lightweight system.) The tracking set transmits the target-location data to the missile's programmer/autopilot unit which makes suitable adjustments to the missile's flight path. ⟡ *Radar* for the techniques involved.

Sentinel. US ABM Project. A ballistic-missile defence programme proposed by the US government in late 1967 and replaced by ⟡ *Safeguard*. Both were based on the same components, but whereas Sentinel was primarily oriented against a 'Chinese' attack (ICBMs only, and coming over the Pole) against 25 selected US cities, Safeguard is intended to provide an all-round defence primarily to defend the ⟡ *Minuteman* silos and bomber bases of the US deterrent forces as well as the Washington command centres (though it has a 'thin' area defence capability against a small or unsophisticated 'Chinese' ICBM strike).

'Serb'. Soviet submarine-launched missile. The Serb, which has been operational since 1964, is a submarine-launched ballistic missile with a range of about 650 miles. First displayed in 1964, it is thought to have a two-stage storable-liquid-fuel rocket motor. It is about 33 ft long and has a base diameter of 5 ft, and it carries a one-megaton warhead. Unlike the earlier ⟡ *Sark*, the Serb can be launched from under water.

Sergeant MGM-29A. US tactical missile. Sergeant is a solid-propellant rocket-powered battlefield ballistic missile designed to replace the Corporal and operational since 1961. Sergeant has alternative nuclear and chemical warheads and advanced ground electronics equipment. After launch Sergeant is set on a ballistic path by an ⟡ *inertial guidance* system. Despite its weight and complexity, the whole system has been made mobile on three large semi-trailers and a single standard truck. Sergeant has a cylindrical all-metal structure with no wings and small cruciform tail surfaces linked to the flight control jet deflectors.

Specifications: Length: 34 ft 6 in. Body diameter: 2 ft 7 in. Wing span: 5 ft 10 in. Firing weight: 10,000 lb. Minimum range: 28 miles. Maximum range: 85 miles.

'Shaddock'. Soviet tactical cruise missile. The Shaddock is powered by turbojet with rocket-assisted take-off. It has a pointed nose-cone, but little else is known about it. Its range is estimated at 250 miles. The Shaddock has been operational since 1957. It is about 40 ft long and has a kiloton-range warhead.

SHAPE. Supreme Headquarters Allied Powers in Europe. SHAPE is one of the two major commands within ⟡ *NATO*, the other being ⟡ *ACLANT*, Allied Command Atlantic. Its HQ is at Casteau in Belgium, and its commander, SACEUR, Supreme Allied Commander Europe – who has always been an American general – controls ACE, Allied Command Europe. Apart from the forces under ACE, SHAPE is also involved in strategic planning through the commitment of British and American strategic nuclear delivery systems to its commander SACEUR. SHAPE is responsible for the defence of all European member country territory except Britain and Portuguese coastal waters.

SHAPE's operational command organization is ACE (Allied Command Europe). ACE controls directly a mixed mobile force consisting of about 7 augmented battalions, a light armoured unit and several fighter-bomber squadrons. It also supervises the NATO air defence system NADGE, which consists of 17 radar-computer centres linked by a communications network. But ACE's main combat forces come under three area commands: AFNORTH, AFCENT, and AFSOUTH.

AFNORTH, Allied Forces Northern Europe, is commanded by a British general, and its HQ is at Kolsaas in Norway. The command, which is responsible for the whole of Norway, Denmark and the Baltic as well as a slice of northern Germany, comprises two German air wings and one division, as well as most of the armed forces of Norway and Denmark. It is the weakest of NATO's area commands and the only one which shares a border with European Russia. *AFCENT*, Allied Forces Central Europe, is headed by a German general, and its HQ is at Brussum in the Netherlands. It is divided into two sub-commands, NORTHAG, Northern Army Group, and CENTAG, Central Army Group. NORTHAG comprises four German divisions, three British divisions and one Canadian brigade, as well as most of the combat-ready units of the Belgian and Dutch armies. These forces are supported by the Second Allied Tactical Air Force. CENTAG is responsible for all of West Germany south of the Göttingen–Liège axis, and it comprises seven German divisions, and the US 7th Army which has some two hundred thousand men. The air element is the Fourth Allied Tactical Air Force, made up of American, German and Canadian units. *AFSOUTH*, Allied Forces Southern Europe, is commanded by an American admiral, and its HQ is in Naples. It is responsible for the defence of Italy, Greece and Turkey and for the security of the sea lanes in the Mediterranean, as well as Turkish coastal waters in the Black Sea. The local members of the alliance contribute 30 divisions, of which 7 are Italian, 9 Greek and 14 Turkish. They also contribute part of their air forces. The American contribution is the US navy's Sixth Fleet, and there are also some British air and naval units. Since the Russian naval expansion in AFSOUTH's area, a surveillance force, MARAIRMED (Maritime Air Forces Mediterranean), has been formed and equipped with American, British and Italian long-range patrol aircraft. In May 1969 an 'Allied On Call Naval Force' was also initiated. At present it includes only a small number of destroyers.

ACE has some 3,000 tactical aircraft of all types based on 150 airfields, as well as some 55 divisions, half of which are armoured or mechanized, as well as ◊ *Pershing*, ◊ *Sergeant* and ◊ *Mace* missiles with nuclear warheads. Its artillery is augmented by ◊ *Honest John* rockets, and there are some 7,000 nuclear warheads in the area, part of which are, however, to be delivered by the aircraft attached to the command. ACE's air-defence intercep-

tors are supplemented by ◊ *Hawk* and ◊ *Nike* surface-to-air missiles.

Sheridan M.551. US light tank. This 15-ton vehicle, which went into production in 1966, represents a new concept in tank design. It is a 'light tank' in the sense that it weighs less than one third as much as the M.60, the current US main battle tank, and its armour is sufficient only to counter heavy machine-gun fire and shell splinters. On the other hand, its armament is at least as good as that of MBTs, unlike previous light tanks, which were as poorly armed as they were armoured, so that their value – apart from economy – rested on their air-transportability. In fact such light tanks as the M.41 'Walker Bulldog' were (at 22·8 tons) far too heavy for the airborne or reconnaissance roles, while having inadequate weaponry to deal with full-weight tanks. The Sheridan has a conventional layout with 4 crew, and its armour is a steel and aluminium composite. It mounts a revolutionary combined howitzer/missile-launcher which can fire the ◊ *Shillelagh* guided missile as well as conventional shells. This weapon is more than adequate for both 'soft' and 'hard' targets. The powerful engine, low ground pressure and swimming ability give it an exceptionally good all-round mobility; but it is still too heavy for its official 'airborne assault' role. Another defect is the poorly shaped turret.

Specifications: Length: 16 ft. Width: 8 ft 2 in. Height 7 ft 6 in. Weight: 15 tons. Engine: turbo-charged diesel 300 bhp. Road speed: 43 mph. Water speed: 4 mph. Road range: 370 miles. Ground pressure: 6·8 lb sq in. Ground clearance: 1 ft 7 in. Armament: 152-mm howitzer/launcher with 20 HEAT rounds and 10 Shillelagh missiles. One 0·5-in and one 0·3-in MG.

◊ *Main Battle Tank* and *Light Tank* for competitive vehicles.

◊ *Armoured Fighting Vehicle* for general subject entry.

Sherman M.4. US medium tank. An American-built medium tank first introduced in 1941 and produced in large numbers in various versions, of which the last was designated M.4A3 and fitted with a 76·2-mm gun. After the Second World War it was supplied to almost all members of NATO and SEATO as well as a number of other countries. It is now obsolete, but is still serving with some Asian armies. The Israeli army uses a large number of converted Shermans fitted with a new 105-mm gun as well as a number of locally built components.

15 **Shillelagh XMGM-51A. US anti-tank missile.** Shillelagh is a lightweight close-support missile which can be fired from a howitzer/launcher also used for conventional shells. This combined weapon is mounted on the ◊ *Sheridan M.551* tank and on development models of the M.60 (◊ *Patton*) and ◊ *MBT-70* ◊ *main battle tanks.* The Shillelagh is powered by a solid-fuel rocket motor and controlled by hot gas jet reaction. Guidance is line-of-sight, where the operator keeps the target in the cross-hairs of an optical sight and a programmer/shaper sends correcting impulses to the missile in flight. The warhead is a HEAT (shaped) charge. Shillelagh is now in large-scale production and heli-borne variants

be launched from all aircraft equipped with the ◊ *Sparrow* missile system, and it has a passive ◊ *radar* homing guidance unit. After the enemy radar has identified itself by sweeping the launching aircraft with its beams, the Shrike is launched, and it then 'rides' enemy radar impulses back to the radar set.

◊ *Missile* (air-to-surface) and *Standard Missile.*

Sidewinder AIM-9 Series. US air-to-air missile. Very simple and cheap air-to-air missiles produced in large quantities and extensively used in combat. The basic version, the Sidewinder 1A (AIM-9B), consists of a rocket, a 10-lb warhead, a proximity fuse and an infra-red homing device. This is effective in fine weather and if fired almost dead astern. Sidewinder 1A is still used by almost all NATO and allied air forces, but in US service it has been replaced by an improved version, the AIM-9C/D Sidewinder 1C. This has higher speeds, better range and improved wider angle ◊ *infrared homing* (AIM-9C) or ◊ *semi-active radar* (AIM-9D) guidance system.

Sidewinder Missile – Specifications:

	Length	Body diameter	Weight	Speed	Range
AIM-9B	9 ft 2 in	5 in	159 lb	Mach 2·5	2 miles
AIM-9D	9 ft 6 in	5 in	185 lb	Mach 2·5	over 2 miles

are under development. It is thought to be the best all-round AT missile, because of its performance and reliability.

Specifications: Length: 3 ft 9 in. Body diameter: 6 in. Control surfaces: small tailplanes which emerge after missile leaves gun/launcher. Weight: 60 lb.

◊ *Missile* for general subject entry.

Shmell. Soviet anti-tank missile. ◊ *Snapper.*

Shrike AGM-45A. US air-to-surface missile. The Shrike is an air-to-surface missile intended for use against ground radars. It weighs about 500 lb and has a 10-mile range. This weapon can

SINS. Ships Inertial Navigation System. ◊ *Inertial Guidance.*

'Skean'. Soviet intermediate-range ballistic missile. A liquid-propellant ballistic missile with a range of about 2,000 miles. It has no tail fins and a blunt nose-cone. Skean has been operational since 1961. It is 75 ft long and has a body diameter of about 8 ft. It is thought to have a one-megaton warhead. Some Skeans at least are launched from ◊ *hardened* silos.

Skyhawk A-4. US attack aircraft. The Skyhawk is a small, simple and cheap subsonic attack aircraft designed to perform tactical bombing missions from aircraft carriers. The Skyhawk, the only naval combat plane which does not need folding wings for storage below decks, is a single-seat low-wing machine, whose overall smallness accentuates the apparent size of the cockpit. The A-4 is powered by a single turbojet, rated at 9,300-lb thrust in the A-4F version. The A-4 series started with the A-4A, which had a 7,700-lb thrust turbojet and very limited avionics; the later A-4B was adapted to carry the ◊ Bullpup air-to-ground missile and had improved bombing and navigation systems. The A-4C is equipped with terrain-clearance (i.e. air-to-ground) radar as well as autopilot and gyro attitude indicators. A more powerful 8,500-lb thrust turbojet, introduced in the A-4E, has been retrofitted to the A, B and C types. The A-4E, with increased range and greater payload and the more powerful 8,500-lb thrust engine, has a maximum weapon load of 8,200 lb. The A-4F has a modified airframe, additional armour to protect the pilot and other sensitive parts, and updated avionics. The maximum payload of the A-4F is 10,000 lb, and it is powered by a 9,300-lb thrust turbojet. A training version, the TA-4F, has also been produced, and various export versions for Australia (A-4G), New Zealand (A-4K), Argentine (A-4B), and Israel (A-4E) have been derived from the basic types.

Specifications (all types): Wing span: 27 ft 6in. Length overall: from 39 ft of the A-4A to 41 ft 3½ in of the A-4F. Height: 15 ft. Gross wing area: 260 sq ft. Weight empty (E): 9,853 lb. Maximum take-off weight: A, B and C, 22,500 lb; A-4E and F, 24,500 lb. Fixed armament: two 20-mm guns in wing roots. Maximum speed at normal take-off weight: from 664 mph of the A to 674 mph of the E. Maximum range: 2,000-plus miles.

◊ *Fighter* for general subject entry.

Small arms. Weapons which can be carried and operated by one man; more generally light infantry weapons as opposed to artillery and armour. In terms of modern military equipment this includes ◊ *sub-machine-guns, rifles,* rifle-calibre ◊ *machine-guns,* ◊ *grenade-launchers,* light-rocket-launchers, most ◊

recoilless weapons, as well as pistols, revolvers and carbines. Before the introduction of recoilless and low-velocity weapons, the distinction between ◊ *artillery* and small arms was based on calibre; everything below 20-mm (light cannon) calibre was 'small arms'. This distinction is obsolete.

SMG. ◊ *Sub-Machine-gun.*

'Snapper' (Shmell). Soviet anti-tank missile. A first-generation wire-guided anti-tank missile with joystick control and no autopilot. Usually used on a triple vehicle mounting, Snapper is tracked by an operator through optical sights which combine periscopic binoculars with an illuminated sight. Remote firing is possible up to 165 ft away. Snappers were captured by the Israeli army during the 1967 war, thus there are specifications: Length: 3 ft 8 in. Diameter: 5 in. Weight: 49 lb. Speed: 201 mph. Minimum range: 1,700 ft. Maximum range: 7,650 ft. Warhead: HEAT 11·5 lb.

Soman. ◊ *Nerve Gases.*

Sonar. A communication and position-finding device used in underwater navigation, target detection, and weapon control. Sonar is an acronym of SOund NAvigation and Ranging, and it is based on the use of acoustic energy in a manner similar to the use of electro-magnetic energy in ◊ *radar.* Both techniques depend on the emission of energy beams, which are 'bounced off' reflecting objects to detect their presence (by the act of reflection) and to measure their range (by timing the return of the beam to the receiving apparatus). In sonar as in radar, the direction of the object is determined by comparing the intensity of the returning beam as the source of the beam is traversed and elevated, the returning beam being most powerful when the receiver is pointing directly at the reflecting object. But sonar operates in a far less favourable medium than radar. Sound travels in water at variable speeds around a median of 1,500 metres per second; significant errors can therefore arise from post-ranging movements of the object observed.

Further, the speed of sound in water varies with the salinity, the temperature and the depth, thus affecting the stability of the time/distance conversion. These factors severely limit the accuracy of sonar ranging. (By contrast, the electromagnetic beams of radar devices move at 300,000,000 metres per second, so that post-ranging errors are insignificant.) The other limiting factor in sonar operation is the long wavelength of acoustic energy (10 cm at least). This sets a maximum ceiling on the resolution (minimum distance between separately detectable points) of sonar devices, and one far inferior to the resolutions available with high-frequency radar. But sonar techniques are still the most useful in ◊ *anti-submarine warfare* (ASW), since the other means available (optical detection and radio techniques) are even less reliable and severely limited in range.

There are two basic forms of sonar, active and passive. Passive sonar depends on the sound produced by the object observed and a simple, direct listening can be employed, though there are usually a transducer (to convert the acoustic energy into electromagnetic energy), an amplifier and a directional device. The advantage of the passive mode in sonar observation is of course the fact that its operation cannot be detected.

Active sonar consists of a transducer, which provides the acoustic energy beam, a transmitting antenna and a receiver-transducer-amplifier which processes the returning beam. As in many types of radar, the sound beam is sent in pulses, so that range can be determined by timing the returning beam, identified by its place in the series of pulses.

Sonar can also be used for underwater communication, including direct voice communication, where the human voice is transmitted as modulated ultrasonic energy by an active sonar device.

In navigation and target detection, a scanning active sonar can be used to provide a realistic visual picture of sound-reflecting objects around the source. The scanning beam is rotated, and the reflected beams show the position of objects in a given plane as echo 'blips'. This is identical to the PPI displays used in radar, and, as in the latter, another

display is required to show the depth/altitude of the object observed, since the PPI-type display shows only range and direction in a given plane.

Sonar is also used to control weapons. Acoustic homing torpedoes are guided by a sonar device which 'locks on' the target, a programmer which converts target position data into directional instructions, and a servo-mechanism which actuates the torpedo's control surfaces. One typical homing torpedo (the US 12-in AST) of this kind has an active life of 13 minutes at 30 knots. Such torpedoes are launched from ordinary deck tubes and aircraft dispensers, but they are also used as the terminal stage of AS missiles such as the ◊ *Asroc*, the ◊ *Ikara* and ◊ *Subroc*.

VDS, Variable Depth Sonar, is fitted on most modern ASW vessels. As the name indicates, the device can be suspended below the hull. Fixed-bow sonar is used for long-range detection/location, and one such device has detected a submarine at ranges in excess of 2,000 miles. (But see below.)

Dipping sonar devices equip DAS/DASH and manned AS helicopters and fixed-wing aircraft; the aircraft flies low and suspends a sonar device at the end of a wire-link which is held just below the surface of the water. The sonar technique is also used in sonabuoys, which use active or passive sonar to detect and locate submarines. Sonabuoys provide ASW data to aircraft which cannot dip sonar devices while hovering overhead, the data being transmitted by coded radio signals.

Apart from the general constraints on the accuracy of sonar mentioned above, there is the problem of living and sound-reflecting organisms in the sea. These tend to concentrate within certain depth ranges, and they provide abundant cover for a submarine hiding amongst them. Equally important is the 'clutter' produced by these organisms, which interferes with the operation of sonar through those depth levels. It is these factors which make submerged systems such as ◊ *Polaris* and ◊ *Poseidon* attractive propositions, in terms of survivability.

Sparrow IIIB (AIM-7E). US air-to-air missile. The Sparrow IIIB is a homing air-to-air missile designed to operate at high altitudes and in all weathers. It currently equips the 'air-superiority' fighters of the USAF and allied air forces, such as the ⬦ *Phantom F-4* and the ⬦ *Starfighter F-104S*. The Sparrow has a cylindrical body and pivoted cruciform wings and tail fins. Propulsion is by a solid-fuel rocket motor and guidance is by ⬦ *semi-active radar homing*.

Specifications: Length: 12 ft. Body diameter: 8 in. Weight: 400 lb. Speed: Mach 2·5. Range: 8 miles plus. Warhead: 60 lb; high explosive.

Spartan. US ABM missile. The Spartan is a three-stage interceptor missile fitted with a megaton-range warhead which is to be deployed as part of the Safeguard ⬦ *ABM* system. The Spartan has been developed from the Nike-Zeus, which achieved the first-ever, intercept of an ICBM in July 1962. It is 55 ft long and fitted with prominent tail and wing control surfaces which are controlled by command-guidance from the MSR unit of the ground system. The warhead, whose yield is estimated at two megatons, is of the minimum-fission type designed to minimize the ⬦ *blackout* effects on friendly radars. The range of the Spartan has been estimated at several hundred miles, and the missile is intended to intercept enemy warheads outside the atmosphere, where the ⬦ *X-ray kill* of its own warhead is effective within a radius of several miles. ⬦ *Sprint* for the other, atmospheric, interceptor of the Safeguard programme. The detonation of the Spartan's warhead would not cause any direct ground damage at the minimum allowed altitude (below which the Spartan warhead would not be detonated). The general ⬦ *fallout* effect would, on the other hand, be limited by the minimum-fission warhead and the high altitude of the detonation. In any case the Spartan's global fallout effects would be a fraction of those of its target: a warhead intended to detonate near the ground.

⬦ *Safeguard* for the Spartan's missions within the US ABM system.

Sprint. US ABM interceptor. The Sprint is a short-range (or 'terminal') interceptor missile which is part of the ⬦ *Safeguard* ⬦ *ABM* system. It has a slant range of 25 miles, and it is therefore intended to intercept enemy warheads in the last phase of their trajectory to target. Since the intercept is to take place well within the atmosphere, the Sprint's warhead has been kept quite small (one or two kilotons) in order to minimize damage to friendly or 'neutral' territory. The range and yield of the Sprint are such that the missile needs an exceedingly high acceleration to reach the warhead in time, as well as great accuracy to achieve the kill. The acceleration is in fact 'far in excess of 200 g', which is unprecedented for any missile; the accuracy cannot be determined, but the Sprint would have to come within one or two hundred feet of an enemy warhead in order to destroy it. (This rough estimate assumes a kill by means of blast and heat effects, but a ⬦ *neutron kill* may result in a larger kill-radius.)

The Sprint is 27 ft long, conical in shape, with a base diameter of 4 ft 5 in.; apart from the warhead there are two stages, both propelled by solid-fuel rocket motors. The Sprint's batteries will be housed in bins with soft plastic covers. Upon firing, the missile is ejected by a chemical explosion which rips the plastic covers; it then hovers for a short time (under one second) while it is being aligned towards the incoming warhead by the missile-control MSR unit. Once the attitude is set, the MSR ignites the Sprint's booster and guides the missile to the target; the more manoeuvrable second stage then takes over for the final intercept.

⬦ *Spartan* for the long-range, extra-atmospheric, interceptor used in the Safeguard programme.

Spy Satellite. ⬦ *Surveillance Satellite.*

Spz. Neu. ⬦ *Schutzenpanzer Gruppe.*

Squadron. A common designation for air-force units. A squadron is not a standard-sized unit, although the term is often used as if it were. A French bomber squadron consists of just four aircraft, while a US fighter squadron can include as many as twenty-five machines. In practice, fighter squadrons tend to number around twenty aircraft in most Western air forces though only about a dozen in British, French and Russian-style air forces.

56 SR-71 (YF-12A) (A-11). US Mach 3 long-range aircraft. A very advanced aircraft originally developed from the A-11 and intended to replace the U-2 long-range reconnaissance aircraft. The A-11 airframe is made largely of titanium and its alloys: this is necessary for sustained Mach 3 speeds. The elongated fuselage has an unusual shape with long lateral fairings; the crew of two is housed in tandem. The A-11 has been developed into the SR-71 (for strategic reconnaissance), which is the only aircraft in the world known to be cruising at Mach 3 over intercontinental ranges. The SR-71 has been operational since 1966 with the US Strategic Air Command, which operates about fifteen of them. It carries a large payload of cameras, air-to-ground radars and other sensors. It can radar-map (with filming) 60,000 square miles of territory in one hour. The two turbojets of 32,500-lb thrust each (with afterburning) consume 8,000 US gallons of fuel per hour. The operational ceiling is over 80,000 ft. The external dimensions are a wing span of 55 ft 7 in; a length of 107 ft 5 in; and an overall height of 18 ft 6 in. A conversion, known as SR-71B, serves as a trainer with an elevated second cockpit. An experimental fighter-interceptor version, known as YF-12A, has also been developed: it retains the same basic airframe and engine but has a slightly shorter fuselage. This version has a large nose radar and other sensors, and bays for eight ◊ *Falcon* AIM-47A missiles (which are 12 ft long). It is not in production, but has set speed (2,070 mph), height and payload records.

◊ *Fighter* for general subject entry.

SRAM AGM-69. US stand-off missile. The SRAM (Short-Range Attack Missile) was being deployed in the course of 1970 with the G and H versions of the ◊ *Stratofortress B-52* bombers of the USAF's Strategic Air Command. As in the case of stand-off missiles, such as the US ◊ *Hound Dog* and the British ◊ *Blue Steel*, the SRAM is carried by the bomber to the target area and then released to

deliver its thermonuclear warhead over the final phase of the attack. The SRAM's primary target is the enemy's ◊ *Air Defence System* to open the way for the bomber. The higher speed and smaller radar cross-section of the SRAM enable it to penetrate air defences too 'thick' for the bomber. Further, with each B-52 carrying six SRAMs, the number of targets presented to enemy air defences complicates the latter's task. The use of this exhaustion tactic is possible with the SRAM because of its small size and weight: in the case of the Blue Steel a bomber (the ◊ *Vulcan 2*) can deliver only a single missile, while even the B-52 can carry only two Hound Dogs. If the projected ◊ *A.M.S.A.* bomber is developed by the United States, it will be equipped with eight SRAMs or their equivalents. The SRAM is powered by a solid-fuel rocket motor and guided by an ◊ *inertial* system. Its range exceeds 60 miles.

SS-11. A one-megaton Soviet ICBM which has been mass-produced since 1966. About 800 were in service by 1971. ◊ *ICBM, Soviet.*

SSBS. French medium-range ballistic missile. The SSBS (Sol-Sol-Balistique-Stratégique) is a silo-launched two-stage solid-fuel ◊ *ballistic missile.* Twenty-seven of these missiles in ◊ *hardened* silo-launchers, together with submarine systems (◊ *Redoutable*), are intended as the second-generation weapons for the French independent deterrent ('Force de Frappe'). The silos are ready, though as yet empty, and these IRBMs are currently planned to be operational in the early seventies.

Specifications: Length: 49 ft 6 in. Body diameter: 4 ft 11 in. A first-stage rocket of 35,275-lb thrust, and a 22,050-lb thrust second stage. Warhead estimated at one quarter of a megaton.

◊ *France, Strategic Offensive Forces* for other French systems.

◊ *Missile* for general entry.

STANAVFORLANT. Standing Naval Force Atlantic. A ◊ *NATO* naval force of four destroyers drawn from various members of the alliance and under the command of SACLANT, the NATO area commander. The stated purpose of STANAVFORLANT is to study the problems and prospects of a multi-national naval force. ▷ MLF for a planned mixed-manned multi-national force.

1   Standard Missile RIM-66A/67A. US ship-to-air missile. A shipborne anti-aircraft ▷ missile being developed to replace both the ▷ Tartar and the ▷ Terrier SAMs. The missile uses ▷ semi-active radar homing guidance, where a large shipborne radar illuminates the target and the missile follows the reflected pulses. An autopilot controls the four tail surfaces. Two versions, the ER (extended range), a two-stage 35 miles plus range missile, and the MR (medium range) dual-thrust version, have been produced. Both have solid-state electronic controls.

fighter is a Mach 2 fighter with very small, very thin, 'supersonic' wings without sweepback. Its development started in 1951, and eventually an interceptor (F-104A) was produced for the USAF Air Defense Command (ADC) which became operational in 1958. But the ADC ordered only a small number since the aircraft developed flight stability problems and their correction by electronic controls restricted manoeuvrability. Some were delivered to Pakistan and, much later, to Jordan. A number of training two-seater versions, designated F-104B, D, DJ and F, were delivered to the US, Japanese and West German air forces. The original version for the ADC had a turbojet rated at 14,800 lb of thrust with afterburning, while a more powerful 15,800-lb engine was mounted on a fighter-bomber version (designated F-104C) for the USAF

Standard Missile – Specifications:

	Length	Weight	Range
ER	26 ft	3,000 lb	35 miles
MR	14 ft	1,300 lb	15 miles

An adaptation of the RIM-66A/67A has been developed as ARM (Anti-Radiation Missile). This is an air-to-surface missile intended to home on radiation emitted by radars; it consists of an improved ▷ Shrike anti-radar warhead plus the RIM-66 propulsion system.

Stand-off Missile. A missile, generally fitted with a ▷ nuclear warhead, which is launched by a bomber and substitutes for the latter in the final phase of the attack. The stand-off missile concept was developed in order to extend the operational life of the conventional high-flying bomber in the face of improving air defences. ▷ the US Hound Dog, the British Blue Steel and the Soviet Kitchen and Kipper.
▷ Missile for general subject entry.

4   Starfighter F-104. US fighter and NATO fighter-bomber. The Star-

Tactical Air Command and for Japan (under the designation F-104J).

This version became the basis of the F-104G, a 'multi-mission' fighter which first flew in 1960 and now equips in various versions the West German, Dutch, Belgian, Italian, Canadian, Spanish, Greek, Turkish, Formosan and Norwegian air forces. The F-104G programme of licensed construction was a private venture vigorously supported by the US military authorities, and it is interesting to note that the F-104G, which was never deployed in quantity by the USAF, has been marketed in larger numbers (1,500+) than any other post-war Western fighter. The basic F-104G has a single turbojet, rated at 15,800 st (with afterburning), a reinforced general structure and extra 4·2-in chord in the wings and modified flaps intended to improve manoeuvrability at low altitudes. A 20-mm multi-barrel cannon is fitted in the fuselage, but there is a centreline store fitting for alternative uses of the payload. Like other versions of the Starfighter, the G has automatic pitch control to deal with the stability problems originally encountered by the USAF ADC. There is a three axis auto-stabilizer system and flap-blowing to reduce the landing run. Apart from the optional gun, the G can carry up to 4,200 lb of bombs, missiles (◊ *Sidewinder* and ◊ *Bullpup* AGM-12B), rockets or extra fuel tanks. The G also carries a lot of electronics: the multi-purpose N.A.S.A.R.R. radar (◊ *Thunderchief F-105*), electronic fire-control, a bombing computer, and alternative navigation systems including ◊ *inertial guidance*; there is also a particularly versatile autopilot. 977 F-104Gs were produced by a European consortium under licence, 179 were built for export by the US licensing company, and 200 were built in Canada under the CF-104 designation. 137 two-seater trainers designated TF-104G were built by the US licensers and another 38 were built in Canada, as the CF-104D. A further version, the F-104S, with a more powerful 17,900-lb thrust turbojet/afterburner, has been developed for Italy, carrying the Sparrow missile as an interceptor.

Specifications (F-104G): Overall length: 54 ft 9 in. Height: 13 ft 6 in. Wing span: 21 ft 11 in. Wing area: 196·1 sq ft. (i.e. very little). Weight empty: 14,082 lb. Maximum take-off weight: 28,779 lb. Maximum level speed at 36,000 ft: Mach 2·2. Maximum diving speed: Mach 2·2. Maximum cruising speed: Mach 0·95. Rate of climb at sea level: 50,000 ft per minute. Ceiling: 58,000 ft. Radius with maximum fuel: 745 miles.

◊ *Fighter* for general entry.

Starlifter C-141. US long-range transport aircraft. This four-jet long-range transport has been operational since 1965 and has given good service in its intended role: fast strategic deployment. Given the logistic load of modern military units, only this kind of aircraft allows significantly large forces to be moved over long ranges without requiring an impracticable and uneconomic number of transport aircraft. The C-141 is immediately distinguished from its civilian counterparts by the high mounted wings.

The four turbofan engines give a combined thrust of 84,000 lb and there is an 'all-weather' landing system as well as a full range of navigation radars and computers. The transport capacity can be used to move 154 troops (or 123 paratroopers) or 5,283 cu ft of freight. The dimensions, although inadequate for the available payload, are sufficiently large to accommodate a ◊ *Minuteman* missile and the US army's ◊ *Sheridan* M.551 tank. The plane can also be used for paradrops of men and equipment – the first jet to have been so used. A rear loading ramp of full cabin size and an integral loading system facilitate fast turn-around, and it can be used to paradrop the Sheridan tank and similar cargoes. 284 C-141s have been delivered to the US air force.

Specifications: Length: 145 ft. Height: 39 ft 3 in. Internal dimensions: 70 ft by 10 ft by 9 ft (the 10 by 9 is the shortcoming). Floor area: 718 sq ft. Wing span: 159 ft 11 in. Gross wing area: 3,228 sq. ft. Weight empty: 133,733 lb. Maximum payload: 70,847 lb. Maximum take-off weight: 316,600 lb. Maximum level speed at 25,000 ft: 571 mph. Economic cruising speed: 495 mph. Take-off run to 50 ft: 5,650 ft. Landing run (maximum weight): 2,240 ft. Range with 31,780-lb payload and reserves: 6,140 miles. Range with maximum payload: 4,080 miles.

◊ *Hercules, Galaxy,* and also Soviet *Cub* and *Cock.*

Sterling. British sub-machine-gun. The Sterling, developed from the Patchett, has replaced the old 'Sten' SMG in British and British-supplied armies. It operates, like the Sten, by direct blowback, but it has much better trigger, safety, sights and general finish. It is accurate to a maximum of 150 yards, with careful aim, and fires well on automatic. The long curved magazine holds 34 rounds of the 9-mm Parabellum pistol cartridge.

Specifications: 9-mm L2A3. Length: 19 in folded, 27 in extended. Weight (loaded): 8 lb. Cyclic rate of fire: 550 rpm.

◊ *Sub-machine-gun* for general entry.

STOL. Short-Take-Off and Landing. The term invariably applies to fixed-wing aircraft. ◊ *Combat Aircraft Part II.*

Stoner (small arms). ◊ *M.16.*

Strategic Air War. War waged by non-nuclear bombing intended to destroy enemy war-making potential by selective strikes against critical targets, typically industrial capacity, transport facilities and energy sources. As between 'Great Powers', this way of playing the game is strictly pre-nuclear but may be used by them against lesser powers as an alternative to nuclear strikes. Small powers typically import war materials rather than produce them, so that the targets for strategic air war are transport facilities, especially 'interfaces' such as harbours.

Strategic Intelligence. The collection, collation, evaluation and dissemination of information relevant to military and other decisions. The purpose of the activity is to support planning and to estimate the capabilities and forecast the intentions of other states. In terms of the confrontation of the major powers, this amounts to a coverage of basic economic, social, demographic, scientific and technological trends as well as their product in terms of the military capabilities deployed and the policies adopted.

Most of the raw data is collected by overt means and consists of published and broadcast materials, as well as the visual and other evidence of plant or equipment (mainly by ◊ *surveillance satellites*). Covert means, including classic espionage, provide a small segment of the total data collected, and it tends to relate to intentions (or policies) rather than to capabilities.

The interpretation of the raw data involves processes such as photo-interpretation, code-breaking, statistical analysis and electronic impulse discrimination, as well as plain linguistic translation. The collation of the usable data aims at concentrating data collected from all sources into subject categories and 'fitting' fragmentary data into meaningful patterns.

Evaluation processes (i.e. thinking) are intended to relate the collated information to policy decisions by analysing the information in terms of the problems at hand, or perhaps to suggest new ones.

Dissemination is what renders strategic intelligence productive. Unless the data and interpretations are made available to those who need them, the whole activity is purposeless (and customers do not necessarily know what they need). Strategic intelligence is therefore concerned with a broad range of foreign affairs regardless of actual states of conflict. Intelligence on any one local manifestation of the opponent, especially in times of armed conflict, is known as ◊ *tactical intelligence.*

Strategy. The art of developing and using military and other resources in order to achieve objectives defined by national policy. The term has become very fashionable and it is used to describe plans, tactics, or intended courses of action whether military or not.

Strategy for a specific party facing a specific contingency is defined as above, but military strategy in general is the body of theoretical knowledge applicable to armed conflict. The conventional distinction between strategy and tactics is that the first deals with the development and deployment of military forces – including the selection of the form of warfare and of the 'battlefield' – whereas the latter deals with the use of armed forces within the context set by strategic decisions.

One of the elements in the conventional distinction is that 'strategic' decisions are thought to be 'political' and so a matter for the national leadership, while tactical decisions are thought to be 'technical', and so a matter for the soldiers. This distinction has been eroded by the increased coherence and sensitivity of modern states and lately even fighter-bomber targeting has been based on 'political' decisions.

'Strategic' with reference to equipment means long-range in geographical terms.

Stratofortress B-52. US strategic bomber series. The B-52, which was first deployed in 1955, has been the main manned bomber in the US strategic offensive forces, and about 500 are now deployed with the USAF's Strategic Air Command. The B-52 is a very large subsonic aircraft with swept wings and very large tail surfaces, powered by eight engines. The B-52 production programme involved 744 aircraft in eight variants which introduced progressive improvements in range, payload and avionics. Structural modifications were introduced to improve low-altitude flight performance and – in the latest G and H types – to convert them into platforms for the ◊ *Hound Dog* stand-off missile. As Soviet air defences improved, B-52s were equipped with more electronic countermeasures (◊ *ECM*) equipment, including the decoy missile ◊ *Quail*. More recently, variants C to F have been reinforced to enhance external payload in their non-nuclear role in Vietnam.

The G and H variants, which have increased fuel capacity and turbofan engines, are to be retained until the mid-seventies carrying the ◊ *SRAM* according to present plans, and this reflects the USAF confidence in the 'penetration aids' of this slow (650 mph) aircraft, while the supersonic Hustler B-58 is being retired.

Specifications: Power plant: 8 turbofans rated at 17,000 lb st each. Length: 157 ft 6 in. Height (at tail): 40 ft 8 in. Wing span: 185 ft. Sweepback: 35 degrees. Gross loaded weight: about 488,000 lb. Maximum payload: 75,000 lb. Maximum level speed: 650 mph. Service ceiling: over 60,000 ft. Maximum range (low payload): 12,000 miles.

◊ *Hustler B-58* and FB-111 (◊ *F-111A*) for other US bombers.

◊ *USSR* and *UK, Strategic Offensive Forces* for competitive types.

Strv.103. Swedish battle tank. The Strv.103, also known as S-tank, is the Swedish counterpart of the main battle tanks (MBTs) now coming into service in other countries. Like these, the S-tank has good armour protection, a powerful gun and good automotive characteristics. It is, however, much lower – very important for battlefield survival – and has an automatic loading device which allows almost twice the usual rate of fire and reduces the crew to three from the usual four. At the same time, the S-tank has rather more armour than most MBTs, while being lighter in weight. All these important advantages derive from the totally new design: there is no turret, and the gun is mounted in a fixed position in the centre of the hull. But unlike other turretless vehicles, such as ◊ *tank destroyers*, the S-tank can rapidly engage targets all

round; the gun is elevated or depressed by altering the pitch of the hull (by means of a new type of hydro-pneumatic suspension) and traversed right–left by turning the whole vehicle. This can be done quickly and accurately because of a very advanced steering/differential system. Turrets are heavy and their use involves vulnerability and design-complexity penalties, and the S-tank is the first vehicle to retain MBT capabilities without having a turret. There are also many other innovations: propulsion by a 240-bhp diesel with a 490-hp gas turbine for peak loads; a gyro-stabilized combined periscope/binocular sight, and ribbed instead of solid-plate armour. These characteristics make the S-tank a very good MBT indeed.

Specifications: Length: 22 ft 8 in. Width: 10 ft 10 in. Height: 7 ft. Weight: 36·5 tons. Road speed: 30 mph. Road range: 200–250 miles. Ground pressure: 12·8 lb sq in. Ground clearance: 1 ft 8 in. Main armament: 105-mm high-velocity gun firing APDS and HEAT rounds. 50 rounds carried, all in automatic loader. Night-vision aids. Secondary weapons: remote-controlled machine-gun on hull roof and two more machine-guns fixed to fire parallel to the main gun. Crew: 3.

◊ *Main Battle Tank* for competitive types.

◊ *Armoured Fighting Vehicle* for general entry.

'Styx'. Soviet surface-to-surface cruise missile. The Styx is a cruise missile which is standard armament on Soviet patrol boats of the *Komar* (2 carried) and ◊ *Osa* (4) classes. It has a liquid-propellant sustainer rocket motor and a detachable solid-fuel booster; it has an aeroplane configuration, with cropped delta wings and three identical tailplanes, with trailing-edge control surfaces. The Styx system consists of a search and acquisition radar with a range of about 15 miles for destroyer-sized targets, and ◊ *IFF* transponder to discriminate between 'friendlies' and 'hostiles' and a simple programmer. This feeds target data to the missile and sends it on an interception course. No corrections can be made to the flight path after blast-off, but the missile carries a ◊ *radar* homing device which is programmed to switch on within a few miles

of the target area in order to make final
flight path adjustments. The Israeli
destroyer *Eilat* was sunk in October 1967
by three Styx missiles fired from 15 miles
away: a fourth Styx fell among the sur-
vivors after the sinking.

Specifications (estimates): Length: 20
ft. Wing span: 8 ft 10 in. Weight: about
3 tons. Speed: Mach 0·9. Range: about
20 miles. Warhead: about 800 lb.

**SU-57-2 (and SU-23-4). Soviet anti-
aircraft tank.** This vehicle (SU stands
for *Samochodnaja Ustanovka*) consists of
a large open-top turret with twin 57-mm
high-velocity guns mounted on the very
good chassis of the ▷ *T-54 tank*. The
mounting makes it primarily suitable
for anti-aircraft use, but the SU-57-2
is also effective against ground targets. As with other AA gun systems, the
SU-57-2 is inadequate against Mach 2
aircraft, and the lack of radar fire control
renders it obsolescent. It is well armoured
and has very good cross-country mobility,
but the men in the turret have no overhead protection, and as in other Soviet
AFVs crew comfort, provisions, and communication gear are primitive.

The more modern SU-23-4 is based on
the chassis of the ▷ *PT-76* tank. It has
four 23-mm cannon in the large turret and
(simplified) radar control. It is replacing
the SU-57-2 in the Soviet army.

Specifications: Length: 20 ft 4 in.
Height: 7 ft 10 in. Width: 10 ft 9 in.
Weight: 35 tons. Engine: 520 hp diesel.
Road speed: 40 mph. Road range: 250
miles. Armament: 2 S-60 57-mm guns
with (very long) 73-calibre barrels. Cyclic
rate of fire per gun: 120 rpm. Maximum
effective range: 8,800 metres. Crew: 6,
with the driver in the hull and 5 men in the
turret.

Sub-machine-gun (SMG). A class of
simple and lightweight automatic weapons
which fire low-energy pistol cartridges.
First introduced during the First World
War in the face of opposition motivated by
their unsuitability for drills and parades,
SMGs were popularized by Chicago
gangsters and actual combat conditions.
Almost all SMGs operate by direct blow-
back. When the cartridge is fired, ex-

pansion gases propel the bullet outwards
and then push back the heavy bolt against
a strong spring; as the bolt moves back it
extracts and ejects the spent cartridge and,
as it retreats over the magazine lip, it
allows the next bullet to be pushed up by
the magazine spring. If the trigger remains squeezed, the powerful spring
behind the bolt pushes it forward and
the process is repeated.

SMG (= pistol) ammunition is low-
powered, and the further loss of power in
the mechanism means that muzzle velocity
is very low so that the maximum accurate
range is far shorter than in the case of
rifles. Early crude SMGs, and later crude
SMGs such as the Sten, the US M3 and
the Soviet PPS, fired when the trigger
was pressed – and sometimes when it was
not. Since the war a new breed of SMGs
with safe triggers, good finish and improved configurations has been introduced, and, with good training, these are
useful at up to 200 yards. The Israeli
▷ *Uzi*, the British ▷ *Sterling* L2A3 and
the Belgian Vigneron represent the new
trend. Commandos, armoured troops, and
security forces use SMGs in all armies,
but the regular infantry of different
countries seem to rate the value of SMGs
very differently. Both the Soviet and the
US forces have tried to abolish them.
The former have adopted as official the
APS pistol with a sinister holster that
doubles as a screw-on stock; the US
Army and Marines have a profusion of
indifferent M3s (grease-guns) and carbines but none issued to standard infantry.
Others, such as the Israelis, issue them
widely.

▷ *Machine-gun* and *Rifle* for other small
arms.

Submarine. The submarine emerged as
a key naval weapon during the First World
War, when battleships were involved only
in inconclusive and marginal encounters,
aircraft carriers did not exist, and other
combat vessels did little more than protect
shipping from submarines. This state of
affairs was, of course, due to the fact that
one power, Britain, effectively controlled
the seas and prevented the operation
of the German surface fleet. The submarine was therefore seen as the weapon
of the power whose surface navy was in-

ferior. And it was an inferior weapon, far slower than its rivals on the surface, and lacking in range, ballistic protection and armament, with only a small gun on the deck and perhaps two dozen inaccurate torpedoes for submerged action. Above all, in its only compensating role, submerged cruise and attack (using battery-fed electric motors), its endurance was very limited, and its speed far inferior to that of surface vessels. Because of this, submarines cruised on the surface, diving only for attack or defence.

The submarine suffered from these drawbacks until nuclear propulsion was introduced, though the snorkel (a small-diameter air-intake tube) improved the speed and endurance considerably at shallow depths. But when nuclear propulsion was introduced, the situation was reversed. The nuclear submarine is faster, under water, than surface vessels, while its range, diving depth and endurance below the surface are limited only by crew resistance factors. The speed of surface vessels is constrained by their length (since the ship must be longer than the waves, crest to crest, which its motion causes), but the speed of the submarine is constrained only by its propulsion power, and nuclear power plants can supply all that is needed.

The movement from the submersile surface vessel, which is what old submarines really were, to the true underwater vehicle has led to major changes in role and design. Instead of the knife-edge prows of old-pattern submarines, designed to achieve the high speeds on the surface, nuclear-powered submarines have whale-like bulbous prows intended for maximum submerged speeds. Most modern non-nuclear submarines are also designed to cruise at high submerged speeds, thus sacrificing performance on the surface, instead of the other way round. In place of small conning towers, intended to reduce the boat's profile and so its visibility when surfaced, modern submarines have large fin-shaped ones, known as 'sails', intended for stability. Modern submarines are also equipped with diving planes attached to the 'sail' or the hull; these are used to manoeuvre the boats aircraft-fashion. Other advances include ◊ *inertial navigation* units for submerged

and very accurate position-finding, ◊ *sonar* homing torpedoes and, for US submarines, the ◊ *Subroc* missile.

Nuclear-powered submarines are used as strategic-missile platforms (◊ *Polaris*) as well as for the hunter-killer and anti-shipping roles. In these missions three months or more of unsupported cruise are sustainable, though hunter-killer (i.e. anti-submarine) and anti-shipping missions do not normally require such extended submerged cruises. Communications do, however, present a problem, since only low frequency (usually VLF) radio can be employed, and its performance is not always satisfactory in the underwater medium. The value of FBMSs (Fleet Ballistic Missile Submarines) as missile platforms arises from the low vulnerability of the modern submarine, which cannot be reliably detected if a low speed (which reduces sonar noise) is maintained and certain precautions taken (such as cruising at certain depths). The available submarine detectors – sonar, magnetic and heat sensors, gas sniff and ripple observations – are all relatively inefficient, though nuclear depth charges have a vast kill-radius, thus reducing the need for accurate fixing of the target.

◊ the following seven entries for comment on different countries' individual types of submarine.

◊ *Anti-submarine Warfare.*

Submarines, British. The British navy 7 operates a force of 32 submarines of all types, as well as three ballistic-missile submarines, entered separately under the class name, ◊ *Resolution.*

The *Valiant* class of nuclear-powered hunter-killers is to include seven vessels according to present plans; the first three are already operational. They are similar to the *Thresher* class of US hunter-killers, but have stern diving planes, instead of the wing-like arrangement at the conning tower, and the *Albacore* whale-like hulls of earlier US nuclear submarines. Their armament consists of six 21-in torpedo tubes for which 24 torpedoes are carried. The single *Dreadnought* the first British nuclear-powered submarine, largely based on the US Skipjack class, introduced the British navy to the range, accommodation, detection devices and sustained submerged travel capabilities of this type of weapon.

The *Oberon* class, completed between 1960 and 1964, is the last class of conventional submarines with diesel-electric propulsion. It features extensive use of glass-fibre laminates, advanced sonar, and

fied by alteration of the conning tower (into the large fin shape now standard) and the extension of the hull. This was cut and a new section was inserted between the two halves. Other additions include the search radar, snorkel and improved sonar. The guns and external torpedo

British Submarines – Specifications:

	Resolution†	Valiant	Oberon	Porpoise	A	T
Displacement, full-load, tons						
surface	n.a.*	n.a.*	2,030	2,030	1,385	1,505
submerged	7,500	4,500	2,410	2,405	1,620	1,700
Length (feet)	425	285	295	295	283	288
No. of torpedo tubes	6	6	8	8	6	6
Engine, nuclear (shp)	15,000	15,000	—	—	—	—
Engine, diesel-electric						
diesel drive (bhp)	—	—	3,680	3,680	4,300	2,500
electric drive (shp)	—	—	6,000	6,000	1,250 hp	1,450
Maximum speed (knots)						
surface	20	n.a.*	12	12	19	15
submerged	25	30	17	17	8	9
Maximum radius at						
10 knots (miles)	60,000	60,000	n.a.*	n.a.*	n.a.*	n.a.*
Crew	141	103	68	71	68	65
Year first of class laid down	1964	1960	1957	1954	1943	1941

* not available.
† for comparison purposes; ◊ also *Resolution*.

long snorkel tube. The *Oberon* class is basically an improved version of the *Porpoise* class.

The *Porpoise* class, the first of post-war construction, completed between 1954 and 1957, is oriented towards 'long-endurance operations', i.e. several days below snorkel depth, and some months of loitering (but not cruise) without outside support. Like the *Oberon*, it has a prominent search radar bulge on the super-structures and eight torpedo tubes.

The 'A' class, laid down in 1944–5, are substantially smaller and armed with six torpedo tubes, but have been retrofitted with the fin-shaped conning towers now universal on submarines. Other modifications include the deletion of the 4-in gun and the addition of snorkels and search radar.

The *T* class, of which some are still operational, were laid down between 1941 and 1944 and have been extensively modi-

tubes have been deleted.

Submarines, French. The French ballistic-missile submarine programme is entered under the class name ◊ *Redoutable*, and the following entry refers to the rest of the fleet.

The French submarine fleet consists of ten *Daphne* class medium-to-small patrol vessels, four small 'coastal' hunter-killers, and six larger ocean-going multi-purpose submarines. There are also two experimental submarines, one built recently as a missile-carrying prototype and the other, a former German U-boat, used for general research.

Daphne class. These were completed between 1964 and 1966, and two more are now under construction. Their hull design reflects a compromise in mission requirement and is suited for medium speeds both on and under the surface. Their torpedo tubes are, at 21·7-in diameter, slightly larger than the tubes fitted on British, US, Soviet and almost all other submarines in the world. This reflects a French predilection for non-standard calibres whenever possible, a state which is generally 'interesting' for a variety of material and other reasons but tends to lead to logistic problems in times of war.

Aréthuse class. Four of these were built between 1955 and 1960. They are quite small and probably very effective for their in-shore hunter-killer role (i.e. anti-submarine-submarine). As usual with such vessels, the submerged speed is greater than the surface speed, and much effort has been put into making them as silent as possible.

Narval class. These ocean-going subs were built as part of the first rearmament programme and have immediate post-war design features, largely based on the most advanced German U-boats of 1945.

▷ *Submarine* for general subject entry.

structed Italian submarine originally laid down a few months before the Armistice of 1943, and four modern submarines now being completed: the *Toti* class.

The first of these boats was launched in March 1967. They are very modern and quite small coastal submarines intended for the 'hunter-killer' role, that is against other submarines. They have a compromise hull giving good submerged speeds and, like most modern submarines of their size, only four torpedo tubes. There is a circular housing in the prow which is to be fitted with a search radar.

Specifications: Displacement: 460 tons standard, 524 tons surface, and 582 submerged. Length: 46·7 metres (153 ft 2 in). Beam: 4·6 metres (15 ft 4 in). Draught: 4 metres (13 ft 1 in). Four 21-in torpedo tubes. 2 diesel and one electric motor. Diesel/electric drive on single shaft giving 2,200 hp. Speed: 14 knots on the surface,

French Submarines – Specifications:

	Daphne	*Aréthuse*	*Narval*
Displacement (tons)			
standard	n.a.*	400	1,200
surface	850	529	1,640
submerged	1,040	650	1,910
Length (feet)	190·3	164	256
Beam (feet)	22·3	19	23·6
Draught (feet)	15·4	12·8	18
Torpedo tubes (all 21·7-in diameter)	12	4	8
Engine power			
surface	1,300 bhp	1,060 bhp	4,000 bhp
submerged	1,600 hp	1,300 hp	5,000 hp
Speed (knots)			
surface	16	16	16
submerged	16	18	18
Radius (nautical miles) at 8 knots with *snorkel*	n.a.*	n.a.*	15,000
Complement	45	45	68

* not available.

Submarines, Italian. The small Italian submarine fleet consists of five modernized US submarines of wartime vintage (three *Balao* and two *Gato* class), one recon-

15 knots submerged. Radius: 3,000 miles at 5 knots. Crew: 18.

There is a project for two diesel/electric submarines of ocean-going size, 1,370

standard tons; but since the *Toti* class was projected in 1956 and the first launch did not occur until 1967, no early completion is anticipated.
   ▷ *Submarine* for general subject entry.

Submarines, Japanese. The Japanese submarine fleet consists of eight, soon to be eleven, modern boats completed since

Submarines, Soviet. The Russian submarine fleet is the largest in the world, and there are about 300 submarines of all types in the Soviet navy, deployed in four fleets – the Baltic (about 75), the Arctic (about 150), the Black Sea (about 40), **4**

from poor habitability; the provisions for ancillary equipment are limited.

Japanese Submarines – Specifications:

	Oshio	Hayashio	Oyashio
Displacement (tons)			
standard	1,600	780	n.a.*
surface	n.a.*	n.a.*	1,130
submerged	2,400 (estimated)	1,200 (estimated)	1,420
Length (feet)	288·7	193·6	258·5
Torpedo tubes			
bow	6	3	4
stern	2	none	none
Engine power			
surface	2,300 bhp	1,350 hp	2,700 hp
submerged	6,300 hp	1,700 hp	5,960 hp
Speed (knots)			
surface	14	11	13
submerged	18	14	19
Radius (miles)	n.a.*	n.a.*	5,000 (at 10 knots)
Complement	80	40	65

* not available.
▷ *Submarine* for general subject entry.

1960 in Japanese yards. In contrast with wartime Japanese designs, these emphasize habitability and safety as well as ample provision for sonar, radar and communications equipment.

Oshio class. Three of these boats were laid down between 1963 and 1965 and are already operational; three more are being built. These are ocean-going, deep-diving multi-purpose types with bow vanes and small fin-shaped conning towers.

Hayashio class. Four of these were completed in 1962–3. They are medium-sized patrol types with three torpedo tubes in the bow. Their manoeuvrability is said to be outstanding.

Oyashio. This was completed in 1960, the first built in Japan since the war. It is very fast for a conventionally powered submarine, but appears to suffer

with the remainder in the Far East. As well as the large number of diesel-electrics, there are about 80 nuclear-powered submarines of three main types.

Missile launchers. E (1 and 2): ocean-going types equipped with nuclear-warhead cruise missiles which can be fired only on the surface. ▷ (*Shaddock*). Suitable for off-shore nuclear bombardment; H: long-range type equipped with three vertical tubes for ballistic missiles (▷ *Sark, Serb*) in the large conning tower (and not in the hull as in the US ▷ *Polaris* types). The short range of the missiles limits their value as strategic weapons, and the small number of missiles reduces their cost-effectiveness *vis-à-vis* the Polaris types. But a Y class of Polaris copies is now in accelerated production, and ten were operational by 1970.

Attack type. N: this is the only nuclear-powered 'hunter-killer' class in the Soviet navy. These are very fast boats equipped with six 21-in torpedo tubes.

The Soviet navy has a larger number of conventional diesel-electric submarines in operation than the US navy, and some of them are equipped with missiles:

J: A modern, medium-sized class with a high surface freeboard. It has four launchers for cruise missiles and four torpedo tubes in the bow.

G: A ballistic-missile carrier with three tubes in the very large conning tower. It is slower than the J class on the surface but faster when submerged, and it has a 22,700-mile surface range. The Chinese navy has a single G class submarine but not the Sark or Serb missiles that go with it. There are six 21-in torpedo tubes. See also Z and W classes.

The main attack types are:

F: The most recent type of non-nuclear submarine, these are large ocean-going boats equipped with eight 21-in torpedo tubes for which 20 torpedoes are carried. One F class submarine has been supplied to India.

Z: This class of ocean-going submarine was in production until 1960, and formed the basis on which the F class was developed. It carries 24 torpedoes which can be fired through eight 21-in tubes, or 40 mines. Some Z class submarines have been converted to carry two launch tubes for ballistic missiles and have a larger crew (85) and a higher speed (16 knots) when submerged. Other Z class submarines have been supplied to satellite countries or converted into radar pickets. The range is estimated at 26,000 miles in optimum surface cruise.

R: Medium-range submarines with six 21-in torpedo tubes, developed from the W class.

Q: A smaller medium-range class, the last of which was completed in 1960. The cruising range is estimated at 7,000 miles, and there are four 21-in torpedo tubes in the bow. These boats are propelled by a single screw and carry fifty tons of fuel oil.

W: The most numerous boats in the fleet, of which about 170 had been built when production ended in 1957. They have a maximum range estimated at 16,500 miles and are armed with four bow and two stern 21-in torpedo tubes. The W class boat can operate as a mine-layer, and 40 mines can be carried. Some W class submarines have been converted to carry cruise missiles and have an unusual ramp/launcher behind the conning tower. Two submarines of this class have been adapted for oceanographic research.

Soviet Submarines – Specifications:

	E-1	E-2	H	N	J
Displacement (tons)					
surface	5,000	4,600	3,500	3,200	1,800
submerged	5,600	5,000	4,100	4,000	2,500
Dimensions (feet)					
length	393·7	385	328	328	328
width	33	33	33	32	27
draught	27	27	25	24	20
Speed (knots)					
surface	n.a.*	n.a.*	n.a.*	n.a.*	19
submerged	22	20	30	30	15
Crew	100	92	90	88	85
Engine type	nuclear	nuclear	nuclear	nuclear	d/e
Engine, bhp diesel	—	—	—	—	n.a.*
hp electric	—	—	—	—	n.a.*

* not available.

MV: An obsolescent class still operational with client-state navies. These boats are very small, carry two 21-in torpedo tubes and have a 4,000-mile range at 10 knots in surface cruise, or 100 miles submerged at 5 knots.

6   Submarines, US. The US navy's submarine fleet consists of about 100 conventional and nuclear attack submarines and 41 ballistic-missile submarines of the *Lafayette, Ethan Allen,* and *George Washington* classes, which are entered separately under ◊ *Polaris.*

Nuclear-powered (SSN) 'attack' or 'hunter-killer' types include the following classes.

Thresher (or *Permit*). The main class of this category, still in production. They have 'tear-drop' hulls and diving planes attached to the sail, so that the assembly has a tailplane appearance. The automated controls operate the vessel for all routine (diving and steering) operations, and this class is supposed to be quieter than other submarines. *Thresher* itself was lost in 1963, and the class has been renamed *Permit.* There are four torpedo tubes set amidships, instead of in the bow, and the outer tubes are intended for the ◊ *Subroc* ASW missile. The maximum submerged speed is stated to be 30 knots but is widely believed to be substantially higher, given the use of advanced drag-reduction methods. The *Thresher* class is divided into several sub-classes but the specifications of only the original and the latest (Narwhal) sub-classes are given below.

Skipjack. An earlier and somewhat smaller class. They have a whale-shaped (so-called *Albacore*) hull and the basic diving plane, propulsion unit and single propeller features retained in the *Thresher* class. There are six bow 21-in torpedo tubes with 24 torpedoes normally carried.

Skate. The first group of nuclear-powered submarines based on the *Nautilus* class, with twin screws, and the *Nautilus*'s bulbous prow. Instead of the conning-tower diving planes of the later types, these have stern diving planes. The original *Skate,* which named the class, cruised a record 120,862 miles on her first nuclear core and was 'refuelled' after 39 months. Noise-reduction modifications have been made. Other nuclear-powered submarines include the *Triton,* the longest and most powerfully engined submarine ever built, with a 'knife-blade' prow for high surface speed, instead of the bulbous *Albacore* and 'tear-drop' prows of the other nuclear-powered types. The *Triton* (a single vessel, not a class) was originally designed as a radar picket, which explains

G	F	Z	R	Q	W	MV
2,350	2,000	1,900	1,100	650	1,030	350
2,800	2,300	2,200	1,600	740	1,118	420
320	300	295	246	185	240	167·3
28	27	26	24	18	22	16
22	19	19	14·5	13	15	12
17·6	20	20	18·5	18	17	13
17	15	15	15	16	15	10
86	70	70	65	40	60	24
d/e	d/e	d/e	d/e	d/e	d/e	d/e
6,000	10,000	10,000	4,000	3,000	4,000	1,000
n.a.*	4,000	3,500	2,500	2,500	2,500	800

the surface-orientation of her hull design. She has three deck levels and the largest complement (170) of any submarine. She has now been converted to a 'hunter-killer' role with four bow and two stern torpedo tubes.

The *Tullibee* (another single vessel) inaugurated the 'tear-drop' hull used in the *Thresher* class and was used as a testbed for hydrophone and sonar devices. It has four 21-in torpedo tubes.

Halibut was originally designed as a platform for cruise missiles, and when these (Regulus 1) were removed, she was converted to a 'hunter-killer' role by the removal of the missile equipment. *Halibut* has four 21-in torpedo tubes.

The *Nautilus* (completed in April 1955) was the first nuclear-powered submarine and a second prototype submarine, the *Seawolf*, was completed two years later. Both are used for experimental purposes.

Non-nuclear submarines include the following attack classes:

Barbel. Three submarines of this class were the last non-nuclear 'hunter-killers'. They have the *Albacore* hull used in the later nuclear types; the conning tower is small and lacks the prominent diving fins of those later types. They have six 21-in tubes forward. The last of this class (*Blueback*) was completed in 1960.

Tang. A class of six submarines laid down in 1949–50 and completed in 1951–2. They have six forward and two aft 21-in torpedo tubes. The experimental (extra-silent) *Darter* is a *Tang* conversion.

Tench. A late wartime class of which about 20 types are still operational. There are six torpedo tubes forward and four aft. Almost all vessels have been retrofitted with extended snorkel and other improvements.

Balao. A wartime class of which about 50 remain operational. Many have been converted to T training types with closed torpedo tubes and no propellers. Others have been retrofitted with improved sonar and snorkels (G-type). Still others, designated F G, have been fitted with new conning towers and extended hulls (by the addition of a new middle section to the two sectioned halves). *Balao* class submarines have the old-style arrangement of six bow and four stern torpedo tubes.

US Submarines – Specifications:

	Thresher*	Skipjack
Displacement (full-load, tons)		
surface	3,750	3,075
submerged	4,300	3,500
Length (feet)	279	252
No. of torpedo tubes	4	6
Engine, nuclear (shp)	15,000	15,000
Engine, diesel-electric		
diesel drive (bhp)	—	—
electric drive (shp)	—	—
Maximum speed (knots)		
surface	20	20
submerged	30	35
Radius (miles at 10 knots)	60,000	60,000
Crew	107	93
Year first of class laid down	1959	1958

* The *Narwhal* sub-class of the *Thresher* class is now in production. This displaces 4,450 tons on the surface and 5,350 tons submerged. The length has been increased to 303 feet, and the power plant uprated to 17,000 shp. Like the main *Thresher* and *Skipjack* classes these boats are equipped with the Subroc missile.

Two *Balao* class submarines, the *Perch* and the *Sealion,* have been converted into troop carriers (for frogmen, commandos and marines) with an amphibian capability; they can carry 184 officers and men apart from the 74-man crew.

Other submarines include the experimental *Dolphin,* the only post-1960 conventionally powered submarine, intended for deep-diving experiments; the two *Grayback* class submarines originally intended as platforms for the now obsolete Regulus cruise missile; two *Sailfish* class submarines, intended as radar pickets but now converted to the attack role; the single *Albacore* used as a hull prototype with a 'tear-drop' configuration and used for testing sonar and other devices; and various *Balao, Tench* and *Gato* class submarines which have been reconverted from the obsolete radar picket role into attack, training or experimental vessels. Two of the smallest submarines built in recent times are the *Marlin* and *Mackerel* T-types, used for training purposes, which have a single torpedo tube and an 18-man crew. The *XI,* a 36-ton (submerged) micro-submarine, and the *Barracuda* are two non-operational types.

The *Nautilus,* laid down in 1952, is 320 feet long and displaces 3,764 tons on the surface and 4,040 tons submerged. The nuclear power plant is rated at 15,000 shp, and, unlike later submarines, it also has a full diesel-electric stand-by plant. The maximum speed on the surface is 20 kts and it can travel at 23 kts below the surface.

Submarines, West German. The West German submarine fleet is being developed on the basis of two classes, a 'coastal' type with a submerged displacement of about 450 tons and a larger 'hunter-killer' type with a submerged displacement of about 1,000 tons. The coastal class, numbered U1 to U24, are handy boats with 'tear-drop' hulls and eight torpedo tubes in the bow. U1 to U3, which were launched between 1961 and 1962, developed various problems, and these led to a modified and heavier design for the subsequent boats. The modifications were retrofitted but not the non-magnetic hulls of U4 to U8. U9 and

Skate	Barbel	Tang	Tench	Balao
2,570	2,150	2,100	1,800	1,816
2,861	2,895	2,400	2,500	2,425
268	219	269	311	312
6	6	8	10	10
6,600	—	—	—	—
—	3,100 hp	4,200	6,500	6,500
	n.a.[†]	3,200	4,610	4,610
20	15	15	20	20
25	25	18	10	10
n.a.[†]	n.a.[†]	25,000 surface	14,000 surface	12,000 surface
95	77	83	82	80
1955	1956	1949	1944	1942

[†] not available.

subsequent boats use special steel for magnetic damping instead of the zinc sheathing used on U4 to U8. Only eleven of these boats were operational by 1970. The 'hunter-killer' type was due to be delivered as of 1967 but the schedule has been affected by the 1966 defence cuts and may be further delayed or altered. These large boats are to have an unspecified number of bow torpedo tubes for homing torpedoes.

There are also two midget submarines used for training and research purposes, which were launched in 1965. The midgets have two torpedo tubes, a maximum crew of six, and a submerged displacement of 150 tons. Two old wartime submarines, the *Wilhelm Bauer* and the *Hecht* (ex U2540 and U2367), are used for weapon trials.

turns upwards and leaves the water. A thrust vectoring system then propels it towards the target area; at a set distance from the target, the depth bomb disconnects and continues a guided but unpowered descent to the target. It impacts in water at supersonic speeds and then sinks, with the nuclear warhead exploding at a pre-set depth. Subroc equips nuclear-powered attack submarines. ◊ *Submarines, US*.

Specifications: Length: 21 ft. Maximum diameter: 1 ft 9 in. Weight: 4,000 lb. Maximum range: 30 miles (estimate).

Subversion. The acquisition of disguised but effective control over a population or group which is supposed to be under the control of some constituted authority. Subversion is the principal

West German Submarines – Specifications:

	Hunter-killer	Coastal
Displacement (tons)		
surface		370
submerged	1,000	450
Length (feet)		142·7
Beam (feet)		15·1
Torpedo tubes		8
Diesel		12,00 bhp
Electric		1,700 bhp
Speed (knots)		
surface		10
submerged		17
Complement	60	21

Subroc UUM-44A. US submarine-to-submarine missile. The Subroc is a very advanced long-range submarine-to-submarine missile system with a nuclear depth-charge warhead. It consists of (a) detection equipment to identify undersea targets, (b) computer equipment to calculate missile intercept courses, and (c) a ballistic missile fired from torpedo tubes which is fitted with a nuclear depth bomb and its own guidance unit. The missile ignites at a safe distance from the submarine and follows the set path controlled by an ◊ *inertial guidance* system; after underwater ignition, Subroc

technique of ◊ *revolutionary war*, but it is used in all forms of political warfare. In Vietnam for example, the agents of subversion extract food and conscripts from a population theoretically governed by the Republic of Vietnam. When this level of subversion is achieved, the constituted authority is in apparent control (since its military and security forces 'hold' the territory), but the population is actually administered by the agents of subversion. Thus in the context of revolutionary war, subversion = covert administration.

The tools of subversion are ◊ *pro-*

paganda and terrorism. Propaganda conditions the population to accept a certain image of the agents of subversion and their aims; terrorism is used to intimidate or eliminate those who can or could lead popular action against the subverters. Its main purpose is to reduce the target group to isolated individuals unwilling to risk their lives to oppose the organized forces of subversion. Once the population is atomized, propaganda plus the mere threat of terrorism is sufficient to control it. The agents of subversion can then covertly collect food and conscripts by simple request instead of clandestinely at gun-point. These resources are then used to supply the guerilla forces, the other arm of revolutionary war.

In order to function effectively the agency of subversion needs (a) a covert network of 'civilians' operating in their home communities and (b) a separate group of clandestines. These may be part of the guerilla forces, but their role is really to link the 'civilians' with the 'guerillas'. The 'civilians' carry out the propaganda action, act as informers for the executioners within the clandestines, and collect combat intelligence for the guerilla forces. Thus terrorism can be used effectively against carefully selected targets (for example, an honest Government official is a better target than a notoriously corrupt one, whose actions are useful for propaganda); at the same time surviving police authorities cannot find the gun-men, who do not belong to the area and who come in it only briefly, to do their job. The more subversion the more food, information and conscripts for the clandestines and guerilas, who can then further reduce the government presence. This facilitates further subversion. . . . Farmers do not like to give their crops and their sons to the guerillas, but they can withold them only if they are effectively protected against the agents of subversion and their clandestine allies. Only ground security and counterterrorism can protect the farmers, and not air power or conventional forces.

Sukhoi-7B. ◊ *Fitter.*

Sukhoi-9. ◊ *Fishpot.*

Sukhoi-11. ◊ *Flagon.*

Surveillance Satellite. An artificial earth satellite equipped with optical and other sensors which is used to collect data primarily intended for intelligence purposes. These differ from weather-observation and resource-identification satellites mainly in terms of the far higher precision of the sensors and in the exclusive recovery of the data by the party who deploys the system (unlike weather-satellite data, which is freely available to all).

There is little doubt that surveillance satellites are the most important by-products of the entire space-exploration effort, both from the point of view of Western defence needs and in terms of the more general ◊ *arms control* interest. Although both the Soviet Union and the United States now deploy surveillance satellites, it is clear that the latter obtains far greater direct benefit from the new technology. The 'intelligence balance of power' is now a great deal closer to equilibrium than it was in pre-satellite days, when Soviet planners knew almost everything about US defence from freely available open sources, while US planners knew only what Soviet planners wanted them to know, except for the limited and unreliable data gathered by covert means.

The high-quality intelligence now available to defence planners favours arms control, since 'missile-gap' and consequent over-reaction phenomena are far less likely than in the days when faulty intelligence – and deliberate deception – combined with the natural prudence of defence planners led to destabilizing deployment policies for strategic weapons. Multilateral arms control by formal argeement is even more favoured, since surveillance by satellites can often, though not always, replace inspection arrangements (which the Soviet Union has consistently refused to accept). On the other hand, satellite mapping and the pin-point location of missile silos also provide the intelligence ingredient for dangerous ◊ *counterforce* deployments aimed at the other side's deterrent forces, but this is more than offset by the stabilizing effects of the technology.

The sensors available on currently operational surveillance satellites include both still and moving optical cameras, multi-spectral cameras (for penetrating clouds, ground cover and deliberate camouflage), TV receivers for direct retransmission, and other electronic receivers which can intercept communication, telemetry and electro-magnetic emanations in general. The data is transmitted both by direct radio link (usually in a coded and/or scrambled form) and by the recovery of film and tape casettes. (The United States air force operates a fleet of aircraft equipped with nets which catch the re-entry capsules ejected from the satellite in orbit.)

The current surveillance satellite effort of the United States is organized into the Integrated Satellite System (ISS), which has replaced the series of partial programmes which apparently started with the SAMOS system first deployed in 1961. Very little information has been published about the precise capabilities of the ISS sensors, but it is known that they can unambiguously identify fixed-site land-based ICBM silos and detect missile firings, including abortive ones. TV pictures of large-scale troop concentrations and movements can be televised directly back to the ground; many kinds of radio and microwave communication can be intercepted, including others' missile telemetry; radar and other electronic apparatus can be identified and located and their capabilities assessed. The sensor and recovery hardware of the satellites is supplemented by a great deal of 'software': automatic scanning techniques for processing the vast amount of data produced; bargaining and technical procedures to allocate expensive satellite 'space' to different intelligence collection requirements; and human evaluation of the data produced by the hardware. (The early phases of satellite surveillance were characterized by gigantic backlogs of unprocessed data.)

Surveillance satellites make easy targets, with their low (100 to 300 miles) and regular orbits. Even if they are manoeuvrable, the isolation of the satellite in space makes it a sitting duck for homing systems, while ordinary ICBMs can readily be reprogrammed to deal with satellite targets of the non-manoeuvrable kind. The Soviet Union has already experimented with satellite-to-satellite kill techniques, and one 'cosmos' destroyed another by conventional high-explosive detonation in one of these tests.

Surveillance satellites have the technical capability of providing earliest (that is, blast-off stage) tactical warning of an impending nuclear attack, but the MIDAS programme, intended to do just that with full coverage of Soviet missile silos, is not believed to have reached operational stage within the ISS. In any case, such warning does not now, and probably will not in the future, serve as a 'trigger' for nuclear retaliatory strikes.

'Swatter'. Soviet anti-tank missile. The Swatter is a wire-guided anti-tank missile which may have terminal homing IR guidance. It is 3 ft 8 in long and has 2 ft 2 in wing span.

Swingfire. British anti-tank missile. The Swingfire is a second-generation wire-guided anti-tank missile specifically designed for vehicle mounting but man-portable. It has an autopilot, which keeps the missile flying in a straight line set by an automatic programme generator built in the ground equipment. Older wire-guided missiles, such as the French SS.11 (♢ Nord) series, have acceleration control in which the operator keeps the missile on target by making repeated corrections with a joystick. He sees a small flare behind the missile and he sees the target; if the two are out of line he makes adjustments by joystick moves carried to the missile by the link wire paid out by the latter. Such repeated corrections need a skilled operator and impose low (subsonic) missile speeds to allow for the changes. Swingfire is automatically gathered on the programmed path, compensating for traverse and elevation differences automatically. The operator, who can be up to 150 ft away, uses velocity control to make final corrections to the flight path. Velocity control allows higher speeds and needs less skilled operators.

Specifications: Missile: 42 in long, HEAT warhead, solid-fuel rocket propulsion, swivelling-jet guidance, four folding wings. Ground equipment: battery-operated programmer, fault detector, tilt detector with periscope and wire connexion. Can be mounted in the smallest vehicles.

T

16 **T-54 (includes T-59, T-55, T-62). Soviet battle tank.** Ever since the appearance of the T-34/76 in 1940 Soviet designers have been ahead in the field of armour, setting trends followed by others rather than competing with existing designs, their basic combination of gun power, armour and mobility in cheap and reliable designs being far superior to Western tanks, at any rate until quite recently. On the other hand, the communication gear, fire control and crew comforts have been consistently inferior.

The T-54, which was introduced in 1955, is of conventional layout, with driver in the hull and a three-man turret crew; the armour is very well used with no angular surfaces or 'shot traps' in the turret and well-sloped frontal plates. The T-54 appeared with a 100-mm gun, larger than those on contemporary Western tanks, and, at the same time, was lighter and far more mobile than its competitors. Only the introduction of 105-mm guns in most Western tanks, firing the superior APDS ammunition, and carrying more of it, has outmatched the T-54's armament. (But ◊ T-62 below.) Other good features of the T-54 include the wading capability (to 18 feet) and the infra-red devices for night vision. Its defects are the cramped turret and driver position, and, more important, the quality of the armour.

The T-54/A (a NATO designation) introduced vertical gyro-stabilization, a bore evacuator and better vision devices.

The T-54/B added a narrow-beam 'shooting' searchlight, better periscopes and an optical range-finder.

The T-59 is a Chinese copy of the T-54, without some of the vision and fire-control devices.

The T-55 is an improved T-54 with minor modifications, including the removal of the turntable on the loader's hatch, a large-diameter tube for deep wading, and better infra-red vision and aiming aids.

The T-62 is the T-54/T-55 replacement, the first Soviet main battle tank, since the T-54 was to be supported by the heavier JS III/T-10 122-mm gun tank. The basic layout and smooth general appearance have been kept, but there is an enlarged hull and turret. This mounts a more powerful, 115-mm, gun which may be equipped with a laser range-finder. The turret ring is larger and moved farther back, probably to accommodate a filtration/ventilation system for protection against fallout and biological or chemical weapons. In spite of these improvements, the T-62 weighs only one ton more than the T-54 series.

The Soviet army received its first T-62s in 1965, but since they have not yet been delivered to the Egyptians, few details are available.

Specifications (T-54 basic): Length: 19 ft 8 in. Width: 10 ft 9 in. Height: 7 ft 11 in. Weight: 36 tons. Engine: water-cooled diesel 580 bph. Road speed: 34 mph. Road range: 220 miles. Ground pressure: 11·8 lb sq in. Ground clearance: 1 ft 5 in. Main gun: 100 mm with 35 rounds of APC ammunition (◊ AP). Secondary weapons: two MGs; one heavy MG on turret roof (this is not present in the T 55). Crew: 4.

◊ *Main Battle Tank* for competitive types.

◊ *Armoured Fighting Vehicles* for general entry.

Tabun. ◊ *Nerve Gases.*

Tactical Intelligence (including Combat Intelligence). The collection and evaluation of information about enemy forces in or on a given terrain or about the terrain itself. 'Tactical intelligence' is therefore limited to a given local and current situation; beyond these limits, the proper term is *strategic intelligence.* 'Combat intelligence' is synonymous but sometimes has the implication of being conducted at the level of single operational units (division or brigade) rather than at the highest level amongst armed forces deployed in a particular theatre (area command).

The main sources of raw data for tactical intelligence are: (a) ◊ *reconnaissance* operations and the reconnaissance 'angle' of combat operations; (b) interrogation of prisoners, deserters and the civilians who live in or have moved to the occupied area (i.e. refugees); (c) clandestine operations and classic covert espionage operations (though these play a marginal role in most circumstances). Tactical intelligence evaluation and dissemination is usually conducted at the level of the area or 'front' command with detached officers in major operational units; each branch of the armed forces represented in the area will also have its complementary (or competitive) HQ. The detached officers are in any case normally members of the branch of the armed forces to which the operational unit belongs. For example, the intelligence structure of the X-country expeditionary force in Y-country or sector will consist of (a) an 'Intelligence Staff' at general HQ level, (b) army, air force and naval intelligence HQ, and (c) intelligence officers with each wing, division and naval squadron.

Tactics. The 'art' of using armed forces in battle; more generally, elements of knowledge or belief applied to decisions over armed forces deployed in a particular setting or 'theatre' determined by a strategic decision. Conventionally, strategy is supposed to cover the development and deployment of forces; tactics covers their use once their composition and location are set. For example, the strategic response to an air defence problem may lead to the development and deployment of x interceptors, y missiles and z radars. The coordination and particular use of x, y and z would reflect tactical decisions. More conventually, if country A goes to war with country B and decides to invade it, the choice of the invasion axis is 'strategy', whereas the mode of the invasion is 'tactics'. Traditionally, 'strategy' has been regarded as an area of decisions made at the highest 'political' level, whereas tactical decisions are the proper concern of 'technicians', i.e. soldiers. In fact, the progressive improvement in communications has meant that battlefield decisions at lower and lower levels can be controlled at long range, and such control is increasingly being exercised by political authorities. In the case of the bombing of North Vietnam, individual targets were selected at the level of the US Presidency: here the distinction strategy/tactics was therefore almost non-existent.

The adjective 'tactical' is loosely used to mean short-range (as of a bomber), small or unimportant (as of a retreat); tactical as applied to nuclear warheads means that they are intended for use against a battlefield manifestation of the enemy, rather than against his homeland. In this sense, 'tactical' is sometimes used as a euphemism, in order to disguise an increase in the level of violence.

For some tactical concepts, ◊ *Indirect Approach*, *Blitzkrieg* and *Defence in Depth*; these cover broad problems in terms of a general posture and are sometimes described as 'grand tactics'. For more specific, or low-operational-level tactical concepts, ◊ *Hedgehog*, 'Fire and Movement', 'Marching Fire'. ◊ *Combat Aircraft, Combat Vessels, Fighter, Armoured Forces* and *Anti-submarine Warfare* for main tactical problem areas and some solutions.

◊ also *Guerilla Warfare, Nuclear Warhead, ECM* and *Penetration Aids* for the tactical end of eminently strategic issues.

Tallinn Line. US term for a Soviet radar-computer-communication network intended to provide early warning and weapon-control against US bombers and their stand-off missiles flying the Polar route to the Soviet Union. Initially many US analysts associated the Tallinn

Line with the Galosh 'anti-missile-missile', with the two forming a ballistic-missile defence system (◊ *ABM*). Later, the Tallinn Line was estimated to be a much less ambitious anti-bomber system. ◊ *PVO Strany* for the Soviet air defence system.

Talos RIM-8E. US shipborne surface-to-air missile. The Talos is a medium-range anti-aircraft shipborne ◊ *missile* with a choice of alternative nuclear or high-explosive warheads, which is currently deployed on US guided-missile light cruisers. The Talos has a cylindrical body with three sets of cruciform wing/fins. The solid-fuel rocket booster weighs about 4,000 lb, and flight is sustained by a ramjet fuelled with a kerosene/naphtha mixture. Guidance is by beam-riding, with semi-active radar homing in the final stages (for which there is a shipborne target-illuminating radar); there is, additionally, a proximity fuse. The Talos can also be used against surface targets. Like the Terrier and the Tartar, the Talos had a long and unsatisfactory development history.

Specifications: Length: 31 ft 3 in. Diameter: 2 ft 6 in. Wing span: 9 ft 6 in. Weight: 7,000 lb. Speed at burn-out: Mach 2·5. Range: 65 miles + (slant).

◊ *Terrier, Tartar* and *Standard* for other US shipborne missiles.

Tank. ◊ *Armoured Fighting Vehicle, Main Battle Tank, Light Tank.*

Tank Destroyer. A specialized ◊ *armoured fighting vehicle* intended to fight tanks and other armoured vehicles. Originally, these turretless tracked and armoured vehicles, with a powerful gun mounted in a frontal casemate instead of a revolving turret, were obtained by converting tank chassis. The turret was removed, and a larger gun was put in a limited traverse position within an armoured frontal box. This reduced cost, increased ammunition-carrying capacity, and made a bigger gun possible for the same weight. Examples of this type include the Soviet wartime SU 100 and JSU 122-mm, still used outside the USSR; and the Swiss G-14 Jagdpanzer. Only the Soviet Union and West Germany

have continued the development of this kind of vehicle: the USSR with light-weight air-transportable vehicles such as the ◊ *ASU 85*, and West Germany with the ◊ *Jagdpanzer Kanone*. But specialized tank destroyers have instead appeared in many different configurations: the US M.50 'Ontos', an 8-ton vehicle with six recoilless rifles; and the various French and Swiss wheeled vehicles with 90-mm guns firing fin-stabilized rounds (i.e. the ◊ *AML*). The availability of efficient anti-tank missiles ideally suited for defensive deployment (i.e. the tank destroyer role) has led to missile-carrying conversions of ◊ *armoured cars*, ◊ *armoured personnel carriers* and ◊ *light tanks*. Only one specially designed vehicle has so far been produced: the West German Jagdpanzer Rakete, this being the long-range counterpart of the 90-mm Jagdpanzer Kanone in defensive anti-tank deployments. Light tanks with powerful weaponry, such as the US ◊ *Sheridan M.551* and the French ◊ *AMX 13*, are also described as tank destroyers.

Tartar RIM-24. US shipborne surface-to-air missile. The Tartar is a small US shipborne anti-aircraft ◊ *missile* with a slant range exceeding ten miles. It is powered by a dual-thrust rocket, with an initial high-thrust acceleration phase and a longer, lower thrust sustainer. Guidance is by active homing. Tartar missiles are deployed on ◊ *destroyer* class vessels and currently equip 31 US navy vessels as well as those of several other countries. Like other early US naval SAM missiles, the Tartar had a long and troubled development history.

Specifications: Length: 15 ft. Body diameter: 13 in. Launching weight: about 1,200 lb. Speed at burn-out: Mach 2·5. Slant range: 10 miles +. Effective heights: 1,000 to 4,000 ft.

◊ *Terrier, Talos* and *Standard* for other US shipborne surface-to-air missiles.

Terrier (Advanced Terrier) RIM-2. US shipborne surface-to-air missile. The Advanced Terrier, an improved version of the original Terrier, whose development was long and unsatisfactory is operational in US ships of the frigate class and above as a short–medium range anti-aircraft ◊ *missile*. Advanced Terrier has low aspect ratio wings and movable tail surfaces. Guidance is by the beam-riding method, and the system is also effective against surface targets. Propulsion is by a two-stage solid-fuel rocket engine. The booster stage weighs about 1,900 lb and has a 16-in diameter, while the 12-in diameter sustainer-warhead stage weighs 1,100 lb. The Terrier is used in association with a search radar and an electronic fire-control system; these are integrated in the Naval Tactical Data System.

Specifications: Length 27 ft. Wing span: 1 ft 8 in. Total weight: 3,000 lb. Range: 20 miles +.

◊ *Talos*, *Tartar* and *Standard* for other US shipborne surface-to-air missiles.

Test Ban Treaty. This Treaty is probably the most important ◊ *arms control* agreement concluded since the introduction of nuclear weapons. It was initially signed by the Soviet Union, the United States and Britain in 1963, but over one hundred countries have since adhered to it, though two nuclear powers, China and France, have refused to sign and continue testing nuclear weapons in the atmosphere. The terms of the Treaty prohibit test detonations of nuclear devices except underground, both outer space and underwater tests being prohibited as well as atmospheric ones. Provisions of the Treaty include the prohibition of atmospheric testing by proxies who are not signatories, i.e. any X-country assistance to a Y-country testing programme is prohibited, on the presumption that test results would be shared by X-country.

Although a large number of underground detonations have been carried out by both the Soviet Union and the United States, the Treaty is a severe constraint to the further development of nuclear weapons by the two countries. Underground testing is less reliable as a source of data for the development of certain nuclear weapons, i.e. those whose kill-effect relies on radiation rather than blast effects. Underground testing is also more expensive and – for the United States at

any rate – raises internal political problems in areas within which the tests take place.

Any signatory power can resume testing within three months of announcing its intention to do so, this 'escape' clause being seen as essential to protect the parties against a rival's technological breakthroughs. But it is generally agreed that the political costs of resuming atmospheric testing would be exceedingly high for any signatory, and especially the two super-powers. Attempts to extend the coverage of the Treaty to include underground testing initially foundered on the key problem of inspection. The Soviet Union refused to allow the deployment of adequate seismic control facilities on Soviet territory, while the United States refused to accept instead the less reliable remote seismic detectors located outside the Soviet Union. ◊ *CTB* for the continuing debate on this issue.

Threshold. Recognizable demarcation lines between levels of violence in conflicts between states. The value of the concept lies in its application to nuclear war, once widely thought of as (automatically) total. A set of thresholds is implicit in the policy of ◊ *controlled response*. This would include at least the following thresholds:

1. Sub-megaton weapons used on the battlefield.
2. Any nuclear weapon used on 'not-homeland' targets.
3. Recognizably selective nuclear strike on homeland.
4. 'Spasm', a term used by Herman Kahn to describe the situation where a state about to die simply uses all the nuclear delivery means at its disposal.

Throw-weight. The total weight which a missile can deliver over a stated range and in a stated trajectory; this depends on the power of the booster-sustainer combination as well as on the intended range and the trajectory chosen. Unlike 'payload', throw-weight includes the total re-entry vehicle (◊ *REV*) plus guidance unit weight.

Thunderbird. British surface-to-air missile. The Thunderbird is an anti-aircraft missile with solid-fuel rocket propulsion and ◊ *semi-active radar homing* guidance. It is a point-defence weapon system for field army use, consisting of the missile, power generators, launching ramp, and search and 'illumination' radars, all of which are air-portable and ground mobile. ◊ *Bloodhound* for the guidance method. The Thunderbird Mk 2 was developed from the Mk 1 by the substitution of a new radar unit in the same basic system.

Specifications: Power plant: 4 solid-fuel rocket-boosters and solid-fuel rocket sustainer. Length: 20 ft 10 in. Body diameter: 1 ft 8 in. Warhead: high-explosive fitted with proximity fuse.

Thunderchief F-105. US fighter-bomber. The Thunderchief is a large single-seat fighter-bomber, specially developed as such, instead of being an adapted fighter-interceptor like most other fighter-bombers. It is a large aircraft with a pronounced bulge in the rear fuselage and swept-forward air-intake ducts. The main production version, the F-105D, has a monopulse doppler radar system (N.A.S.A.R.R.), which provides 'all-weather' coverage for air search, automatic tracking, ground-mapping and a limited degree of terrain avoidance.

Fixed armament consists of a single multi-barrel 20-mm gun with 1,029 rounds of ammunition, and alternative external loads include a wide range of missiles, bombs, napalm and fuel tanks and rocket packs. Though air-to-air missiles can be carried (and the Thunderchief can 'look after itself' in air-to-air combat), air-to-ground weapons are standard, and up to sixteen 750-lb bombs can be carried on under-wing pylons. The Thunderchief F-105D has been retrofitted with armour-plating and provisions for the ◊ *Walleye* guided weapon. They have been widely used in the Vietnam war, but production ceased in 1963.

The speed performance of this plane is very good for a fighter-bomber of 1954 prototype vintage: Mach 1·1 at sea level and Mach 2·1 at 36,000 ft. It has a maximum range of 2,070 miles and a 32,000 ft-per-minute rate of climb at sea level. The maximum take-off weight is 52,546 lb.

◊ *Fighter* for general subject entry.

Tigercat. ◊ *Seacat.*

Titan II LGM-25C. US intercontinental ballistic missile. A complex, heavy and expensive liquid-fuelled ballistic missile which first became operational in 1963 and of which about 54 are still operational. Titan II is an improved version of the Atlas and Titan I generation of ICBMs, which have been replaced by the cheaper and more reliable ◊ *Minuteman* series. Titan II is silo-launched (hardened against 100 psi of blast) and was the first US ICBM to navigate exclusively on ◊ *inertial guidance,* dispensing with the radio commands required in earlier systems. Its motor was also an advance on earlier liquid fuels in that it is storable, so that count-down was cut to about one minute; and it can be fired from its rest position within its silo.

Titan II weighs 360,000 lb, has a 9,000-mile range, and carries a large warhead exceeding five megatons in power.

◊ *US, Strategic Offensive Forces* for other US strategic missile systems.

TNT and other high explosives. Trinitrotoluene or TNT, $CH_3C_6H_2(NO_2)_3$ was first developed in Germany and there first tested in 1890. It was extensively used during the First World War as a bursting charge for artillery shells. It takes the form of non-volatile colourless crystals which melt above 75–80° C but which do not ignite below 250° C. Since TNT does not detonate on impact, a triggering explosion (or detonator) is required. It is this remarkable stability which has led to the extensive use of TNT, since in terms of explosive power it is equivalent to only 65 per cent of the equivalent weight of blasting gelatine.

Amatol is a mixture of 60 per cent TNT and 40 per cent ammonium nitrate which was extensively used during the Second World War as a bomb and shell filling. Its explosive power is greater than that of TNT since the oxygen balance is better.

PETN (Pentaerythritoltetranitrate) $C(CH_2NO_3)_4$ is rather more powerful than Amatol or TNT but its use has been somewhat limited owing to its high cost of manufacture.

RDX, also known as Cyclonite, is another wartime explosive. Like TNT and PETN, RDX is a white crystalline solid, whose formal name is Cyclotrimethylene-trinitramine, $C_3H_6O_6N_6$. It is still widely used.

Torpedo. A marine weapon consisting of an underwater vehicle, a warhead, and a control mechanism, which can be a homing unit, an impact or proximity device or a time fuse. Initially developed as a surface-to-surface weapon, the torpedo has evolved into a diversified class of weapons capable of submerged launching or targeting – or both. The propulsion mechanism usually consists of an electric motor which, with the associated power storage, takes up most of the torpedo's cylindrical structure. The warhead is generally a high-explosive one, though nuclear warheads are available for some systems. Torpedoes were initially equipped with time or impact fuses, and therefore required precise linear aiming. Magnetic fuses were introduced during the Second World War but performance improved much more significantly with the introduction of acoustic homing units. These use an active or passive ◊ *sonar* to detect the target, a programming unit to compute path corrections, and an autopilot which actuates the control gear. Wire-guided torpedoes, such as the US Mk 39, are controlled by command impulses to target and require either target visibility or the sonar tracking of both torpedo and target. Torpedoes now equip ◊ *submarines*, surface vessels, and aircraft as well as ballistic rockets (the US ◊ *ASROC*), and are often dual-purpose in that they can be used against both surface vessels and submarines. The introduction of fast nuclear submarines has resulted in the development of deep-diving high-speed torpedoes, such as the US Mk 46 Mod.1; this last is reported to have solid-propellant driven turbines, instead of the more common electric motors. ◊ *Subroc* represents a new approach in weapons launched from submarines: it emerges on the surface, flies to the target as a rocket and then drops a (nuclear) depth charge which is inertially guided to the target area.

Total War. A war in which at least one party perceives a threat to its survival and in which all available weapons are used and the distinction between 'military' and civilian targets is almost completely ignored. It is noteworthy that even Germany in 1944–5 refrained from using all available weapons, for example nerve gases, and from other rational measures, such as shooting unproductive POWs. In practice, restraint is exercised for sub-rational reasons.

The phrase is now in literary use, while professionals use ◊ *central war*, for a direct confrontation between the nuclear superpowers, and ◊ *general war* for an all-weapon conflict between them.

TOW MGM-71A. US anti-tank 69
missile. TOW (Tube-launched, Optically-tracked, Wire-guided) is an anti-tank missile package, consisting of a missile, tube launcher, a simple tripod, an optical sight unit, and an electronic guidance unit. The whole, with one missile, weighs 200 lb. Unlike many other anti-tank missiles, TOW is not guided by joystick control. Instead, the missile is kept on target by keeping the latter in the telescopic sight of the optical unit; an electronic programmer converts movements in the sight into impulses sent down two wires to the missile. This has two stages, the first to project the missile to a safe distance from the operators and the second for the rest of the flight. TOW has scored kills on tank-type targets moving at 30 mph and more than a mile away. Apart from this, what makes the TOW a second-generation missile is its short minimum range: 70 ft, or so. The maximum range is about 1·5 miles. A helicopter-launched version has also been developed.

Tracking (or Narrow-beam) Radar. A radar intended to acquire and follow targets. This is done by means of an antenna which emits narrow or 'pencil' beams which oscillate around the target, so that the latter cannot evade the beam by abrupt changes in its flight path. When this operates as intended, the radar is said to be 'locked on' the target. Such radars are used in ◊ *air defence systems* in conjunction with computing gear (used to predict the flight path of the target). Anti-aircraft guns are usually controlled by tracking

radar, where the radar beam follows the target and the gun is slaved to it.
◊ *MSR* for an advanced multi-beam radar application.

59   Transall C-160. Franco–German military transport aircraft. The Transall is the product of a joint Franco–German private venture which later acquired formal government backing. It is a twin-turboprop high-wing, and there are provisions for fitting two under-wing turbojets to reduce the length of the airstrip required. Like other modern military transports, the Transall is equipped with a wide range of navigation and communication avionics. The original prototype (C-160A) flew in 1965, and the two production versions, the C-160D and F, intended for the German and French air forces, have been operational since February 1968. Apart from the four-man crew the Transall can carry 93 troops or 61 to 81 paratroops, or 62 litter cases. In freight configurations, loads up to 35,270 lb can be carried, while single loads of up to 17,640 lb can be air-dropped. The cabin floor area is 584 sq ft, and the volume is 4,072 cu ft; and there is a rear ramp loading door.

110 Transalls have been ordered for the Luftwaffe and 50 for the French air force, and these have been built in both countries under a production sharing agreement.

Specifications: Length: 105 ft 2 in. Height: 38 ft 5 in. Wing span: 131 ft 3 in. Gross wing area: 1,722 sq ft. Weight empty: 61,095 lb. Maximum payload: 35,270 lb. Maximum level speed at 14,760 ft: 333 mph. Take-off run at maximum take-off weight: 1,970 ft. Range with 17,640 lb: 3,013 miles. Range with maximum payload: 1,070 miles.

Trigger. A strategic theory used to justify small national 'independent' nuclear deterrent forces. The basic 'trigger' argument is that a nuclear delivery force too small in itself to deter a super-power can do this by provoking the release of another super-power's retaliatory forces. While the usual 'equalizer' argument for an independent deterrent is that it frees its owner of the need for an alliance with a super-power (since any nuclear strike, however small, is supposed to be unacceptable to an enemy), the 'trigger' argument runs rather the other way, since it is a means of ensuring that such an alliance becomes operative. Apart from cataclysmic visions of general war resulting automatically from any nuclear strike on either super-power, a subtler – and more realistic – version of a 'trigger's' role has been proposed. The sequence of events is that a small country allied to super-power *A* would protect a vital interest against super-power *B* by threatening it with a strike which would pre-empt enough of *B*'s nuclear delivery forces to make *B* vulnerable to *A*'s attack (and therefore make such an attack profitable to *A*, a prospect intended to deter *B* from violating the small power's vital interest). The trouble is that forces such as the French and British nuclear deterrents are simply too small to make a dent in the delivery forces of either super-power, thus vitiating the argument.

◊ *Catalytic War* for a more vicious use of small nuclear forces.

Tripwire. A military force in itself too weak to resist successfully, but nonetheless deployed against, a potential threat in order to trigger off intervention from outside the area. The concept is used in the continuing debate on ◊ *European security*. The idea is to replace US ground forces in Europe with a thin screen of American soldiers which, if attacked by the Soviets, would trigger off an American nuclear strike against the Soviet Union. The proposal has been ridiculed by advocating the use of a single American soldier who would interpose himself against a Soviet land offensive. In practice, a tripwire strategy is an adaptation of the old form of ◊ *massive retaliation* designed to achieve security at minimum cost.

Another form of Tripwire was the implicit strategy of the United Nations Emergency Force which until May 1967 patrolled the Israel–Egypt armistice lines. Though far from a token force, it was a 'tripwire' in the context of the two heavily-armed contenders (but it was envisaged as sufficiently strong to arrest a conflict until Great Power or United Nations intervention would freeze it once more).

Trojan FV 432. British armoured personnel carrier. The Trojan equips the infantry element of the UK armoured brigades and is being generally issued as a tracked combat vehicle to 'first line' units. Like the US ⟡ *M.113*, it is a transport rather than a true combat vehicle, since its frontal armour is thin, there is no weapon turret, and the crew cannot fight from within it. This 'battle taxi' layout derives from Anglo–American infantry doctrine which envisages the use of APCs to move troops up to the combat zone but then requires them to leave the vehicle and fight on foot. The 'mechanized infantry' concept favoured by almost all other armies emphasizes the use of APCs for the assault, willingly sacrificing aimed fire for the sake of speed and momentum. The Trojan is very similar to the M.113 but weighs about four tons more, so that it cannot float without preparation; the preparation includes the fitting of a 'flotation screen', which is vulnerable, expensive and inconvenient in combat. The Trojan is also almost three times as expensive as the M113, due to short pro-duction runs and a more complex chassis. The engine is a stressed cylinder diesel of the 'multi-fuel' variety, which gives a good road speed and range. The high silhouette and the fully enclosed crew compartments are serious disadvantages, because while the Trojan is highly visible to the enemy the combat crew inside is 'blind' and cannot study the tactical situation before having to meet it. (Other APCs have ports or scopes or both.)

Specifications: Length: 16 ft 9 in. Width: 9 ft 9 in. Height: 7 ft 2 in. Weight, laden: 15 tons. Engine power: 240 gross bhp; diesel 'multi-fuel'. Maximum road speed: 32 mph. Road range: 360 miles. Ground pressure: 11·3 lb sq in. Ground clearance: 1 ft 4 in. Driver and crew: 12.

⟡ *Armoured Personnel Carrier* for general subject entry.

Tu-16. Soviet bomber. ⟡ *Badger.*

Tu-20. Soviet bomber. ⟡ *Bear.*

Tu-22. Soviet bomber. ⟡ *Blinder.*

U

UK, Strategic Defensive Forces. British strategic defensive forces are part of the RAF Strike Command. The defensive portion of Strike Command consists of: (a) a radar-computer-communication system, linked to ◊ *B M E W S*; (b) a force of 70 ◊ *Lightning* fighter-interceptors and up to 100 ◊ *Phantom F-4*s for area defence against strategic bombers; (c) ◊ *Tigercat*, and ◊ *Bloodhound 2* surface-to-air missiles.

For comparison purposes, ◊ *N O R A D*, *C A F D A*, *P V O Strany* and *China, Strategic Defensive Forces.* ◊ *Air Defence System.* For the potential opposition, ◊ *U S S R, Strategic Offensive Forces*, and the same for *U S, France* and *China.*

UK, Strategic Offensive Forces. British nuclear-capable delivery systems include:

(a) 50 medium bombers (◊ *Vulcan 2*s) equipped with ◊ *Blue Steel* missiles, supported by: 24 tanker Victor 1s, 12 ◊ *Victor 2* strategic reconnaissance aircraft, as well as Canberra photo-reconnaissance aircraft. This force no longer has a primarily strategic role.

(b) 4 nuclear submarines with 16 ◊ *Polaris* missiles each, of which 3 are currently operational.

These forces have been formally committed to ◊ *N A T O* and are under the planning control of SACEUR. Their targeting is coordinated through SACEUR with the US Joint Strategic Planning System, with HQ at Omaha, Nebraska. These arrangements do not rule out the use of this force outside the context of N A T O or U S strategic decision-making. Though Blue Steel significantly enhances the capabilities of this force, the V-system's chances of penetrating Soviet air defences are not very high, and the Polaris element is small enough (if used independently of U S missiles) to be neutralized even by a limited anti-ballistic missile defence system (◊ *P V O Strany*). This force would, however, be of some significance against France, or other even less likely antagonists.

For comparison purposes ◊ *U S, Strategic Offensive Forces* and the same for *U S S R, France* and *China.*

US, Strategic Offensive Forces. The strategic forces of the United States, ◊ *Minuteman* and ◊ *Titan II* ICBMs, ◊ *Polaris* – ◊ *Poseidon* missile-firing submarines, and the long-range bomber force, are intended to implement the stated policies of ◊ *controlled response* and ◊ *assured destruction.* The weapons and the associated command and control facilities are to a greater or lesser extent ◊ *hardened* or otherwise protected against nuclear attack, since a deterrent can deter only if it can cause unacceptable damage to any combination of enemies after absorbing a nuclear attack – and do so reliably in the face of engineering and enemy-action hazards. The realization that survivability is the key requirement, as opposed to, say, the gross size of the strategic forces, came well after the deployment of the first generation of deterrent forces (B-36 and B-47 bombers), and it was due to the path-breaking Rand Corporation study on the basing of strategic bombers (R-266, formally published in 1954). Survivability is a far more subtle matter than it seems, and it is a dynamic one, linked to the evolution of others' strategic forces, rather than a question of attaining certain engineering standards. (◊ *Assured Destruction.*)

The Joint Strategic Planning System, with headquarters at Omaha, Nebraska, coordinates the retaliatory forces under the direction of the national command centres, the locus of the overall political decisions made by the President and his associates. ⟡ *NATO* is formally linked with the system through its commander in chief, SACEUR, who is supposed to assume control of US *Polaris* submarines in the Atlantic, as well as the British ones, in the event of a war which involved Europe directly. The forces maintained under the revised 1969–73 strategic programmes consist of about 300 B-52 ⟡ *Stratofortress* bombers (G and H versions) equipped with ⟡ *Hound Dog* and *SRAM* stand-off missiles; about 150 older C and F B-52s not adapted for low-level penetration and armed with free-fall bombs; a declining force of B-58 ⟡ *Hustlers* which is being phased out; and a small (30–40) FB-111 (⟡ *F-111A*) force which is now being deployed. Unless the A.M.S.A. project results in an actual deployment programme, manned bombers will cease to play a significant role in the strategic context by the mid-seventies. This bomber force is supported by a network of ground facilities, including dispersal airfields, about 500 KC-135 tankers and 15 ⟡ *SR-71* strategic reconnaissance aircraft.

The main retaliatory forces are, however, the ICBMs. The Minuteman missiles, whose number has been frozen at 1,000, consists of three variants, including Minuteman 3 which is fitted with MIRV warheads and which is being phased in to replace the 650 Minuteman 1s originally deployed. There are also some 54 Titan II, older liquid-fuel (storable) ICBMs, which are to be phased out by the mid-seventies or earlier.

The US navy operates the more survivable but costly ballistic-missile submarine force. There are 41 of these, of which no more than 25 are on station at any one time, each fitted with 16 missiles, now Polaris A3s. 31 of these submarines are to be converted to the Poseidon missile, which is to be fitted with a MIRV payload. There appear to be command and control problems with this force, which reduce its flexibility from the point of view of strategic planning.

⟡ *USSR, Strategic Offensive Forces, ICBM, Soviet* and, for the likely op-

position, *PVO Strany*. ⟡ also *NORAD* and *Safeguard* for the strategic defensive forces of the United States.

USSR, Strategic Offensive Forces. Soviet strategic deployment policies have gone through three main stages since the introduction of strategic missile systems: the 1958–62 period of systematic deception (⟡ *Missile Gap*), when the deployment of a very small force was masked by a veil of secrecy and by propaganda claims of superiority over the United States; the 1963–6 period of intense development efforts intended to prepare the ground for a major expansion of Soviet strategic forces; and the years since, when the Soviet Union has been moving to a net superiority in land-based systems and towards some sort of parity in submarine and other systems.

The ICBM force is now larger than the American one in terms of both the number of launchers and – by a wide margin – in the total payload deliverable (⟡ *ICBM, Soviet* for details). Further, the crude and vulnerable ICBMs of the early sixties have been superseded by a number of very costly and advanced systems including the SS-9 ⟡ *Scarp* and the first Soviet solid-fuel ICBM, the ⟡ *Savage*. The Soviet Union now deploys a force of some 48 missile-carrying submarines, of which 18 are nuclear-powered, but these are inferior to the US ⟡ *Polaris* ⟡ *Poseidon* system, both in the number of missiles carried and in their range; but the new Y-class submarines are being produced at the rate of four per year. These nuclear-powered vessels carry sixteen missiles each whose capabilities are similar to those of the early Polaris (⟡ *Sawfly*). When these come on station as of 1971, they will add some 64 launchers per year to the present force of about 160 ⟡ *Sark* and ⟡ *Serb* SLBMs carried in a variety of submarines.

The bomber component, on the other hand, is and will remain inferior, with some 100 ⟡ *Bison* and 100 ⟡ *Bear* heavy bombers, of which about 150 are available for strategic bombardment. This inferiority is of course of limited importance considering the uncertain effectiveness of strategic bombers in the present state of air defences (⟡ *A.M.S.A.*).

Unlike the United States, the Soviet Union has worthwhile strategic targets at medium and intermediate ranges from its own territory. For these it has available a force of 600 ◊ *Badger* medium bombers and some 150 supersonic ◊ *Blinder* jet bombers, as well as some 700 medium and intermediate range ballistic missiles, ◊ *Sandal* and ◊ *Skean* being the most important types. (The mobile ◊ *Scamp* may be coming into limited deployment as an IRBM.)

◊ *US, Strategic Offensive Forces* as well as the same for *China, France* and the *UK*. For the presumed opposition, ◊ *NORAD* and *Safeguard*.

Uzi. The Israeli army leadership received its introduction to things military in the pre-independence period of conflict with the British colonial power and the Arab feudal leadership. During this underground period, the supply of small arms was, as usual in such cases, a critical problem, and though some arms were imported covertly, local clandestine manufacture became important during the final pre-independence period. Inevitably only simple weapons could be manufactured: mortars (◊ *Artillery*) and ◊ *sub-machine-guns*, both of which can be fashioned out of standard tubing. Instead of treating these as remedial weapons to be abandoned as soon as better equipment became available, every effort was made to extract the maximum benefits from these weapons by developing specialized training and tactics. Thus in the early fifties, when the equipment of the Israeli defence forces started to resemble that of a 'real' army, sophisticated versions of these weapons were manufactured in Israel, while most other items are imported. The mortars are particularly well designed, but it is the Uzi sub-machine-gun which has found a wide international clientele and is generally considered to be the best weapon of its class. Several versions, with wooden, metal, full and skeleton stocks are used, all built around the same basic weapon. This has a very safe 'safe' (it can be fired only when the grip is squeezed as well as the trigger), an internal barrel easily removable for maintenance, and a simple but effective sight arrangement. It weighs 8·9 lb (loaded) and is 17·9 to 25·2 in long, depending on whether the stock is folded or not. The effective range is about 100 metres, and the stability is such that it can be fired in a single hand-grip, like a pistol. It has a rate of fire of 550 rpm and has optional 25–32 round magazines.

V

V750 VK. ⟡ *Guideline.*

VDS. ⟡ *Sonar.*

Vela. An American system of satellite sensors which monitors nuclear detonations: the Vela S for surface ones and Vela H for space detonations.

Victor B.2. British medium bomber. A developed version of the Victor bomber (the first version, the B.1, became operational in 1958). The Victor is one of the three original 'V-bombers', the others being the (scrapped) Valiant and the still operational ⟡ *Vulcan.* But now the Victor 2 is used for reconnaissance while the Victor 1 is a tanker. The Victor 2 is a large (200,000-lb) swept-wing aircraft with a nose compartment for the crew of five. It carries electronic countermeasures equipment in the tail cone and elsewhere, as well as anti-radar chaff 'window' dispensers on the trailing edge of each outer wing. The Victor is transonic in cruising flight (though it has exceeded Mach 1 in a shallow dive). In non-nuclear missions the Victor can deliver up to 35,000 lb of bombs, and it has actually been used for high-altitude bombing. The Victor BK.1 and 1A are tanker conversions of the original B.1 bomber, and the B(SR).2 is a strategic reconnaissance aircraft for radar-mapping, day and night photography, and in-flight first analysis.

Specifications: Power plant: four 20,000 lb static thrust turbofans. Length: 114 ft 11 in. Height: 30 ft. Wing span: 120 ft. Gross wing area: 2,597 sq ft. Cruising speed: Mach 0·92. Maximum cruising height: 55,000 ft. Combat radius: Hi-lo, 1,725 miles; Hi-hi, 2,300 miles.

⟡ *Bomber* for general subject entry.

Viggen. (SAAB-37) AJ/SK/S/JA/37. 47 **Swedish Mach 2+ fighter.** The Viggen, now in the advanced development stage, is a very sophisticated multi-purpose fighter of original design, part of a weapon system known as System 37. It has a unique configuration, with a canard foreplane and a large delta wing. This, and the power plant combination, allows the Viggen to exceed Mach 2, cruise economically at low level, and, at the same time, use short runways. (Enlarged road sections about 1,700 ft will be used as dispersal airfields.) The avionics are equally advanced, with a single digital computer integrating the navigation and attack functions, as well as a selective display device which projects critical data on the windscreen. Four versions are currently being delivered:

AJ 37: Primarily a ground-attack version with some interceptor capability. The AJ 37 is equipped with an optional combination of air-to-surface missiles, bombs, mines and guns.

JA 37: Primarily an interceptor with some ground-attack capability.

There is also a reconnaissance version (S 37) and a two-seat training version (SK 37).

The first prototype of the Viggen flew in 1967, and the first operational versions (AJs) became available in 1970.

Specifications: Power plant: 26,450 lb st with afterburning. Length: 53 ft 5 in. Height: 18 ft 4 in. Wing span: 34 ft 9 in. Combat take-off weight (AJ): 35,275 lb. Maximum level speed: exceeds Mach 2. Time to 36,000 ft: 2 minutes.

⟡ *Fighter* for general subject entry.

Vigilant. British anti-tank missile. 68 The Vigilant is a lightweight wire-guided missile system operational since 1963. It consists of a 3 ft 6 in long missile, a carry box which doubles as a launcher, and a hand-held guidance set with a pocket battery; the missile itself weighs 31 lb, and the whole system 52 lb. Vigilant is guided by electrical impulses sent down a thin wire, which the missile unrolls in flight; these impulses originate from the optical line-of-sight command set. This can be held by the operator at up to 70 yds away from the missile launcher. Vigilant has velocity control, with a twin-gyro autopilot, which makes guidance a good deal easier than on earlier missiles such as the SS.11 or Entac (◊ *Nord*). The warhead is a 13·2-lb HEAT charge, which can penetrate practically all tank armour. Minimum range is 200 yds (at which range the warhead is armed), and it has a 5,280 ft maximum range. The speed is higher than that of many comparable missiles, at 348 mph; the flap controls assure good manoeuvrability at high speeds.

Vigilante A-5 (and RA-5). US naval Mach 2 bomber. The Vigilante is an all-weather two-seat twin-jet which was initially intended as a bomber to be operated from attack carriers as part of the US strategic nuclear forces. The A-5A became operational in this role in 1962, and nuclear weapons were carried in an internal tunnel and released by rearward projection attached to (empty) fuel tanks – previously used in flight. This manner of release and the attached tanks facilitated accurate delivery, since the bomb's aerodynamic properties were improved by the fuel tanks. External weapons were also carried in under-wing racks. An extended range version, the A-5B, was built in 1962 with extra internal fuel and an improved boundary layer control. As the strategic offensive mission was withdrawn from US attack carriers, the A-5A and the A-5Bs have been converted into a reconnaissance version, the RA-5C. This retains the bomb-carrying equipment of the A-5s but mounts vertical, oblique and horizon-to-horizon cameras, radio sensors, side-looking radar, infra-red sensors, TV cameras with light intensifications, and ◊ *ECM* equipment. The data so obtained are fed for analysis to the carrier-based or land units of the IOIS, Integrated Operational Intelligence System, of which

the RA-5C is the airborne part. The RA-5C first became operational in 1964, and new ones are still being produced.

Specifications: Length: 75 ft 10 in. Height: 19 ft 4 in. Wing span: 53 ft. Gross wing area: 700 sq ft. Maximum take-off weight: 80,000 lb. Maximum level speed at 40,000 ft: Mach 2·1. Maximum level speed at sea level: 685 mph. Service ceiling: 64,000 ft. Normal range: 2,650 miles.

Vijayanta (Vickers 37-ton). Main battle tank. A British-developed tank assembled locally for the Indian army. This is a British attempt at a light and economical main battle tank, and it incorporates the engine of the ◊ *Chieftain* and the 105-mm gun of the ◊ *Centurion*. The layout is conventional, with a driver in the hull and a three-man crew in the turret – the latter being welded because of the lack of casting facilities in India. Partly as a result of this, this tank is poorly shaped from the ballistic point of view and it presents many vulnerable vertical surfaces. Good features include the high power-to-weight ratio, though its mobility is limited by the ground clearance, which is too small, and the ground pressure, which is too high. Crew comfort is also reported to be good, as is the gun, which is superior to that available on Chinese T-59 (◊ *T-54*) and ◊ *Patton* M.48 tanks which equip India's two major enemies.

Specifications: Length: 23 ft 11 in. Width: 10 ft 5 in. Height: 8 ft. Weight: 37·5 tons. Engine: diesel 'multi-fuel' 700 bhp water-cooled. Road speed: 30–5 mph. Road range: 220 miles. Ground pressure: 12·8 lb sq in. Ground clearance: 1 ft 4 in. Main armament: 105-mm high-velocity gun with 44 rounds of APDS and HESH ammunition. Ranging machine-gun. Secondary weapons: smoke dischargers, coaxial MG and turret-top MG. Crew: 4.
   ◊ *Main Battle Tank MBT* for competitive types.
   ◊ *Armoured Fighting Vehicle* for general subject entry.

Voodoo F-101B. US fighter. The F-101B fighter-interceptor is a large, long-range, twin-jet, two-seat fighter with a top speed of Mach 1·85. It is now being phased out from its main operational

home, the USAF Air Defense Command
of the ◊ *NORAD* US–Canadian air
defence organization. It remains in service
with the RCAF. Like the ◊ *Delta Dart
F-106*, the Voodoo is equipped with three
◊ *Falcon* and two ◊ *Genie* air-to-air
weapons and fitted with a so-called 'all-
weather' radar detection and fire-control
system.

Specifications: Power plant: 2 turbojets
rated at 30,000 lb each. Length: 67 ft 4 in.
Height (at tail): 18 ft. Wing span: 39 ft
7 in. Sweepback: 35°. Combat radius:
over 700 miles.

VTOL. Vertical Take-Off and Landing.
The term includes helicopters but is more
often applied to fixed-wing aeroplanes.
◊ *Combat Aircraft Part II.*

Vulcan B.2. British medium bomber.
The Vulcan, of which the first (Mk 1)
version became operational in 1957, is one
of the three 'V-bomber' types which were
originally deployed as the British 'independent deterrent'. The Vulcan B.2 can
be recognized by the very large delta wing
whose roots contain the four turbojets.
With a maximum take-off weight of about
200,000 lb the Vulcan can carry a substantial weapon and electronics payload,
and there is a five-man crew. This includes an electronics man who controls the

electronic countermeasures (◊ *ECM*)
equipment in the bulged tail cone and
elsewhere. The Mk 2 was designed to
carry the ◊ *Blue Steel* supersonic stand-off
missile, which is credited with a range of
200 miles; this and the low-level flight
modifications in 1967 have increased the
versatility of the Vulcan as a nuclear delivery weapon. The Vulcan could deliver
a 15–20 megaton payload in high-level
flight over the target, or, more safely, a
smaller weapon-load with the Blue Steel
while keeping clear of the 'hotter' defence
areas. Alternatively, it can deliver a smaller
weapon-load in low-level flight (about
1,000 ft) to target by using glide bombs to
avoid exposure to weapon effects at the
moment of explosion. The Vulcan has
already been used for non-nuclear
bombing: in this role they can deliver up
to 54,000 lb of bombs.

Specifications: Power plant: 4 turbojets
of 20,000 lb thrust each. Length: 99 ft 11 in.
Height: 27 ft 2 in. Wing span: 111 ft.
Gross wing area: 3,964 sq ft. Maximum
cruising speed: Mach 0·94 at 55,000 ft.
Combat radius: Hi-lo, 1,725 miles; hi-hi,
2,300 miles. Radar: nose-cone air-to-air
and doppler air-to-ground.

◊ *Bomber* for general subject entry.
◊ *Victor* for the other surviving V-
bomber.

VX. ◊ *Nerve Gases.*

W

Walleye GW 1/o. US guided glide bomb. The Walleye is a maximum precision air-to-ground weapon intended for use against hard targets such as bridges. It is a 1,100-lb glide bomb with crop-delta cruciform wings with attached control surfaces. It carries a TV camera which is focused on target by the launch aircraft pilot. Once launched, the guidance system homes on target with no further assistance from the pilot. Power supply is provided by a ram air turbine; this is sufficient for both the guidance and the hydraulic control energy requirements. Walleye has been operational since early 1969. It is 11 ft 3 in long and has a body diameter of 1 ft 3 in and a wing span of 3 ft 9 in.

War: Four Views. A form of international relations in which organized violence is used in addition to other instruments of policy; more generally, the use of force between groups which do not belong to the same law-enforcement system. In order to protect their interests and extend their sphere of influence, states seek to manipulate the policies of other states; this is done by influencing the costs and benefits of particular policies as perceived by them. This manipulation is carried out by means of a whole battery of instruments: formal diplomacy, the modulation of trade, financial and travel flows, ◊ *propaganda*, ◊ *political warfare*, ◊ *economic warfare*, threats of war, and finally war itself. War and the threat of war are therefore ordinary instruments of international relations in the absence of extra-terrestrial or other supra-national regulation and rule-enforcement.

This view of war is essentially the classical view of war as defined by von Clausewitz: 'War . . . is an act of violence intended to compel our opponent to fulfil our will . . . ', where 'War is not merely a political act, but also a real political instrument, a continuation of political commerce.' There is a prevalent but mistaken confusion of von Clausewitz's concept of war with the concept of ◊ *total war*, but his view refers to something less than total war with the weaponry now available: 'The result of war is never absolute . . . '; indeed, the very idea of a political objective always implies some limitation, since the total destruction of the system cannot be an objective of those who operate within it.

But total destruction may well be an objective if the purpose of war is idealistic, that is, when war becomes a 'crusade'. While the classical view of the state at war sees it as maximizing costs and benefits, including its net power position, the 'crusade' is fought against opponents seen as intolerable to the deity or to stated ideals. Then, no compromise is possible, and the only choice that can be allowed to the enemy is either 'unconditional surrender' or total annihilation as a political entity. While compromise, and even co-operation against erstwhile allies, is inherent in the war-as-political-instrument, it becomes very difficult or impossible once a war is allowed to deteriorate into a conflict about ideals, or when absolute ultimate objectives are set.

In terms of the weaponry available today, no war between nuclear powers can be both 'political' and total, although the threat of ◊ *central war* remains a useful, indeed essential, instrument of policy. But this does not mean that politics between the nuclear powers becomes an endless game of bluff: the threats are either partial or conditional. ◊ *Assured Destruction, Deterrence,* and *Controlled*

Response as opposed to *Brinkmanship*,
Massive Retaliation and *Local War*.
How does the Clausewitzian state approach disarmament negotiations? Firstly,
since war is an instrument and not an
ideal, it is replaceable by other instruments, partly or wholly. One of these
instruments, or rather one of the applications of the diplomatic/propaganda
instruments, is disarmament, and negotiations about it. Secondly, the object
remains the same – protection and
extension of interests and influence. Thus
disarmament agreements are merely part
of the overall strategic posture. This implies no cynicism: the parties may well
aspire to a relaxation of the threat of
(nuclear) war; it is only that this relaxation cannot be traded for reduced security.
This classical view of war as a permanent manifestation of social life in the context of a world of nation-states is disputed
by many. Three rival conceptions of war
are presented here.
1. The Marxist–Leninist doctrine of
war sees it as a projection of the class
struggle in the international arena:
capitalist classes in control of the state use
war as an instrument in their perpetual
struggle for more labour, more raw
materials and more markets. While their
own working class is deceived or forced to
fight for a supposedly national cause, the
goals of war are not national but those of
the ruling capitalist class: the exploitation
of further (foreign) populations. Apart
from such market-competitive wars there
are also 'diversionary' wars, intended to
stifle revolutionary sentiments by divert-
ing them into nationalist channels.
War between 'capitalist' and 'socialist'
states is seen as a more direct form of the
class struggle. Both sides use the resources of their respective states to further
class interests. This explanation of the
causes of war implies that wars between
'socialist' states are by definition impossible. Since war is an instrument of
class struggle, and since there can be no
struggle within the one universal 'working
class', it follows that there can be no war
between them either. The Marxist–
Leninist view of war therefore sees it as a
temporary phenomenon associated with
the existence of waning capitalist class rule
and other historically declining political

phenomena, as opposed to the permanent
phenomenon implied in the Clausewitzian
viewpoint.
But the central assumption of the Marxist–Leninist doctrine, that in 'socialist'
countries the state becomes an instrument
of the 'working class', has been contradic-
ted by experience. Classes do not rule, do
not make political decisions, do not conduct diplomacy or war, and the political
leadership which does rule has power-
accumulation interests. War is a major
tool for the accumulation and projection
of power internationally, and no state can
renounce it; the political leadership of
'socialist' countries is as likely to engage
in war as any other; wars between
'socialist' countries are as likely as war
across the 'class-barrier'.
2. The non-violent resistance/pacifist
view is that war is unnecessary as an
instrument of policy. No political leadership should be allowed to engage in war;
no political leadership should develop and
maintain armed force organizations.
National defence can be adequately assured by means of non-violent resistance
against an aggressor, which will nullify
the expected benefits of invasion and in-
duce a withdrawal. Peaceful propaganda
and demonstrations to demoralize the
invading troops, and passive resistance by
bureaucrats to prevent effective administration by the army of occupation, were
some of the means which were employed
in the non-violent resistance which the
Czechs opposed to the Soviet-led ◊ *Warsaw Pact* invasion of 1968; earlier methods
used against the British in India and what
was then called Palestine included the
non-payment of taxes and human bar-
ricades to prevent enemy movements. But
no account is taken of aerial bombing,
naval blockades and other forms of
military force which do not require personnel on the ground. Further, it is assumed that the invader can always be
neutralized, but this is so only when his
own values or self-image deter the use of
terror to enforce obedience. Non-violent
resistance can work only when the
'invaders' operate within a restricted code
of behaviour; against determined 'invaders' who freely use violence and other
means of compulsion, it cannot work.
Thus the non-violent resistance/pacifist

alternative to war is very limited in its application.

There is of course pacifism in the absolute, that is a unilateral rejection of the use of violence without conditions or alternative means of defence, but this has a very limited appeal to those concerned with life, as opposed to after-life.

3. The 'peace-research'/conflict-resolution concept of war sees it as an event, rather than as an instrument of the state. Wars are seen as the outcome of political forces and structural facts which 'generate' conflict, rather than the result of conscious rational decisions. While the classical-Clausewitzian view sees war as a chosen instrument of politics willingly used (though of course this may result from a misreading of the data), here war 'happens' because a given situation is inherently unstable.

Since wars are 'events', peace-researchers seek to uncover the causes of these events; once the causes are known the illness can be cured. Thus models of the international system of nation-states are built up, and behavioural parameters are fed in to observe the outcome, the intention being to discover stable systems which do not generate wars. Much of this resembles economic model-building and, as in the latter, an equilibrium is presumed, and is also assumed to be attainable. The ultimate aim of peace-research is conflict-resolution by a 'balancing' institution constructed for the purpose, rather like a powerful UN which would operate on the causes of conflict rather than on the conflicts themselves. Neither the empirical researchers nor the model-builders of this school have yet isolated the 'causes' of war; no credible institutional framework for conflict-resolution has yet been suggested.

As long as humans are organized into state entities, as long as these entities are constructed as vehicles of power-accumulation, and as long as war remains an efficient means of power-accumulation, there will always be one party who will initiate war. If there is one party initiating war other parties will be forced to engage in it too. Thus an 'anti-war system' must prevent all states from using war all of the time in order to succeed; the invention of such a system does not appear probable,

and there is no evidence to indicate that it is possible.

Warsaw Pact. A multilateral military alliance between the Soviet Union, Albania, Bulgaria, Czechoslovakia, East Germany, Hungary, Poland and Romania. The Pact was signed in May 1955 to supplement the bilateral treaties which exist between all members except for Albania (which has none) and Romania (which does not have a bilateral treaty with East Germany). The Warsaw Treaty Organization, which is supposed to implement the Pact, has a High Command and a parallel Political Consultative Committee, consisting of Party Secretaries, Heads of Governments, and the Foreign and Defence Ministers. Officers are in Moscow, and both the Commander-in-Chief and the Chief-of-Staff have always been Soviet officers. A Joint Secretariat and Permanent Commission are the continuing bodies, and their senior posts are also filled by Soviet personnel. Marshal Yakubovski is the current Commander-in-Chief, and as before the post is held in *ex-officio* association with that of Soviet First Deputy Minister of Defence.

Known operational arrangements include a northern group of forces with HQ in Legnica, Poland; a southern group with HQ in Tököl, Hungary; and a group of Soviet forces in Germany, with control over all operational East German divisions, with its HQ in Wunsdorf, near Berlin. Nuclear warheads are thought to be in exclusive Soviet control, though battlefield nuclear-capable missiles have been delivered to other member countries. Albania has not participated in Pact activities since 1961, and Romania appears to be disengaging from Pact activities.

Total peacetime Warsaw Pact forces now amount to about 1,300,000 ground troops, 19,000 battle tanks, 4,300 fighters (ground-attack and interceptors), and about 800 other combat aircraft.

The entry on ◊ *European Security* mentions some of the problems faced by the Warsaw Pact's model and presumed opposition, ◊ *NATO*. But the strains on the Warsaw Pact have, if anything, been even more severe than those experienced by NATO, as shown by the 1968 invasion of Czechoslovakia and by the diplomacy by military manoeuvres which characterized Soviet–Romanian relations in the same year. In particular, the super-power-to-European-ally relationship embodied in both NATO and the Warsaw Pact is complicated in the latter case by the visible backwardness – social, economic and political – of the 'greater ally'. A *de facto* monopoly of major strategic decisions produces greater strains in the relationship between lesser powers and the super-power. This is particularly the case when the latter's status is based exclusively on the deployment of long-range nuclear delivery systems and ground forces, and does not reflect either global military power, or a general economic and technological superiority, both of which the USA has and the Soviet Union clearly lacks.

Weapon. A man-made object or an *objet trouvé* which is intended as a means of killing or incapacitating humans or of destroying other objects.

'Strategic weapons' are those capable of striking at the 'homeland' of an opponent, that is, at his population and industrial centres. This is as opposed to 'tactical weapons', which are presumed to be usable only against a battlefield manifestation of the enemy. More generally, the former are long-range and the latter short-range weapons, but the distinction has no precise meaning outside a specific context. 'Conventional weapons' is an unfortunate expression for all weapons other than biological, chemical or nuclear ones. 'Defensive weapons' is a useful category: most of one's weapons

are defensive, while those of the opposition are invariably 'offensive'. 'Weapons of mass destruction' is a preferred Soviet expression for nuclear, biological and chemical weaponry. Western usage is ABC (Atomic, Biological, Chemical) in the USA and France, and CBR (Chemical, Biological, Radiological) in the UK. Nuclear weapons are either fission (A- or atomic) or fusion (H– or thermonuclear). 'Special weapons' is a euphemism for chemical or biological weapons.

The following categories are listed and classified further in the text: *Biological Warfare, Chemical Warfare, Nuclear Weapon, Small Arms* (all portable infantry weapons), *Artillery* (all full-recoil tube weapons whose calibre exceeds 20 mm, or 0·6 in), *Armoured Fighting Vehicles, Grenades, Combat Aircraft* (all manned flying objects designed to carry weaponry or ancillary equipment), *Combat Vessels* (all surface and sub-surface water vehicles with built-in weaponry), *Missiles* (all unmanned but guided flying objects with built-in weapons or ancillary equipment).

Western European Union (WEU). Formed by treaty between Belgium, France, West Germany, Luxembourg, Italy, the Netherlands and the United Kingdom. The WEU succeeded the Brussels Treaty Organization in May 1955, when Italy and West Germany joined the other powers. The WEU is a small-scale and heavily watered-down version of NATO, and its main practical importance derives from the fact that the British commitment to deploy military forces on the Rhine was made as part of the WEU package deal.

X

X-ray Kill. Much of the energy of megaton-range *nuclear warheads* takes the form of X-rays. This weapon effect has been adapted in the *Spartan* interceptor missiles of the *Safeguard* *ABM* programme. The X-ray effect can destroy an enemy warhead at long ranges outside the atmosphere by evaporating the surface layer of the latter's heat shield at such high velocity that the shock waves generated damage the warhead's mechanical structure. Since X-rays are absorbed by air, this mechanism is most effective at higher and especially extra-atmospheric altitudes, where the powerful minimum-fission warhead can be used to defeat incoming warheads with little damage to the ground environment. Shielding against the X-ray effect is possible but costly in payload.

Y

YAK-25. ⇨ *Flashlight.*

YAK-28. ⇨ *Firebar.*

YAK-42. ⇨ *Fiddler.*

Cross-reference Index of Aircraft, Missiles and Armoured Vehicles Designations

Aircraft: United States

US military aircraft are designated by a letter-number code which defines their mission, configuration and status by letters and the model by numbers. The basic configurations are:

A for Attack, that is aircraft intended for strikes against surface targets and not primarily to fight other aircraft
- A-4 ◊ *Skyhawk*
- A-5 ◊ *Vigilante*
- A-6 ◊ *Intruder*
- A-7 ◊ *Corsair*

F for Fighters, that is aircraft which can fight other aircraft
- F-4 ◊ *Phantom*
- F-5 ◊ *Freedom Fighter*
- F-8 ◊ *Crusader*
- F-101 ◊ *Voodoo*
- F-104 ◊ *Starfighter*
- F-105 ◊ *Thunderchief*
- F-106 ◊ *Delta Dart*
- F-111 ◊ *F-111* (this aircraft has no popular name)

B for Bomber, that is aircraft intended to deliver large payloads at long ranges
- B-52 ◊ *Stratofortress*
- B-58 ◊ *Hustler*
- B-1A ◊ *A.M.S.A.*

C for Transport
- C-5 ◊ *Galaxy*
- C-130 ◊ *Hercules*
- C-141 ◊ *Starlifter*

In addition to the configuration letters there may be a first letter which defines the specific mission of an aircraft version. The letter E means that the aircraft has been adapted for electronic warfare. The letter R stands for a reconnaissance version, T for a training version, and Y for a prototype. In each case the entry can be identified by the letter which precedes the model designation (a number separated by a dash).

TA-4 equals A-4 ◊ under *A* (attack aircraft)

YF-105 equals F-105 ◊ under *F* (fighters)

RF-8 equals F-8 ◊ under *F* (fighters)

EA-6 equals A-6 ◊ under *A* (attack aircraft)

The above is a drastically simplified version of the full coding system, but it should cover all the entries in this book except for:
- YF-12A ◊ *SR-71*
- FB-111 ◊ *F-111*
- AH-56A ◊ *Cheyenne*
- AH-1G ◊ *Hueycobra*

Aircraft: Soviet Union

Soviet aircraft have been entered in the text under their NATO code-names. These are imaginative but follow a simple system:

- B for Bombers
- F for Fighters
- C for Transports
- M for Trainers (including trainer versions of fighting planes)
- H for Helicopters

Often, though not always, the design bureau which developed the aircraft is known in the West. Both the full designation and the conventional abbreviations are listed below:

Antonov
- An-12 ◊ *Cub*
- An-22 ◊ *Cock*

Kamov
Ka-20 equals 'Harp', but like all other general-purpose helicopters it is entered under the general heading *Helicopter*

Mikoyan and Gurevich
Mig-21 ⬦ *Fishbed*
Mig-23 ⬦ *Foxbat*

Mil
Mil Mi-4 Hound (⬦ *Helicopter*)
Mil Mi-6 Hook (⬦ *Helicopter*)
Mil Mi-10 Harke (⬦ *Helicopter*)

Myasishchev
Mya-4 ⬦ *Bison*

Sukhoi
Su-7 ⬦ *Fitter*
Su-9 ⬦ *Fishpot*
Su-11 ⬦ *Flagon*

Tupolev
Tu-16 ⬦ *Badger*
Tu-20 ⬦ *Bear*
Tu-22 ⬦ *Blinder*
Tu-28 ⬦ *Fiddler*

Yakovlev
Yak-25 ⬦ *Flashlight*
Yak-28 ⬦ *Firebar* (includes the 'Brewer')

Aircraft: Other Countries

British, French, Italian, Indian and Swedish aircraft are entered in the text under their popular names. There are no widely known official designations.

Missiles and Rockets: United States

As in the case of manned aircraft, US missiles are designated by a complex letter-number code. Except for prototypes of various sorts, the first letter defines the 'launch environment', the second the nature of the intended target, and the third is either M for guided missiles or R for rockets not guided after launch. The numbers and letters which follow the dash define the specific model of missile.

A equals air-launched
AIM stands for Air-launched, Intercept (aerial), Missile

AIM-4 series ⬦ *Falcon*
AIM-7 series ⬦ *Sparrow*
AIM-9 series ⬦ *Sidewinder*
AIM-26B ⬦ *Falcon*
AIM-54A ⬦ *Phoenix*

AIR stands for Air-launched, Intercept, Rocket
AIR-2A ⬦ *Genie*

AGM stands for Air-launched, surface- (or Ground-) attack Missile
AGM-12 ⬦ *Bullpup*
AGM-28B ⬦ *Hound Dog*
AGM-45A ⬦ *Shrike*
AGM-69A ⬦ *SRAM*
AGM-78 ⬦ *Standard* ARM

B equals multiple launch environments
BGM stands for multiple launch environment, Ground-attack Missile
BGM-71A ⬦ *TOW*

C equals a 'Coffin', i.e. horizontal protected storage
CGM stands for Coffin Ground-attack Missile
CGM-13B ⬦ *Mace*
CIM stands for Coffin Intercept Missile
CIM-10B ⬦ *Bomarc*

F equals man-launched
FIM stands for man-launched Intercept Missile
FIM-43A ⬦ *Redeye*

L equals silo-launched
LGM stands for silo-launched Ground-attack Missile
LGM-25C ⬦ *Titan*
LGM-30 series ⬦ *Minuteman*
LIM stands for silo-launched Intercept Missile
(X) LIM-49A ⬦ *Spartan*

M equals mobile-launched
MGM stands for Mobile-launched Ground-attack Missile
MGM-29A ⬦ *Sergeant*
MGM-31A ⬦ *Pershing*
(X) MGM-51C ⬦ *Shillelagh*
(X) MGM-52A ⬦ *Lance*
MGR stands for Mobile-launched Ground-attack Rocket
MGR-1B ⬦ *Honest John*
MGR-3A ⬦ *Little John*

MIM stands for Mobile-launched Intercept Missile
   MIM-14B ◊ *Nike-Hercules*
   MIM-23A ◊ *Hawk*
   MIM-72A ◊ *Chaparral*

R equals ship-launched
RIM stands for ship-launched Intercept Missile
   RIM-2F ◊ *Terrier*
   RIM-8G ◊ *Talos*
   RIM-24B ◊ *Tartar*
   (Y) RIM-66A ◊ *Standard*
   (Y) RIM-76A ◊ *Standard*
RUR stands for ship-launched Underwater attack Rocket
   RUR-5A ◊ *Asroc*

U equals underwater-launched
UGM stands for Underwater-launched Ground-attack Missile
   UGM-27 series ◊ *Polaris*
   UGM ◊ *Poseidon*
UUM stands for Underwater-launched Underwater attack Missile
   UUM-44A ◊ *Subroc*

Missiles and Rockets: Soviet Union

Soviet missiles and rockets have been entered under their NATO code names. In addition to these there is a letter-number system which is widely used in the case of only a few missiles. The actual Soviet designations are not usually known. The NATO naming code is as follows:

A for air-to-air missiles
   Alkali
   Anab
   Ash
   Atoll
   Awl

G for surface-to-air missiles
   Galosh
   Ganef (SAM-4)
   Goa (SAM-3)
   Griffon (SAM-5)
   Guideline (SAM-2)
   Guild (SAM-1)

K for air-to-surface missiles
   Kangaroo
   Kelt

   Kennel
   Kipper
   Kitchen

S stands for all surface-to-surface missiles
Land-based ICBMs with nuclear warheads:
   Sasin SS-8
   Savage SS-13
   Scarp SS-9
   Scrag
   SS-11 (no NATO designation)
Land-based long-range (but not USSR-to-USA) with nuclear warheads:
   Sandal SS-4
   Scamp
   Scrooge SS-X2
   Skean SS-5
Other nuclear-capable missiles:
   Samlet SSCD-1 ◊ *Kennel*
   Scud series SS-1b/1c
   Shaddock SSC-1
Submarine - launched nuclear - warhead missiles:
   Sark
   Sawfly (SS-N-6)
   Serb
Ship-launched non-nuclear missile:
   Styx
Anti-tank missiles:
   Sagger
   Snapper Russian (popular) name *Shmell*
   Swatter

Rockets
'Artillery' rockets with non-nuclear warheads are entered under ◊ *Artillery, Soviet*. Nuclear-capable rockets are given the NATO designation FROG (Free Rocket Over Ground). ◊ *Frog.*

Missiles: United Kingdom

British missile designations are not usually published, and the well-known popular names are used in the text

Air-to-Air
   Firestreak
   Red Top

Air-to-Ground (Nuclear-warhead)
   Blue Steel

Surface-to-Air
   Bloodhound 2

Armoured Fighting Vehicles:
Soviet Union

Russian fighting vehicle designations are
based on a descriptive system which is
usually initialled.

Awiadesantny Samochodnaja Ustanovka
(air-transportable self-propelled gun)
ASU-57 ⬧ *ASU*
ASU-85 ⬧ *ASU*

Broneje Transporter (personnel carrier)
The letter P indicates that the vehicle is
amphibious, while the letter B denotes the
presence of a weapon turret.
⬧ *BIR* for:
    BTR-40
    BTR-50 P
    BTR-60 P
    BTR-152
    M-1967

Plawajustschij Tank (amphibious tank)
PT-76
PT-85 ⬧ *PT-76*

Tank
⬧ *T-54* for:
    T-54
    T-55
    T-62
    T-59

(Z) Senitny Samochodnaja Ustanovka
(anti-aircraft self-propelled gun)
ZSU-23-4 ⬧ *SU-57-2*
ZSU-57-2 ⬧ *SU-57-2*

Armoured Fighting Vehicles: Britain

All British fighting vehicles are designated
by the letters FV followed by the model
number. A further set of letters denotes
special-purpose conversions of the basic
model.
    FV 432 ⬧ *Trojan*
    FV 433 ⬧ *Abbot*
    FV 601 ⬧ *Saladin*
    FV 603 ⬧ *Saracen*
    FV 701, FV 703, FV 704, FV 711 ⬧
Ferret
    FV 4017 ⬧ *Centurion*
    FV 4201 ⬧ *Chieftain*

The numbers of two new vehicles, *Fox*
and *Scorpion*, are not yet known.
    A British commercial tank, the Vickers
37-ton, is entered under its Indian name ⬧
Vijayanta.

Armoured Fighting Vehicles: France

French fighting vehicle designations are
based on three main sets of initials: AML,
EBR and AMX.

Automitrailleuse légère (armoured car,
light)
    ⬧ *AML* for AML 245
            AML S.530
            AML VTT
            AML NA-2

Engin blindé de reconnaissance (armoured
car, heavy)
    ⬧*EBR* for EBR.75
            EBR.90

Atelier de construction d'Issy-les-Moulin-
eaux (defence plant)
    ⬧ *AMX 13* for AMX 13 A/B/D (light
tanks and tank destroyers)
    ⬧ *AMX 30* Char de Combat (main
battle tank)
    ⬧ *AMX-VTT* for AMX VTP Mod.
56 (armoured personnel carrier)
    ⬧ *Artillery, French* (self-propelled) for
OB 105/AMX 105
OB 155/AMX 155

Armoured Fighting Vehicles:
West Germany

The word Panzer, abbreviated as Pz.,
denotes each designation:
    ⬧ *Jagdpanzer* for JPZ-4-5
            Jagdpanzer Kanone
            Jagdpanzer Rakete
            M-1966
    ⬧ *Schutzenpanzer Gruppe* for
        Spz. (Neu)
        Schutzenpanzer Neu
        Marder Spz. Neu
    ⬧ *Leopard* for Standardpanzer
    ⬧ *MBT-70* for KPz. 70
            Kampfpanzer 70

Armoured Fighting Vehicles:
Other Countries

India
 ◊ *Vijayanta* for Vickers 37-ton

Sweden
 ◊ *Strv. 103* for S-tank
      Stridsvagn 103
Hungary
 ◊ *F U G 65/66* for Felderito Uszo Gep-
kocsi

1. *Moskva*, Soviet missile-equipped helicopter carrier

2. *Enterprise*, US aircraft carrier

3. *Eagle*, British aircraft carrier

4. Soviet W class submarine

5. *Plunger*, US *Thresher/Permit* class submarine

6. *Ethan Allen*, US Polaris-type submarine

7. *Resolution*, British ballistic-missile submarine

8. Soviet Kashin class guided-missile destroyer

9. *Varyag*, Soviet Kynda class guided-missile destroyer.
Note the two quadruple mounts for Shaddock missiles

10. Soviet Kresta class guided-missile destroyer

11. Soviet Sverdlov class cruiser

12. Italian cruiser, *Vittorio Veneto*

13. US cruiser, *Long Beach*. Note the phased array radar on the large box-like superstructure

14. *Fife*, British County class guided-missile destroyer

15. US Patton M.60 A1E1 battle tank, being loaded with
a Shillelagh missile

16. Soviet T54 tanks

17. West German Leopard Standardpanzer

18. French AMX 30 battle tank

19. British Centurion tank

20. British Chieftain tank

21. British Vickers Main Battle Tank Mark I

22. Swedish Strv. 103 or S tank

23. British Scorpion light tank

24. Soviet PT-76 light tanks

25. British Trojan (AFV 432) armoured personnel carrier

26. British Saladin armoured car

27. British Ferret light armoured car

28. French AML armoured car

29. French AMX VTT armoured personnel carrier

30. Soviet BTR-60(PB) armoured personnel carriers

31. British Abbot (FV 433) self-propelled 105-mm gun

32. US M.107 self-propelled 175-mm gun

33. US M.109 self-propelled 155-mm howitzer with the British army in Germany

34. West German Jagdpanzer Kanone tank destroyer

35. Crusader F-8, US naval fighter

36. Phantom F-4, US fighter, just after launching
Sparrow missile

37. Delta Dart F-106, US interceptor-fighter

38. Corsair II A-7D, US attack fighter

39. Skyhawk A-4E, US attack aircraft

40. Intruder A-6A, US naval attack aircraft

41. Lightning Mk 6, British fighter, carrying Red Top
missile

42. Jaguar SO6, Anglo–French fighter-bomber

43. Harrier, British VTOL fighter-bomber

44. Starfighter F-104G, US-designed fighter-bomber

45. Fiat G 91, Italian fighter-bomber, shown here in
West German service

46. Mirage V, French attack fighter modification of the
Mirage III

47. Viggen (SAAB-37), Swedish fighter

48. Draken J.35F, Swedish fighter

49. 'Fishbed', Mig-21 Soviet fighters carrying Atoll missiles

50. Hustler B-58, US medium bomber

51. FB-111A, US medium bomber

52. 'Bear' Tu-20, Soviet long-range bomber/reconnaissance aircraft,
shadowed here by a British Phantom

53. 'Bison' Mya-4, Soviet long-range bomber

54. Stratofortress B-52, US long-range bomber

55. Vulcan, British medium bomber

56. SR-71, US long-range reconnaissance aircraft

57. Galaxy C-5A, US long-range transport

58. Hercules C-130E, US military transport

59. Transall C-160, Franco-German military transport

60. 'Cock' An-22, Soviet military transport

61. Hueycobra (AH-1G) attack helicopter

62. MIL Harke Mi-10K, Soviet flying crane helicopter

63. Falcon AIM-4D, US air-to-air missile, being fitted to Phantom 4-F

64. Sparrow, US air-to-air missile, here seen in British service

65. Sidewinder AIM-9D, US air-to-air missile

66. Red Top, British air-to-air missile, attached to a
Lightning fighter

67. Matra R.530, French air-to-air missile, with infra-red
guidance unit

68. Vigilant, British anti-tank missile

69. TOW MGM-71A, US anti-tank missile

70. Swingfire, British anti-tank missile

71. Nord SS.11, French anti-tank missile

73. Crotale, French anti-aircraft system

72. Chaparral, US mobile anti-aircraft missile system

74. Rapier, British anti-aircraft missile system

75. Redeye XMIM-43A, US anti-aircraft missile

76. Thunderbird Mk 2, British surface-to-air missile

77. Hawk MIM-23A, US surface-to-air missile

78. Ganef, Soviet surface-to-air missile

79. 'Guideline' SA-2 (centre), Soviet surface-to-air missile

80. Bloodhound Mk 2, British surface-to-air missile

81. Standard RIM-66A/67A, US ship-to-air missile

82. Styx, Soviet surface-to-surface cruise missile

83. Sea Dart CF-299, British surface-to-air shipborne missile

84. Seacat, British surface-to-air shipborne missile

85. Ikara, Australian/British anti-submarine missile

86. Martel, Anglo–French air-to-surface missile. This picture shows how the system's TV camera relays the target to the crew of the launching aircraft

87. Kennel, Soviet anti-shipping missile

88. Blue Steel, British stand-off missile

89. Hound Dog AGM-28, US stand-off missiles, being
fitted to B-52 bomber

90. Pershing MGM-31A, US tactical missile, here shown
in West German service

91. Honest John MGR-1, US unguided surface-to-
surface missile

92. Frog 7, Soviet surface-to-surface unguided missile

93. Scud, Soviet tactical missile

94. Minuteman 3, US intercontinental ballistic missile

95. Polaris A3, US submarine-launched ballistic missile

96. Poseidon, US submarine-launched ballistic missile.
Note the large MIRV dispenser